AF737025

ENVIRONMENTAL ENDOCRINE TOXICANTS

Biology, Effects, and Management

ENVIRONMENTAL ENDOCRINE TOXICANTS

Biology, Effects, and Management

Edited by

Younis Ahmad Hajam, PhD

Rouf Ahmad Bhat, PhD

Khalid Rehman Hakeem, PhD

Rajesh Kumar, PhD

AAP | APPLE ACADEMIC PRESS

First edition published 2023

Apple Academic Press Inc.
1265 Goldenrod Circle, NE,
Palm Bay, FL 32905 USA

760 Laurentian Drive, Unit 19,
Burlington, ON L7N 0A4, CANADA

CRC Press
6000 Broken Sound Parkway NW,
Suite 300, Boca Raton, FL 33487-2742 USA

4 Park Square, Milton Park,
Abingdon, Oxon, OX14 4RN UK

© 2023 by Apple Academic Press, Inc.

Apple Academic Press exclusively co-publishes with CRC Press, an imprint of Taylor & Francis Group, LLC

Reasonable efforts have been made to publish reliable data and information, but the authors, editors, and publisher cannot assume responsibility for the validity of all materials or the consequences of their use. The authors, editors, and publishers have attempted to trace the copyright holders of all material reproduced in this publication and apologize to copyright holders if permission to publish in this form has not been obtained. If any copyright material has not been acknowledged, please write and let us know so we may rectify in any future reprint.

Except as permitted under U.S. Copyright Law, no part of this book may be reprinted, reproduced, transmitted, or utilized in any form by any electronic, mechanical, or other means, now known or hereafter invented, including photocopying, microfilming, and recording, or in any information storage or retrieval system, without written permission from the publishers.

For permission to photocopy or use material electronically from this work, access www.copyright.com or contact the Copyright Clearance Center, Inc. (CCC), 222 Rosewood Drive, Danvers, MA 01923, 978-750-8400. For works that are not available on CCC please contact mpkbookspermissions@tandf.co.uk

Trademark notice: Product or corporate names may be trademarks or registered trademarks and are used only for identification and explanation without intent to infringe.

Library and Archives Canada Cataloguing in Publication

Title: Environmental endocrine toxicants : biology, effects, and management / edited by Younis Ahmad Hajam, PhD [and three others].
Names: Hajam, Younis Ahmad, editor.
Description: First edition. | Includes bibliographical references and index.
Identifiers: Canadiana (print) 20230156436 | Canadiana (ebook) 20230156509 | ISBN 9781774912102 (hardcover) | ISBN 9781774912119 (softcover) | ISBN 9781003336426 (ebook)
Subjects: LCSH: Endocrine disrupting chemicals. | LCSH: Endocrine disrupting chemicals—Physiological effect. | LCSH: Endocrine disrupting chemicals—Health aspects. | LCSH: Endocrine toxicology.
Classification: LCC RA1270.E53 E58 2023 | DDC 615.9/2—dc23

Library of Congress Cataloging-in-Publication Data

CIP data on file with US Library of Congress

ISBN: 978-1-77491-210-2 (hbk)
ISBN: 978-1-77491-211-9 (pbk)
ISBN: 978-1-00333-642-6 (ebk)

Dedication

This book is dedicated to my beloved mother, *Hajrah Begum*, who left this world so early. To me, she was the whole world. She was the most loving, humble, compassionate, understanding, family-orientated woman. I am going to miss you so much, *Mom*. I pray to Allah (SWT) to bless your soul and grant you, Jannatul-Firdous (the highest level of heaven).

'O Allah, forgive and have mercy upon her, excuse her and pardon her, and make honorable her reception. Expand her entry, cleanse her with water, snow, and ice, and purify her of sin as a white robe is purified of filth. Exchange her home for a better home. Admit her into the Garden; protect her from the punishment of the grave and the torment of the Fire.'

—Prof. (Dr.) Khalid Rehman Hakeem

About the Editors

Younis Ahmad Hajam, PhD
Assistant Professor, Department of Life Sciences and Allied Health Sciences, Sant Baba Bhag Singh University, Padhiana, Jalandhar, Punjab, India

Younis Ahmad Hajam, PhD, holds a doctorate in Zoology from Guru Ghasidas Vishwavidayalaya (A Central University), Bilaspur, Chhattisgarh, India. He is currently working as an Assistant Professor in the Department of Life Sciences and Allied Health Sciences, Sant Baba Bhag Singh University, Padhiana, Jalandhar, Punjab, India. Formerly Dr. Hajam had worked as Assistant Professor and Head of the Department of Zoology, Career Point University, Hamirpur, Himachal Pradesh, India. He has more than four years of experience in teaching and seven years of research experience. His thrust areas of research are animal physiology, applied zoology, nutraceuticals, and biodiversity conservation. He is teaching undergraduate and postgraduate and doctorate students. Dr. Hajam has published 36 research papers in leading national and international journals, 30 book chapters and four in national proceedings. Dr. Hajam has published four books as editor with Taylor and Francis, etc. He has guided five MSc students under the thesis mode and 31 MSc students under the project mode. Currently Dr. Hajam is guiding four PhD students and two MSc students for their dissertation/thesis work. He has completed one major research project (Co-PI), three projects under a HP Chief Minister Start-Up scheme. He has also served as a Junior Research Fellow (JRF) in Zoology at Guru Ghasidas Central University, Bilaspur, Chhattisgarh, India. He has presented his research work in more than 37 national and international conferences; Dr Younis Ahmad has attended more than 30 national and international conferences, and five workshops. He has organized various conferences, seminar, workshops, and webinar of at the national and international levels. He has received a national award for his contributions in biological science. He has also completed one self-sustaining project titled "Establishment of

Vermicompost Unit under Clean Green Campus." He has successfully organized seven national and two international academic events/conferences by collaborating with various government agencies, such as PPCB, AICT, IIC DEST, HIMCOSTE, and NABARD, and has delivered more than 15 invited talks/extension lectures in different institutions. He is an editorial member of various international journals and is also member of the Asian Council of Science Editors.

Rouf Ahmad Bhat, PhD
Department of School Education,
Government of Jammu and Kashmir, India

Rouf Ahmad Bhat, PhD, is presently working in the Department of School Education at the Government of Jammu and Kashmir. He pursued his doctorate at Sher-e-Kashmir University of Agricultural Sciences and Technology, Kashmir (Division of Environmental Science). Dr. Bhat has been teaching graduate and postgraduate students of Environmental Sciences for the past three years. He is an author of more than 50 research articles (h-index 22; i-index 33; total citation >1450) and 40 book chapters and has published more than 30 books with international publishers (Springer, Elsevier, CRC Press, Taylor and Francis, Apple Academic Press, John Wiley, IGI Global). He has specialization in limnology, toxicology, phytochemistry, and phytoremediation. Dr. Bhat has presented and participated at numerous state, national, and international conferences, seminars, workshops, and symposiums. In addition, he has worked as an associate environmental expert in a World Bank-funded Flood Recovery Project and also as as environmental support staff in Asian Development Bank (ADB)-funded development projects. He has received many awards, appreciations, and recognitions for his services to the science of water testing and air and noise analysis. He has served as an editorial board member and reviewer of reputed international journals. Dr. Bhat is still writing and experimenting with diverse capacities of plants for use in aquatic pollution remediation.

Khalid Rehman Hakeem, PhD

Professor, King Abdulaziz University, Jeddah, Saudi Arabia

Khalid Rehman Hakeem, PhD, is a Professor at King Abdulaziz University, Jeddah, Saudi Arabia. He is currently the Head of the Plant Biotechnology Research Unit) at Princess Dr. NajlaaBint Saud Al Saud Center for Distinguished Research in Biotech, King Abdulaziz University, Jeddah. After completing his PhD (Botany; specialization in Plant Ecophysiology and Molecular Biology) from Jamia Hamdard, New Delhi, India, in 2011, he initially worked as Assistant Professor at the University of Kashmir, Srinagar. Later he joined the University of Putra Malaysia, Selangor, Malaysia, and worked there as Postdoctorate Fellow in 2012 and Fellow Researcher (Associate Professor) from 2013–2016, respectively. He joined King Abdulaziz University in August 2016 as Associate Professor and was promoted to Professor in 2019.

Dr. Hakeem has more than 14 years of teaching and research experience in plant eco- physiology, biotechnology, molecular biology, medicinal plant research, plant-microbe-soil interactions as well as in environmental studies. He is the recipient of several fellowships at both national and international levels. He was recently elected as a Fellow of the Royal Society of Biology, London. He has served as the visiting scientist at Fatih University, Istanbul, Turkey, as well as at Jinan University, Guangzhou, China. Currently, he is involved with a number of international research projects with different government organizations. To date, Prof. Khalid has authored and edited more than 75 books with international publishers, including Springer Nature, Academic Press (Elsevier), CRC Press, etc. He also has to his credit more than 155 research publications in peer-reviewed international journals and 65 book chapters in edited volumes with international publishers. At present, he is serving as an editorial board member and reviewer for several high-impact international scientific journals from Elsevier, Springer Nature, Taylor and Francis, Cambridge Press, and John Wiley Publishers. He is included on the advisory board of Cambridge Scholars Publishing, UK. He is also the Fellow of the Plantae Group of the American Society of Plant Biologists, member of the World Academy of Science, member of the International Society for Development and Sustainability, Japan, and a member of the Asian Federation of Biotechnology, Korea.

Rajesh Kumar, PhD
Assistant Professor of Zoology,
Department of Biosciences,
Himachal Pradesh University, Shimla,
Himachal Pradesh, India

Rajesh Kumar, PhD, is currently working as an Assistant Professor in the Department of Biosciences, Himachal Pradesh University, Shimla (Himachal Pradesh) India. He has more than nine years of experience in teaching and research. His thrust areas of research are animal physiology, applied zoology, nutraceuticals, and biodiversity conservation. During his PhD, he was awarded a Junior Research Fellowship by the University Grants Commission under the meritorious students in science scheme. Dr. Kumar has successfully completed two major research projects and two CM startup projects in the area of honeybee physiology and nutrition and waste management, sponsored by the Government of Himachal Pradesh. He has also completed one self-sustaining project entitled "Establishment of Vermicompost Unit under Clean Green Campus." Dr. Kumar has published about 100 research articles and 25 book chapters in indexed journals/with publishers such as Elsevier, Springer, Scopus, WoS, CRC Press, AAP, etc.; 10 research papers in full-length proceedings, five books (CRC Press, Springer), four monographs, 12 awards, and 80 abstracts by participating in more than 75 national/international conferences. He has successfully supervised 2 PhD, 1 MPhil, and 43 MSc students for their research work, and currently three PhD candidates are working under his supervision. Dr. Kumar has successfully organized 16 national-level academic events/conferences by collaborating with various government agencies, like NITI Aayog, DEST, HIMCOSTE, NABARD, etc., and has delivered more than 10 invited talks/extension lectures at different institutions. He is life/fellow member of various prestigious scientific agencies, including ISC, HBSRA, STRA, ISZS, etc. He is also a reviewer for several journals of national and international repute. Dr. Kumar has more than 200+ Google citations, h-index–10, i10 index-10, Vidwan score 9.7/10. He did his master's and PhD in Zoology at Jiwaji University, Gwalior (M.P.), India.

Contents

Contributors

Dinesh Kumar Agrawala
Botanical Survey of India, CGO Complex, 3rd MSO Building, 5th and 6th Floor DF Block, Sector-I, Salt Lake, Kolkata, West Bengal–700064, India

Ritika Anthal
Department of Zoology, University of Jammu, Jammu–180006, Jammu and Kashmir, India

Bindushree Bhaghel
School of Studies in Life Science, Pt. Ravishankar Shukla University, Raipur, Chhattisgarh, India, E-mail: bshri06@gmail.com, https://orcid.org/0000-0002-1026-0864

Musadiq Hussain Bhat
Central National Herbarium, Botanical Survey of India, Howrah, Kolkata, West Bengal–711103, India, E-mail: musadiqali131@gmail.com

Rouf Ahmad Bhat
Sher-e-Kashmir University of Agricultural Sciences and Technology, Kashmir, Jammu and Kashmir, India

Diksha
Division Zoology, Department of Biosciences, Career Point University, Hamirpur, Himachal Pradesh–176041, India

Arup Giri
Baba Mastnath University, Asthal Bohar, Rohtak, Haryana–124021, India, E-mail: arupsatadal@gmail.com

Younis Ahmad Hajam
Department of Life Sciences and Allied Health Sciences, Sant Baba Bhag Singh University, Padhiana, Jalandhar, Punjab, India, E-mail: younismajeed64@gmail.com

Ayesha Hassan
Department of Environmental Sciences, Faculty of Natural Sciences, University of Okara, Okara–56130, Pakistan

Muhammad Ibrahim
Department of Environmental Sciences, Faculty of Natural Sciences, University of Okara, Okara–56130, Pakistan

Neha Jain
Department of Zoology, Guru Ghasidas Central University, Bilaspur, Chhattisgarh–495009, India, E-mail: drnehajain2016@gmail.com, https://orcid.org/0000-0002-9994-468X

Raman Jasrotia
Department of Zoology, University of Jammu, Jammu–180006, Jammu and Kashmir, India, E-mail: ramanjasrotia18@gmail.com

Rajesh Kumar
Department of Biosciences, Himachal Pradesh University, Shimla, Himachal Pradesh–171005, India, E-mail: drkumar83@rediffmail.com

Preeti Kumari
DRDO–Defense Institute of High-Altitude Research (DIHAR), Leh, Ladakh UT–194101, Jammu and Kashmir, India

Shailja Kumari
Department of Biosciences, Division Zoology, Career Point University, Hamirpur–176041, Himachal Pradesh, India, E-mail: shailjakhajuria@yahoo.in

Prem Lata
Department of Biosciences, Division Zoology, Career Point University, Hamirpur, Himachal Pradesh–176041, India, E-mail: premlata1617@gmail.com

Tawseef Ahmad Mir
Center of Research for Ethnobotany, GMS College, Jiwaji University, Gwalior, Madhya Pradesh–474009, India

Seema Rai
Department of Zoology, Guru Ghasidas Vishwavidayalaya, Bilaspur, Chhattisgarh–495001, India

Mohd. Salim Reshi
Department of Zoology, School of Biosciences and Biotechnology, Baba Ghulam Shah Badshah University, Rajouri, Jammu and Kashmir–185234, India, E-mail: reshisalim@gmail.com

Rijupalika Roy
Central National Herbarium, Botanical Survey of India, Howrah, Kolkata, West Bengal–711103, India

Aishwarya Sharma
Department of Biosciences, Division of Zoology, Career Point University, Hamirpur–176041, Himachal Pradesh, India

Yogesh Singh
DRDO–Defense Institute of High-Altitude Research (DIHAR), Leh, Ladakh UT–194101, Jammu and Kashmir, India

Sonika
Department of Biosciences, Division of Zoology, Career Point University, Hamirpur, Himachal Pradesh–176041, India

Shahnaz Tabassum
Department of Zoology, University of Jammu, Jammu – 180006, Jammu and Kashmir, India

Uruj Tahir
Department of Environmental Sciences, Faculty of Natural Sciences, University of Okara, Okara–56130, Pakistan, E-mail: urujtahirjavaid@gmail.com, https://orcid.org/0000-0002-5393-5278

Richa Tikariha
Assistant Professor, Department of Zoology, Government D.B Girls PG College, Raipur, Chhattisgarh, India, E-mail: richamehta3@gmail.com

Abbreviations

3-MC	3-methylcholanthrene
ABP	androgen binding proteins
ABS	aberration assay
ACTH	adrenocorticotropic hormone
ADH	anti-diuretic hormone
ADHD	attention deficit hyperactivity disorder
AFP	α-fetoprotein
AGD	anogenital distance
AhR	aryl hydrocarbon receptor
AP	anterior pituitary
AR	androgen receptor
ARNT	AhR nuclear translocator
BaP	benzo[a]pyrene
BAT	brown adipose tissue
BBB	blood-brain barrier
BBP	butylbenzylphthalate
BFRs	brominated flame retardants
BMI	body mass index
BMR	basal metabolic rate
BPA	bisphenol-A
CAD	coronary artery disease
cAMP	cyclic adenosine monophosphate
CAS	chemical abstract service
CBP	CREB binding protein
Cd	cadmium
CDC	Center for Disease Control
CHD	coronary heart disease
CHLs	chlordanes
CICAD	concise international chemical assessment document
Cl	chlorine
CMV	cytomegalovirus
CNS	central nervous system
CREDO	cluster of research into endocrine disruption in Europe
CRH	corticotropic-releasing-hormones

CVD	cardiovascular disorders
DBCP	1,2-dibromo-3-chloropropane
DBP	dibutyl phthalate
DDD	dichlorodiphenyldichloroethane
DDE	dichloro diphenyl dichloroethylene
DDT	dichloro biphenyl trichloroethane
DEHP	di-(2-ethylhexyl) phthalate
DEP	diethyl phthalate
DES	diethylstilbestrol
DIT	di-iodo-tyrosine
DL	dioxin-like
DNMT3	DNA cytosine-5-methyltransferase gene
DOC	dissolve organic carbon
EA	endocrine activity
EDCs	endocrine-disrupting chemicals
EDs	endocrine disruptors
EDSP	endocrine disruption screening program
EE	ethinylestradiol
EPA	Environmental Protection Agency
ER	endocrine regulation
ER	estrogen receptors
ERβ	estradiol receptor β/estrogen receptor β
ESRRA	estrogen-related receptor [ERR] alpha
ETs	endocrine toxicants
FIPA	familial isolated pituitary adenomas
FLX	fluoxetine
FSH	follicle-stimulating hormones
GD	gestation day
GE	genotoxic effects
GH	growth hormone
GHRH	growth-hormone-releasing hormone
GI	gastrointestinal tract
GLUT	glucose transporter
GnRH	gonadotropin-releasing hormone
GP	glycogen phosphorylase
GSH	glutathione
HbA1c	hemoglobin A1c
HCB	hexa-chlorobenzene
HCHs	hexachlorocyclohexane

hGH	human growth hormone
HMS	homosalate
HPA	hypothalamic-pituitary-adrenal
Hpcal1	hippocalcin1
HPG	hypothalamic-pituitary-gonadal
HPO	hypothalamic-pituitary-ovarian
HPT	hypothalamic-pituitary-thyroid
HPV	high productive volume
HRV	heart rate variability
II-P	infrapyramidal
IL	interleukin
IR	insulin resistance
KC	key characteristics
LH	luteinizing hormone
LIF	leukemia inhibitory factor
MDA	malondialdehyde
MDC	metabolism disrupting chemicals
MEHP	mono-(2-ethylhexyl) phthalate
$MeSO_2$-DDE	methyl sulfonyl-DDE
MetS	metabolic syndrome
miR	miRNA
miRNAs	microRNAs
MN	micronucleus test
MSH	melanocyte-stimulating hormone
NAFLD	nonalcoholic fatty liver disease
NCDs	non-communicable diseases
N-EDCs	natural endocrine-disrupting chemicals
N-EDs	natural endocrine disruptors
NHANES	National Health and Nutrition Examination Survey
NIEHS	National Institute of Environmental Health Science
NK	natural killer
NMDR	nonmonotonic dose-response
NOAEL	no observable adverse level
NP	nonylphenol
NPCSs	national priority contaminated sites
NPs	nonylphenol
NSAID	non-steroidal anti-inflammatory drugs
Nsbp1	nucleosome binding protein-1
OC	organic carbon

OCPs	organochlorine pesticides
OHCS	organohalogen contaminants
OH-PCBs	hydroxylated polychlorinated biphenyls
OMC	octyl methoxycinnamate
PABA	4-aminobenzoic acid
PAHs	polycyclic aromatic hydrocarbons
PBBs	polybrominated biphenyls
PBDE	polybrominated diphenyl ethers
PC	pyruvate carboxylase
PCB	polychlorinated biphenyl
PCDDs	polychlorinated dibenzo-p-dioxins
PCDFs	polychlorinated dibenzofurans
PCOS	polycystic ovary syndrome
PCPs	personal care products
PDBE	penta and decabromodiphenyl ether
p-ERK	phosphorylated extracellular signal-regulated protein kinase
PFAs	polyfluoroalkylated
PFCs	perfluorinated compounds
PFNA	perfluorononanoate
PFOA	perfluorooctanoic acid
PFOS	perfluorooctane sulfonate
PHAHs	polyhalogenated aromatic hydrocarbons
PHAs	polyaromatic hydrocarbons
PK	pyruvate kinase
PKC	protein kinase C
PND	postnatal day
POCs	persistent organochlorine compounds
POPs	persistent organic pollutants
PPAR	peroxisome proliferator-activated receptor
PPAR-α	peroxisome proliferator-activated receptor-α
PPARγ	peroxisome proliferator activated receptor-γ
PRH	prolactin-releasing hormone
PRL	prolactin
PTH	parathyroid hormone
PVC	polyvinyl chloride
PVN	paraventricular nucleus
RD	reproductive endocrine
RH	rhesus
ROS	reactive oxygen species

RXR	retinoid X receptor
SAL	*Salmonella* mutagenicity assay
SCD	Sertoli cell dysfunction
SCE	sister chromatid exchange
S-EDCs	synthetic endocrine-disrupting chemicals
S-EDs	synthetic endocrine disruptors
SERM	selective estrogen-receptor modulator
SHBG	steroid hormone-binding protein
SOD	superoxide dismutase
SON	supraoptic nucleus
SRC	steroid receptor co-activator
SSRI	selective serotonin reuptake inhibitor
StAR	steroidogenic acute regulatory
T2DM	type-2 diabetes mellitus
T3	tri-iodothyronine
T4	tetra-iodothyronine
T4	thyroxine
TA	testicular architecture
TBBPA	tetrabromobisphenol-A
TBT	tributyltin chloride
TCDD	2,3,7,8-tetrachlorodibenzo-p-dioxin
TCS	triclosan
TDS	testicular dysgenesis syndrome
TG	thyroid gland
TGFβ	transforming growth factor-beta
TH	thyroid hormones
THS	thyroid stimulating hormones
TIPED	tiered protocol for endocrine disruption
TNF α	tumor necrosis factor
TPO	thyroid peroxidase
TR	thyroid receptors
TRE	thyroid response element
TRH	thyrotropin-releasing hormone
TSH	thyroid stimulating hormone
TTR	transthyretin
UDP	uridine diphosphate glucose
UDPGT	uridine diphosphate glucuronosyl transferase
USFDA	US Food and Drug Administration
VP	vasopressin

WAT	white adipose tissue
WHO	World Health Organization
XREs	xenobiotic response elements
α-MSH	α-melanocyte stimulating hormone
γ-HCH	γ-hexachlorocyclohexane

Preface

Endocrine-disrupting constituents (EDCs) are present in different ecosystems, including food items, pesticides, and waste materials. EDCs cause health complications and metabolic, endocrine, and reproductive disorders by altering the synthesis of hormones and their metabolism and affecting normal physiological functions. Hence, they modulate the physiology and biochemistry of all the organs by altering homeostasis. Endocrine society proposed a statement that evidenced that endocrine disruptors (EDs) affect reproductive functions, lead to various types of cancers (breast and prostate), and alter the neuroendocrine system and endocrine system. Moreover, EDCs include a diverse group of chemicals such as "organochlorinated pesticides" and industrial chemicals, plastics, their derivatives, etc.

The literature summarized in this book emphasizes sources of endocrine toxicants (ETs) and their impact on human physiology. The basic requirement of a living being is a constant supply of foodstuffs, a clean environment, and safe clean-up of wastes from the environment. However, the natural environment has deteriorated and degraded, and become polluted due to the continuous and persistent use of synthetic chemicals, fertilizers, plastic items, and their derivatives. Therefore, the aim of this book is to explain the effects of endocrine toxicants that are mostly associated with human populations and their control measures.

This book provides recent and updated literature for budding researchers and wide-ranging knowledge regarding the interference of synthetic and other endocrine toxicants in environments. Moreover, it provides support and a simple anthology for undergraduate and postgraduate students to understand biochemical and physiological actions and the mechanism of EDCs on the body and its system.

Special gratitude to Apple Academic Press for their copious teamwork in publishing this volume.

—*Editors*

CHAPTER 1

Chemical Nature and Mechanism of Endocrine Disruptors (EDs): A Critical Review

RICHA TIKARIHA

Assistant Professor, Department of Zoology, Government D.B Girls PG College, Raipur, Chhattisgarh, India, E-mail: richamehta3@gmail.com

ABSTRACT

The endocrine system of an animal/human plays a vital role in the physiological, growth, reproduction, and biochemical aspects of animals. Any alteration in the functioning of this system causes disease or death of the organism. There are many emerging pollutants that have been detected and monitored in the past few decades because of their universal consumption from different sources. Among these emitted environmental pollutants, some of them directly or indirectly interfere with the functioning of the endocrine system. These chemicals are regarded as endocrine disturbers (EDs). In the present context, EDs are reported to be a key area of discussion across the global scientific community due to their estrogen effect, persistence, neurodevelopment delay, and biomagnifications. The discovery of EDs has opened a new direction to toxicology. The interference and influence of EDs on animals and human beings led to the discovery of different EDs 25 years ago, as an emerging category of toxic agents and diethylstilbestrol (DES) is acknowledged as proof of the concept. Ingestion from food or water is reported to be the main route of exposure to EDs. Due to the long half-life of EDs, they can stay in the environment for years, thus affecting the organisms continuously. EDs correspond to a wide variety of chemical classes; they can be synthetic (industrial lubricants and solvents, pesticides, plastics,

Environmental Endocrine Toxicants: Biology, Effects, and Management. Younis Ahmad Hajam, Rouf Ahmad Bhat, Khalid Rehman Hakeem, & Rajesh Kumar (Eds.)
© 2023 Apple Academic Press, Inc. Co-published with CRC Press (Taylor & Francis)

plasticizers, drugs, cosmetics, food, and preservatives) or natural (chemicals present in the environment like phytoestrogens, etc.). The action mechanism of EDs is very difficult to predict as many of these EDs act on the animals/human by interfering with the receptors of estrogen, androgen, thyroid, and hydrocarbon. The EDs can also influence hormone secretion, transport, or synthesis. In addition, EDs also interference with the epigenetic mechanism of an organism. Further elite investigations are required in improving the knowledge about the structure, metabolites, and functioning of EDs.

1.1 INTRODUCTION

The endocrine system of humans and other animals plays a vital role in the regulation of different physiological functions of the body (Table 1.1) and in balancing the stability of internal environment (Marty et al., 2018; Kliene and Rossmanith, 2016). The endocrine system of an organism consists of many glands that synthesize and secretes hormones into the circulatory system, and these hormones control the structure and functioning of cells, tissue, or organ. These hormones are secreted in very small amounts and regulate the overall growth, development, metabolism, reproduction, immunity, and behavior of animals, including human beings. The endocrine system regulates its functioning through a specific pathway, i.e., hormone signaling and activation of specific receptors. Once this pathway is disturbed, endocrine functioning is altered. The alteration in the endocrine physiological functioning is caused by many substances or chemicals. Such chemicals are reported to be endocrine disruptors (EDs).

The endocrine society has defined EDs as; "an exogenous (non-natural) chemical, or mixture of chemicals, that interfere with any respect of hormone action." Recently WHO has defined EDCs as; "These are the substances or a mixture (natural or synthetic) that alter the function(s) of the endocrine system and consequently cause adverse health effects in an intact organism, or progeny, or (sub)population." The EDs bear some properties that are potentiating with their hazards. The EDs are a large group of emerging chemicals found in the environment at low concentrations, but due their potential persistence and range of effects, they have become an emerging concept for toxicologists.

The interference and influence of EDs on animals and human beings led to the discovery of different EDs 25 years ago, as an emerging category of toxic agents and diethylstilbestrol (DES) is acknowledged as proof of the concept. It was pre-described from about 1940 by its induction of cell carcinoma

TABLE 1.1 Endocrine Hormones and Their Major Functions

Endocrine Gland	Hormone Released	Major Function of Hormone
Hypothalamus	Corticotropin	• The release of adrenocorticotropic hormone is stimulated from the pituitary.
	Gonadotropin	• The release of luteinizing hormone and follicle-stimulating hormone is stimulated from the pituitary.
	Thyrotropin	• The release of thyroid stimulating hormone is stimulated from the pituitary.
	Growth hormone	• The secretion of growth hormone (GH) is stimulated from the pituitary.
	Somatostatin	• Release of growth hormones is inhibited from the pituitary.
	Dopamine	• Release of growth hormones is inhibited from the pituitary.
Anterior pituitary gland	ACTH	• Secretion of hormones from the adrenal cortex is stimulated.
	LH	• Synthesis of sex hormones (in both genders) is stimulated.
	FSH	• In women, stimulates follicle development; in men, stimulates sperm production.
	TSH	• Secretion of thyroid hormone is stimulated.
	GH	• Enhances the growth and development of body.
	Prolactin	• Milk production is controlled.
Posterior pituitary gland	Vasopressin	• The level of water and electrolyte is controlled in the body.
	Oxytocin	• The contraction of the uterine is promoted during labor, and Milk ejection is activated in nursing women.
Adrenal cortex	Cortisol	• Carbohydrate, lipid, and protein metabolism is controlled.
	Aldosterone	• The level of water and electrolyte is controlled in the body.
Testis	Testosterone	• The development of male reproductive organs is stimulated, Stimulating sperm production.
Ovaries	Estrogen	• The development of female reproductive organs is stimulated.
	Progesterone	• Uterus is prepared for pregnancy; Mammary glands are prepared for lactation.

TABLE 1.1 *(Continued)*

Endocrine Gland	Hormone Released	Major Function of Hormone
Thyroid hormone	Thyroid hormone	• The metabolic processes are controlled in the cell.
	Calcitonin	• Calcium metabolism is controlled.
Parathyroid hormone	Parathyroid hormone (PTH)	• Calcium metabolism is controlled.
Pancreas	Insulin	• Carbohydrate metabolism is controlled (decreases sugar levels in blood).
	Glucagon	• Carbohydrate metabolism is controlled (Increase sugar levels in blood).

among young women (Smith, 1948). Like DES, many other chemicals which disturb the endocrine system were reported and till today, thousands of chemicals are included in the category of EDs. In the environment, several classes of chemicals like antibodies, non-steroidal anti-inflammatory drugs (NSAID), anti-depressants, contraceptive pills, alkyl phenol, dioxins, industrial, and household chemicals, preservatives, detergents, cosmetics, polychlorinated biphenyls (PCBs), phthalates, and products of personal care which share the properties of EDs (Christensen et al., 2006; Caliman and Gavrilesu, 2009; Pal et al., 2010; Hamid et al., 2021).

These synthetic or natural EDs have the property to interfere the normal endocrine functioning through removing/binding to the receptors of the endogenous hormones, disturbing various processes like hormone secretion, storage, release, metabolism, and transport (Yang et al., 2015; Kabir et al., 2015). EDs produce toxic effects on the growth and development, reproduction, neurological, immune, biochemical, and physiology of humans and other animals. Further, EDs show different responses at various stages of life, which depend on the hormone concentrations (Welshons et al., 2003; Vandenberg et al., 2002). The different exposure routes of human to EDs are given in Table 1.2. Considering the above discussion herein, a slight comprehensive description of nature, action mechanism, and effects of EDs on humans and other animals was provided in this review.

1.2 HISTORICAL PERSPECTIVES OF ENDOCRINE DISRUPTORS (EDS)

The existence and effects of EDs is not a new phenomenon. The scientific community across the globe has acknowledged that there are certain chemicals that are capable of interfering with the growth, development, and functioning of an individual's body (Colborn et al., 1992). In early 1949, it was revealed that the crop dusters who were handling DDT frequently showed a reduction in their sperm counts. It was found that when rats and birds were exposed to DDT, they produced estrogenic response in their reproductive tracts (Bitman et al., 1970). The use of synthetic estrogen (DES) by women (especially pregnant) extensively from 1940s to 1971 to avoid miscarriage resulted in infertility in women, and it also increased the rates of vaginal clear cell adenocarcinomas in their daughters (Colbom et al., 1993). Further, it was pointed out that these chemicals (EDs) affect the reproduction process of animals, including the hatching success of birds. Rachel Carsons book entitled "Silent Springs" was published in 1962 lauds the alarms warning regarding the contamination of environment by different chemicals especially DDT which might cause

TABLE 1.2 Endocrine Disruptor (ED) Routes of Exposures in Humans with Examples

SL. No.	How Human Gets Exposed to EDs	Source of EDs	Examples of EDs
1.	Through contaminated food/water consumption through oral route	Pesticides, industrial wastes, leaching of chemicals, beverage containers.	BPA, PCBs, phthalates, DDT, dioxins, chlorpyrifos, perfluorinated compounds.
3.	Inhalation or through dermal contact	Wood keepers, paints, varnishes, flame retardants, pesticides.	Turpentine, resin, BFR, chlorpyrifos, DDT, pyrethroids.
5.	Intravenous	Intravenous tubing.	Phthalates.
6.	Skin application	Cosmetics, personal care products, perfumes, deodorants, Sunscreens, medications, insecticides.	Phthalates, insect repellants, parabens, triclosan.
7.	Transfer from placenta biologically	EDs present in the maternal body due to previous/present exposure.	Includes the EDs which has the ability to cross the placenta.
8.	Transfer through mother's milk biologically	EDs present in the maternal body due to previous/present exposure.	Includes the EDs which are detected in mother's milk.

reduction in the bird population due to failure of reproduction caused by these toxic chemicals. Exposure to these toxic environmental chemicals is lifelong. Furthermore, these environmental toxicants became a potential debate among scientists across the globe. The identification of EDs came through a deep research of investigation, discoveries, and hardworking of the world scientific community. After working with various EDs chemicals on cell lines, it was observed that these chemicals conserve their ability to produce response. These EDs induced responses were used to show the endocrine active compounds, which are the primary target of EDs during its exposure to animals and human beings. After investigating the effects of EDs on animals, humans, and birds by the world scientific community, the complete scenario of EDs was brought in front of the people. In 2002, WHO defined EDs as; "these are the substances or mixtures (natural or synthetic) that alter the function(s) of the endocrine system and consequently causes adverse health effects in an intact organism, or progeny, or (sub)population(s)." Now the world scientific community till today is updating the people about the cause, effect, and control from these toxic chemicals.

1.3 EDS IMPACT ON THE REGULATORY PROCESSES IN THE BODY

The exogenous or endogenous chemicals which have the ability to antagonize or mimic the effect of endocrine hormones cause severe influence on growth, reproduction, physiology, and genome of an organism. EDs interact with the endocrine system and imbalance the homeostasis of an organism. Since, endocrine system releases specific hormones, which in turn triggers stimuli in specific target cells. It's well known that all biological developments of an organism unfold a sequence of events, which are totally controlled by biochemical signaling mechanism. Across these biochemical signaling, the most important is the signaling mechanism of hormones. The most important among these signaling mechanisms are hormones.

The interference by the EDs to the signaling pathway disturbs the endocrine functioning and thus having adverse effects on the growth and development of an individual organism. Moreover, these potential EDs while interacting with the cell components gives rise to the synthesis of reactive oxygen species (ROS). These ROS disturb the immune system and indirectly the functioning of the endocrine system. Overdoses of EDs cause over toxicity like death of cells, neurotoxicity, genotoxicity, histopathological alterations, and reproductive disorders.

1.4 BIOACCUMULATION OF ENDOCRINE DISRUPTORS (EDS)

It has been revealed that the major route of EDs exposure to the animals and humans is via dust inhalation, ingestion, and dermal exposure. Once EDs reaches to the target cell, tissue, or organ it starts to accumulate in these target places. Further, the EDs accumulate in fatty tissues (bioaccumulate) of animals and human beings. After accumulation of EDs in the tissues, their levels reach higher than the levels present in the environment. There are many EDs which persist in the tissues for a very long period to be degraded and thus retain in the body of an organism for many years. In the present context, the level of EDs in the body of animals and human beings have reached millions of times higher than the natural hormone levels in the body. The bioaccumulation of EDs can be classified in two different ways, which are discussed in subsections.

1.4.1 CHEMICAL MIXTURES

When more than two EDs combine, they are called the chemical mixtures. Human and animals are exposed to these chemical mixtures, and exerts very much toxic effects that they produce individually. The action of these chemical mixtures adds up together and cause cumulative effects. Reports suggest that the effects of these chemical mixtures can be addictive.

1.4.2 BINDING PROTEINS

As we know that above 90% of the naturally circulating estrogen in the body of an animal or human becomes bound to special binding proteins in the blood. Only the remaining free estrogen is able to diffuse into the cells and exert biological effects. However, EDs don't bind significantly to the binding proteins, potentially leaving up to 100% of these chemicals circulating in the blood free to enter cells and exert effects.

1.5 KEY CHARACTERISTICS (KC) OF ENDOCRINE DISRUPTORS (EDs)

Endocrine disruptors (EDs) are exogenous chemicals that antagonize and mimic the natural hormone level by changing the rate of synthesis, transport, metabolism, expression, and action at target receptors (Park et al., 2020). EDs are lined with developmental, reproductive, brain, immune, neurological, and other problems. As they alter the biochemical, physiological, and ROS status of a cell, tissue, or organ in humans and other animals. Almost all EDs are lipophilic, thus the water solubility is lower and accumulate in adipose tissue (Fingler et al., 1992; Abbasy et al., 1999). The EDs are less toxic individually, but in the mixture, they can be more toxic as they alter one another in an additive, adverse or synergetic way. One of the peculiar property of EDs is that even though they are present in the environment in low concentrations, but still, they can cause severe effects. The large half-life of the EDs makes their persistence in the environment more, and thus has potential impacts. The response of EDs shows a disparate response with age, size, sex, organ, tissue, or cell depending on the physiological stress created by the ED on hormone activity (Welshons et al., 2003; Vandenberg et al., 2013). Although in defining a chemical as an ED, the important key is the identification of their intrinsic hazards. On the basis of the rate and type of damage caused by a chemical to a cell, tissue, or organ, it can be further classified as ED or not. Regulatory agencies use various approaches in identification of EDs (Andreson et al., 2012; OECD, 2018; NIHS, 2015; IPCS, 2002; EHHA, 2011; USEPA, 2014). Keeping in view only a few characteristics of chemicals is not appropriate for the identification of EDs. But it's better to identify some key characteristics (KC) to identify an ED.

Some peculiar properties of EDs were identified by reporting the common features of action and regulation of hormones which are independent in number of hormonal effects across the life cycle of an organism. This suggests that there exist features that specify the action of chemicals by which action and regulations of hormones is altered (Michele et al., 2020). Michele et al. (2020) has identified 10 KCs on the basis of evaluation from the existing scientific literature. These 10 KCs of EDs (Table 1.3) defined by Michele et al. (2020) are as:

- Activation or interaction with hormone receptors;
- Receptors of hormones are antagonized;
- The hormone receptor expression is altered;
- Signal transduction is altered in hormone responsive cells;
- Modification in hormone producing/responsive cells occurs through epigenetic mechanism;
- Hormone synthesis is altered;
- Hormone transport is altered across cell membrane;
- The distribution or circulation of hormones is altered;

TABLE 1.3 10 Key Characteristics (KCs) of Endocrine Disruptors and Their Mechanistic Examples

Key Characteristics	Mechanistic Evidences of ED	Key Characteristics	Mechanistic Evidences of ED
• **KC1:** Activation or interaction with hormone receptors	Binding or agonism of hormone receptors	• **KC6:** Hormone synthesis is altered	Expression or activity of enzymes or substrates in hormone synthesis
• **KC2:** Receptors of hormones are antagonized	Antagonism of nuclear or cell surface hormone receptors	• **KC7:** Hormone transport is altered across cell membrane	Intracellular transport, vesicle dynamics or cellular secretion
• **KC3:** The hormone receptor expression is altered	Abundance, distribution, and degradation of hormone receptors	• **KC8:** The distribution or circulation of hormones is altered.	Blood protein expression and binding capacity, blood levels of pro-hormones and hormones
• **KC4:** Signal transduction is altered in hormone responsive cells	Abundance of post-translational modifications, cofactors, transcription factors and transcripts, and activity of associated enzymes	• **KC9:** Hormone metabolism is altered	Inactivation, breakdown, recycling, clearance, excretion, or elimination of hormones
• **KC5:** Modification in hormone producing/ responsive cells occurs through epigenetic mechanism	Chromatin modifications, DNA methylation and non-coding RNA expression	• **KC10:** The fate of hormone producing/responsive cells is altered	Atrophy, hyperplasia, hypertrophy, differentiation, migration, proliferation, or apoptosis

- Hormone metabolism is altered;
- The fate of hormone producing/responsive cells is altered.

The above-mentioned KCs reflect current knowledge and will probably evolve over time with new scientific discovery.

1.6 ENDOCRINE DISRUPTORS (EDS): THE CHEMICAL NATURE

Different groups of EDs are extremely heterogeneous. EDs have a long half-life, persistent, low water solubility, and extremely high lipid soluble. A chemical may have endocrine disrupting effects at a lower dose than the currently accepted no observable adverse level (NOAEL). The EDs are classified in two categories: (i) synthetic endocrine disruptors (S-EDs); and (ii) natural endocrine disruptors (N-EDs). Theoretically, both S-EDs, and natural (endogenous and exogenous) endocrine disruptors (N-EDs) can interact with endocrine receptors and disturb hormonal balance:

1. **Synthetic Endocrine Disruptors (S-EDs):** These are the EDs which are produced by human activities. They include:
   i. Industrial lubricants and solvents, and there by-products like polybrominated biphenyls (PCBs), polybrominated biphenyls ether (PBDE), and doxins, e.g., 2,3,7,8-tetrachlorodibenzo-p-dioxin (TCDD), decabromodiphenylethane (DBPDE).
   ii. **Plastics:** It includes bisphenols, e.g., BPA, BPA, BPF, BPG, BPS, etc.
   iii. **Plasticizers:** It includes the phthalate products.
   iv. **Pesticides:** It includes different insecticides, fungicides, herbicides, weedicides, plant growth hormones (GHs).
   v. S-EDs also include different drugs like DES, ethinylestradiol (EE) as well as NSAID and acetaminophen.
2. **Natural Endocrine Disruptors (N-EDs):** These EDs occur naturally and include, phytoestrogens.

1.7 ENDOCRINE DISRUPTORS (EDS): MECHANISM

To understand how EDs disturb the endocrine system, it is essential to understand that how natural hormones work in the body of an animal or human. Every hormone of endocrine system has a unique chemical composition and three-dimensional shapes. But, a hormone always has a corresponding receptor located on the target cells. The receptor shape of a hormone remains always complementary to its hormone; in such a way that one key (hormone) is specific to a lock (receptor). A specific tissue/organ or cell responds to a hormone due to the presence of its receptors on the target cells and receptor activation by hormone binding. There are many factors which show the ability of a hormone to activate its respective receptor, for example, how much hormone is synthesized, secreted, transported, and how potently a hormone can activate its receptor. These attributes are fundamental to the normal pathway of hormone signaling. EDs have the potential to interfere with any or all of these steps.

Due to the complexity of the endocrine system, the action mechanism of EDs is very difficult to predict (De Coster et al., 2012). But the researchers across the world have made time to time efforts to report the possible mechanism of EDs. However, till date, it has been revealed that the EDs can cause damage to the functioning of the endocrine system by any possible mechanism.

- EDs can alter the signaling of the nuclear receptor (Altenburger et al., 2012);
- EDs can act through non-steroidal receptors, enzymatic pathways, transcription coactivators involved in the steroid biosynthesis and/or metabolism (WHO, 2013);
- EDs have the ability to mimic the natural hormone, thus causing the body of an animal or human being to respond to the (Hamid et al., 2021);
- EDs can cause epigenetic effects like chromatin remodeling, chromosomal aberration, DNA methylation and histone modification (Manikkam et al., 2013);
- EDs can block the effects of hormones from certain receptors;
- EDs directly or indirectly increase or decrease overall production of hormones.

1.8 CONCLUSION

The endocrine system plays a vital role in physiological, biochemical, growth, development, and reproduction processes of a body. The impairment in any of the function of the endocrine system can lead to disease or even death of the respective organism. Many emerging pollutants have been detected and monitored from past few decades because of their universal consumption from different sources. Across all these pollutants, the chemicals which interfere the endocrine hormone functioning are called EDs. This chapter has provided the evidence regarding to EDs, their KCs, chemical nature and basic mechanisms of action. EDs have received global attention from past few decades due to their estrogen effect, persistence, toxicity, and bioaccumulation. The alteration of the endocrine functioning by EDs causes alteration in physiology, biochemistry, histopathology, reproduction, growth, and development of cells, tissue, or organs of humans and other organisms. Thus, it's necessary to remove these toxic substances from the environment. Further research is required to improve knowledge about EDs and their metabolites, mechanism, characteristics, identification, and their possible influence. The Endocrine system of an animal/human play a vital role in physiological, growth, and reproduction, and biochemical aspects of animals any alteration in the functioning of this system causes disease or death of the organism. EDs are classified as the chemical which have the ability to interfere with the hormone signaling pathway of the endocrine system. EDs have received global attention over the past few decades due to their level of toxicity to humans and other animals. The group of EDs are heterogeneous which includes different industrial lubricants, solvents, plastics, pesticides, drugs, and a few natural chemicals. The characteristics of EDs have been classified into 10 KCs on the basis of the action mechanism of EDs on endocrine hormones. EDs causes modification in the signal receptor(s) by following two different pathways, i.e., (a) EDs can mimic hormone binding and thus form a complex hormone receptor; and (b) EDs have the ability to block the specific protein that have the ability to control the homeostasis of a cell.

KEYWORDS

- **diethylstilbestrol**
- **endocrine disruptors**
- **growth hormone**
- **key characteristics**
- **parathyroid hormone**
- **reactive oxygen species**

REFERENCES

Abbassy, M. S., Ibrahim, H. Z., & Abu El-Amayem, M. M., (1999). Occurrence of pesticides and polychlorinated biphenyls in water of the Nile River at the estuaries of Rosetta and Damietta branches, North of Delta, Egypt. *J. Environ. Sci. Health Part B-Pesticides Food Contam. Agric. Wastes, 34*, 255–267. doi: 10.1080/03601239909373196.

Altenburger, R., Scholz, S., Schmitt-Jansen, M., Busch, W., & Escher, B. I., (2012). Mixture toxicity revisited from a toxicogenomic perspective. *Environ. Sci. Technol., 46*, 2508–2522.

Andersson, N., et al., (2018). Guidance for the identification of endocrine disruptors in the context of regulations (EU) No 528/2012 and (EC) No 1107/2009. *EFSA J., 16*, e05311.

Caliman, F. A., & Gavrilescu, M., (2009). Pharmaceuticals, personal care products and endocrine disrupting agents in the environment: A review. *Clean Soil Air Water, 37*, 277–303.

Christensen, A. M., Ingerslev, F., & Baun, A., (2006). Ecotoxicity of mixtures of antibiotics used in aquacultures. *Environ. Toxicol. Chem., 25*, 2208–2215.

De Coster, S., & Van, L. N., (2012). Endocrine-disrupting chemicals, associated disorders and mechanisms of action. *J. Environ. Public Health,* 713696. doi: 10.1155/2012/713696.

Environmental Health Hazard Assessment, (2011). *Green Chemistry Hazard Traits, Section 69403.4 Endocrine Toxicity*. California Code of Regulations, Division 4.5, Title 22, Chapter 54. OEHHA. https://oehha.ca.gov/media/downloads/risk-assessment/gcisor121710.pdf (accessed on 30 August 2022).

Fingler, S., Drevenkar, V., Tkalcevic, B., & Smit, Z., (1992). Levels of polychlorinated-biphenyls organochlorine pesticides, and chlorophenols in the Kupa River water and in drinking waters from different areas of Croatia. *Bullet. Environ. Contamin. Toxicol., 49*(6), 805–812. doi: 10.1007/bf00203151.

Hamid, N., Junaid, M., & De-Sheng, P., (2021). Combined toxicity of endocrine-disrupting chemicals: A review. *Ecotoxicol. Environ. Safety., 215*, 112136. doi.org/10.1016/j.ecoenv.2021.112136.

International Program on Chemical Safety, (2002). *Global Assessment of the State of the Science of Endocrine Disruptors: An Assessment Prepared by an Expert Group on Behalf of the World Health Organization*, Nations Environment Program, WHO/PCS/ EDC/02.2 (World Health Organization, 2002).

Japan Ministry of Health, (2015). *The Endocrine Disruptor Page.* NIHS http://www.nihs.go.jp/edc/english/actions/index.htm.

Kabir, E. R., Rahman, M. S., & Rahman, I., (2015). A review on endocrine disruptors and their possible impacts on human health. *Environ. Toxicol. Pharmacol., 40*, 241–258.

Kleine, B., & Rossmanith, W. G., (2016). *Hormones and the Endocrine System, Textbook of Endocrinology*. Springer International Publishing, Heidelberg, Germany.

Manikkam, M., Tracey, R., Guerrero-Bosagna, C., & Skinner, M. K., (2013). Plastics derived endocrine disruptors (BPA, DEHP and DBP) induce epigenetic transgenerational inheritance of obesity, reproductive disease and sperm epimutations. *PLoS One, 8*, e55387.

Marty, M. S., Borgert, C., Coady, K., Green, R., Levine, S. L., Mihaich, E., Ortego, L., et al., (2020). Distinguishing between endocrine disruption and non-specific effects on endocrine systems. *Regul. Toxicol. Pharmacol., 99*, 142–158.

Michele, A. La-Merrill, et al., (2020). Consensus on the key characteristics of endocrine-disrupting chemicals as a basis for hazard identification. *Nat. Rev. Endocrinol., 16*, 45–57.

OECD, (2018). *Revised Guidance Document 150 on Standardized Test Guidelines for Evaluating Chemicals for Endocrine Disruption.* OECD Series on testing and assessment https://doi.org/10.1787/ 9789264304741-en (accessed on 30 August 2022).

Pal, A., Gin, K. Y. H., Lin, K. Y. C., & Reinhard, M., (2010). Impacts of emerging organic contaminants on freshwater resources, review of recent occurrences, sources, fate and effects. *Sci. Total Environ., 408*, 6062–6069.

Park, C., Song, H., Choi, J., Sim, S., Kojima, H., Park, J., Iida, M., & Lee, Y., (2020). The mixture effects of bisphenol derivatives on estrogen receptor and androgen receptor. *Environ. Pollut., 260*, 114036.

Smith, O. W., (1948). Diethylstilbestrol in the prevention and treatment of complications of pregnancy, *Am. J. Obst. Gynecol., 56*, 821–826.

UNEP & WHO, (2013). *United Nations Environment Program and the World Health Organization, 2013, State of the Science of Endocrine Disrupting Chemicals-2012.* Available at: https://apps.who.int/iris/bitstream/handle/10665/78102/WHO_HSE_PHE_IHE_2013.1_eng.pdf (accessed on 26 October 2022).

US Environmental Protection Agency, (2014). *Framework for Human Health Risk Assessment to Inform Decision Making.* US EPA https://www.epa.gov/risk/risk-assessment-guidance (accessed on 26 October 2022).

Vandenberg, L. N., Colborn, T., Hayes, T. B., Heindel, J. J., Jacobs, D. R., Lee, D. H., Myers, J. P., et al., (2013). Regulatory decisions on endocrine disrupting chemicals should be based on the principles of endocrinology. *Reprod. Toxicol., 38*, 1–15. doi: 10.1016/j.reprotox.2013.02.002.

Welshons, W. V., Thayer, K. A., Judy, B. M., Taylor, J. A., Curran, E. M., & VomSaal, F. S., (2003). Large effects from small exposures. Mechanisms for endocrine—disrupting chemicals with estrogenic activity. *Environ. Health Perspect., 111*(8), 994–1006. doi: 10.1289/ehp.5494.

Yang, O., Kim, H. L., Weon, J. I., & Seo, Y. R., (2015). Endocrine-disrupting chemicals, review of toxicological mechanisms using molecular pathway analysis. *J. Cancer Prev., 20*, 12–24.

CHAPTER 2

Sources of Endocrine Toxicants with Their Biological Implications

AISHWARYA SHARMA and SHAILJA KUMARI

Department of Biosciences, Division of Zoology, Career Point University, Hamirpur–176041, Himachal Pradesh, India,
E-mail: shailjakhajuria@yahoo.in (S. Kumari)

ABSTRACT

An endocrine toxicant is "an exogenous agent that interferes with synthesis, secretion, transport, metabolism, binding action, or elimination of natural bloodborne hormones that are present in the body and are responsible for homeostasis, reproduction, and developmental process." People are exposed to these harmful substances through contaminated water, air, food, skin care products, pesticides, insecticides, fungicides, industrial chemicals, etc. Accumulation of endocrine toxicants (ETs) in humans increases the risk of infertility, diabetes, cancer, cardiovascular diseases, thyroid diseases, kidney disorders, neurological abnormalities, obesity, and other health issues. Exposure to these toxicants at early life interferes with developmental programming, thus increasing the risks of developing different diseases later in adult life. ETs are of great concern these days because exposure to ETs had caused more than 13 million deaths every year, as per data of World Health Organization (WHO). Various guidelines have been laid down to minimize the exposure to ETs by Endocrine Society, WHO, United Nations Environment Program, and from experts in this field. Basel, Rotterdam, and Stockholm Conventions have given hope that all countries will work together to combat the exposure to these harmful chemicals.

Environmental Endocrine Toxicants: Biology, Effects, and Management. Younis Ahmad Hajam, Rouf Ahmad Bhat, Khalid Rehman Hakeem, & Rajesh Kumar (Eds.)
© 2023 Apple Academic Press, Inc. Co-published with CRC Press (Taylor & Francis)

2.1 INTRODUCTION

Endocrine toxicants (ETs) are of great concern these days. ETs are substances which alter the pathways of biosynthesis, metabolism, or action mechanisms of hormones (Diamanti-Kandarakis et al., 2009). The Environmental Protection Agency (EPA) defined endocrine toxicant to be "an exogenous agent that interferes with synthesis, secretion, transport, metabolism, binding action, or elimination of natural bloodborne hormones that are present in the body and are responsible for homeostasis, reproduction, and developmental process" (Diamanti-Kandarakis et al., 2009). Disruption caused by ETs result in health complications like infertility, diabetes, cancer, etc. (Diamanti-Kandarakis et al., 2009). It was estimated by World Health Organization (WHO) that about 13 million deaths occurred because of exposures in the environment every year. Environmental causes have led to about one third mortality as well as morbidity cases affecting various underdeveloped and developing countries (WHO, 2006). Normal growth and development are affected by the exposures to ETs. Some popular environmental chemicals are described as EDCs or ETs, and few of these affect various functions related to reproduction because of estrogenic as well as anti-androgenic properties exhibited by them. Endocrine disruptors (EDs) are identified group of molecules which are known to be highly heterogeneous, including synthetic chemicals which are employed as solvents or lubricants used at industries along with their byproducts. Phytoestrogens like genistein and coumestrol are the natural chemicals present in animal and human food acting as EDs. Moreover, these substances show less binding affinity towards ERs and are consumed widely. They also act as the constituents of infant formula (Figure 2.1) (Kuiper et al., 1998; Dickerson and Gore, 2007).

2.2 VARIOUS ENDOCRINE DISRUPTING CHEMICAL (EDCS) SOURCES PRESENT IN THE ENVIRONMENT

There are various sources that lead to the exposure of ETs varying widely around the world. Decades ago, some EDCs were declared to be banned and with regard to this migrating people provide a study model describing the exposure based on contamination of real as well as new milieus. Several historical examples of spills and PCB contamination, including dioxins, show a causal and direct relation among chemical as well as endocrine manifestation or dysfunction in fauna, including animals and human beings. Although, such type of exposure does not represent persistent exposure of indoor as well outdoor contaminants and chemicals. Contamination caused due to a vast range of chemicals used in industries leads to the characterization of industrial areas. These chemicals leach and percolate into the soil and affect groundwater. Such harmful substances may enter into the food chains and get accumulated at larger levels like humans, eagles, and other animals present at higher trophic levels. Contaminated water, air, food, or soil leads to such exposures, and people involved in work related to pesticides, insecticides, fungicides, or industrial chemicals are mainly at huge risk of the exposure, which gives rise to abnormality of reproductive or endocrine glands. EDCs are present in plastics (PVC), epoxy resins, medical devices such as surgical implants, sealants, and pharmaceuticals especially the pill coatings, supplements, and other products such as nail polish, lotions, soaps, hair sprays, etc., toys like clay, waxes, or paints. It is also used in printing inks, construction materials, lubricating oils, textiles, electronic components and in food packaging (Bradley et al., 2013; Mikołajewska et al., 2015; Erickson and Kaley, 2011).

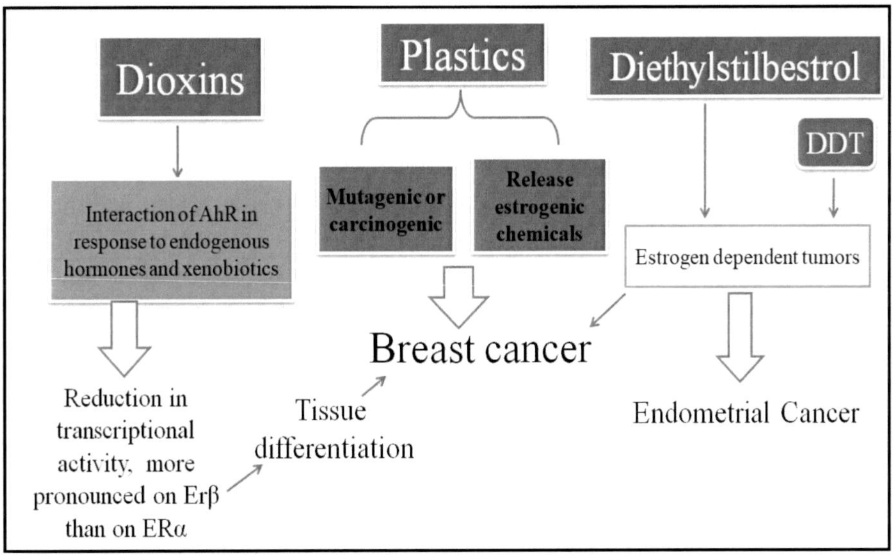

FIGURE 2.1 Different endocrine-disrupting chemicals and their harmful effects.

2.3 CLASSIFICATION OF ENDOCRINE TOXICANTS (ETS)

ETs can be classified based on their sources and modes of action as per toxicological view.

2.3.1 CLASSIFICATION OF EDCS BASED ON THEIR SOURCES

EDCs are classified into two types based on their sources which are as follows (Tavares et al., 2016; Zawatski and Lee, 2013):

1. **Natural:** Examples are genistein, phytoestrogen, and coumestrol.
2. **Synthetic:** Examples phthalates, PCBs (polychlorinated biphenyls), DDT (dichlorodiphenyltrichloroethane), PBBs (polybrominated biphenyls), DES (diethylstilbestrol), vinclozolin, and dioxins.

2.4 CLASSIFICATION OF EDCs BASED ON THEIR MODE OF ACTION

EDCs are classified into five types based on their mode of action, which are as follows (Whaley et al., 2001; Lee et al., 2013):

- EDCs affecting reproductive system;
- EDCs affecting pancreas;
- EDCs affecting thyroid;
- EDCs affecting central nervous system; and
- EDCs affecting other systems.

There are various significant effects of EDCs on hormones and some of them are as follows. They may cause an increase or decrease in hormone production as these substances may mimic or cause overstimulation of naturally occurring hormones like thyroid, androgen, and estrogen. They also affect transport of hormones as well as metabolism. They may also lead to the agonistic and antagonistic effects influencing hormone target cells while binding to the receptor or blocking the functioning of endogenous hormones, i.e., by behaving like an anti-estrogen or anti-androgen. They can cause miscommunication between physiology and environment of organisms as they result in alteration homeostatic systems.

2.5 SYNTHETIC ENDOCRINE TOXICANTS (ETs), ALONG WITH THEIR USES AND EXPOSURE

A Greek word 'plasticos' give rise to word plastic which is a material has an ability of getting molded into various shapes. First plasticizer was synthesized in 1860. With an increase in the global population along with progression in technology, plastics are widely in use as they have various applications in different fields (Benjamin et al., 2015).

2.5.1 PHTHALATES

The alkyl aryl and dialkyl esters of phthalic acid, which are employed to bring an increased flexibility in the plastic materials are known as Phthalates. They are synthesized by the reaction occurring between phthalic anhydride and alcohol as straight chain or branching [(C1/C2 (Methanol) up to C13 (tridecyl alcohol)] and are mainly used to soften rigid plastics as well as polymers. The most abundant derivative of phthalate di(2-ethylhex- yl) phthalate (DEHP) is used as 1% to 40% by weight in plastics and about 90% of DEHP is employed in softening PVC (Erkekoglu and Kocer-Gumusel, 2016). In 1930s, Phthalates were synthesized for the first time. In Japan, DEHP was first synthesized in 1933 whereas at United States it was synthesized in 1939 at a commercial scale. In 1949, DEHP was first used at US and it has known to be the most versatile used phthalate derivative of 20th century. The phthalates consumption was reported to be 3.25 million tons whereas consumption of DEHP was reported to be 2.1 million tons in year 1999 and it was estimated by the European Commission that about 1 million tons of DEHP was used in the year 2000 (Towae et al., 1992). Items constituting of PVC are known to get migrated into the food, air, soil, and water by Phthalates and it thus resulted into human exposure (Waring and Harris, 2005). Human blood as well as urine was used as a biological fluid in evaluating the phthalates exposure with regard to various studies conducted so far. It was revealed in a study that humans have a widespread or ubiquitous exposure towards phthalates, mainly in case of industrialized countries (Durmaz et al., 2010; Frederiksen et al., 2014; Serrano et al., 2014; Johns et al., 2015). Inhalation at the workplace is of much concern because phthalates like DEHP mainly have very less vapor pressure. Among the general population, diet as well as eating habits are the essential sources of exposure caused due to phthalates. Oils and fats may lead to high phthalate exposures in case they are contaminated (Meek and Chan, 1994; Wormuth et al., 2006).

Blood storage bags and blood transfusion equipment employed at the time of blood transfusions and hemodialysis involving use of dialysis bags contribute as the major sources of medical exposure (Labow et al., 1986; Shintani, 1985; Mettang et al., 1999; Nässberger et al., 1987). Phthalates are classified into two groups based on the presence of carbon atoms at their respective alcoholic chains. They exhibit various applications, and toxicological properties along with the classification of phthalates having higher molecular weight as well as phthalates having low molecular weight. Use of such phthalates mostly depends upon their molecular weight. Phthalates having higher molecular weights are employed in the construction materials along with various PVC items like toys, food, clothes, and biomedical equipment whereas phthalates having lower molecular weight are used in cosmetics, color fixatives, textiles, and pharmaceuticals (Heudorf et al., 2007).

2.5.1.1 BIOLOGICAL IMPLICATIONS OF PHTHALATES

2.5.1.1.1 Epigenetic Effect

Many studies revealed that exposure due to EDC would be caused due to mechanisms like epigenetic factors that may further lead to the harmful and cumulative effects influencing coming generations in the future. Epigenetic marks are known to introduce induced alterations in gene expression, which may further persist throughout the life, causing permanent changes leading to adverse effects on health like immune disorders, infertility, diabetes, and cancers. Phthalates are regarded as the epigenetic carcinogens due to their peroxisome proliferators effects. Changes caused due to DNA methylation at CpG site closer to the promoter region of the gene, along with various modifications in histone proteins, along with alteration in the non-coding expression of RNA, including microRNA result into the *in utero* as well as neonatal exposure towards the phthalates like DEHP, MEHP, BBP, DBP, and MBP (Chan and Mu, 2005).

Demethylation occurs when MCF-7 or human breast cancer cells are treated with BBP and as a result ERα associated with promoter along with CpG, indicates that alteration in ERα mRNA expression by the BBP can cause induction in methylation of DNA at the specific promoter regions of the gene. DNA methylation along with various DNA methyltransferase expressions in the testis of mice were induced as a consequence of maternal exposure to DEHP and fetal testis is regarded as a major target with respect to DEHP. DEHP may also result in a syndrome known as testicular dysgenesis, which is caused because of reduction in expression of the hormone (INSL3) similar to insulin along with the production of testosterone (Wu et al., 2010).

A transient exposure towards the mixture containing plastic in case of gestating female rat is trans-generationally known to promote an early onset of puberty while decreasing the size of pool in ovarian follicles in case of the female pups especially at the time of embryonic sex determination. While in case of male pups, sperm promoter regions were also observed and transgenerationally the spermatogenic cell apoptosis was also known to be affected along with DNA methylation of F3 generation (Manikkam et al., 2012).

2.5.1.1.2 Carcinogenicity of Phthalates

Gene and protein expressions can be altered by Phthalates which are also called as peroxisome proliferators, and such ability would also lead to the promotion of hepatic carcinogenesis in the case of rodents (Huber et al., 1996). Literature studies show that phthalates are responsible for an increase in oxidative stress mainly in the liver of rodents before the induction of proximal oxidases. After the treatment with DEHP, Kupffer cells are considered as the potential sources of the oxidants present in the liver of the rodents (Rusyn et al., 2001, 2006). An increase in oxidative stress is a consequence of molecular events which may show interaction with other pathways which are known to be activated by the proliferation of peroxisomes inside the rodent's liver (Rusyn et al., 2001, 2006). DEHP may induce ROS production and this may lead to an increase in oxidative stress at the cellular level *in vivo* and *in vitro* conditions (Rusyn et al., 2006). In 1980, several studies showed hepatocarcinogenic effects caused by DEHP occur as a consequence of its peroxisome proliferation effect. DEHP was classified as group 3 carcinogen in the year 2000 by the IARC, and proliferation of peroxisome has still not been marked in hepatocyte cultures of humans, which are kept much exposed towards DEHP (International Agency for Research on Cancer, 2000). Thus, the mechanism by which DEHP leads to an increase in hepatocellular tumors in case of rats and mice, which might not be applicable in case of humans.

The exposure towards various phthalate esters, mainly towards DBP which may result into a high proliferation in the breast cancer cell lines of humans (Blom et al., 1998), which is explained by the phthalate potency as an xenoestrogenic impact, however phthalates are generally distinguished as anti-androgens instead of xenoestrogens (Fisher, 2004).

2.5.1.1.3 Reproductive Toxicity Caused by Phthalates

Phthalate generally affect the male reproductive system, and the reproductive toxicity potential is being highlighted by the various studies which are carried out recently on phthalates *in vivo* and *in vitro*, respectively. It was proposed in a TDS hypothesis that the extent of the male reproductive disorders like hypospadias, cryptorchidism, infertility, and testicular cancer occur due to prevalence of TDS symptoms which may be due to result of interrupted development of the embryo. Phthalates are known to be the most underlying causes in relation to decline in the male fertility along with an increase in TDS cases since last decade. Fetal exposures towards DEHP and DBP cause reduction in anogenital distance (AGD) in case of the rodents (Sharpe and Skakkebaek, 2008). Leydig cells as well as Sertoli cells are known to be the main targets with response to the phthalates in testis and many researchers say that the interruption in the functioning of these cells is caused by the DEHP. A single dose of MEHP administration (400 mg/kg body weight) into Wistar rats is highly toxic to Sertoli cells as it may lead to detachment of germ cell (Dalgaard et al., 2000). After the single dose (2,000 mg/kg) administration of MEHP, Histopathological interruptions along with cytoplasmic distributions were reported in vimentin present in the Sertoli cells of fisher rats who are 28 days old. TUNEL positive cells and vimentin disruption cells have a correlation among them when studied in the case of treated mice (Tay et al., 2007). DEHP exposure was also determined to cause disruption along with collapse in the filaments of vimentin along with induced apoptotic death in the case of the germ cells (Erkekoglu et al., 2012). A decrease in the production of testosterone in the case of the rodents was observed along with certain reprotoxic effects related to their anti-androgenic potential. This occurred as a result of phthalate exposure, especially DEHP (Jones et al., 1993; Erkekoglu et al., 2011). DEHP causes abnormal sperm production along with a decrease in the count of sperm production and motility on its administration into the 10-week-old rats with 1,000 ppm dosage for consecutive 10 days. Apart from this DEHP also causes a decrease in the levels of testosterone, LH, and FSH (Erkekoglu et al., 2011). Oxidative stress was also induced in testis of rat by DEHP followed by a decrease in the GSH and redox ratio of GSSG which marks an increase in the peroxidation of lipids along with a decrease in GPx4 activity (Erkekoglu et al., 2011).

2.5.1.1.4 Hepatotoxic Effects of Phthalates

In the case of mice and rats, DEHP, along with other types of phthalates like DEHA and DINP. Such substances being hepatocarcinogenic are also known to spread carcinomas and adenomas in relation to the hepatic cells (Astill et al., 1996; Kaufmann et al., 2002). There are various molecular events that underline the potential of such hepatocarcinogenic substances. The property of proliferation of peroxisomes, genotoxicity along with consequences of epigenetic effects reflect the mechanisms which are mostly studied (Erkekoglu et al., 2014).

2.5.1.1.5 Bisphenol-A (BPA)

Bisphenol-A (BPA) can be employed for hardening plastics and for manufacturing of polycarbonate plastics along with epoxy resins. It is known to have a great abundance in the environment and is thus regarded as either EDC or Xenoestrogen (Vandenberg et al., 2007; Flint et al., 2012). Earlier, BPA were classified to be weak environmental estrogens but according to the new studies BPA exerts divergent effect on the cell physiology as well as tissues at very low concentrations (Iso et al., 2006). They are capable of binding to the nuclear as well as nonnuclear receptors (Wetherill et al., 2007). BPA metabolites are mostly known to have stronger estrogenic activities as compared with the main compound (Alonso-Magdalena et al., 2012). BPA binds with ARs and this can be related to prostate stimulation known to cause cancer. At the certain concentrations, BPA causes modulation of androgen-dependent cell proliferation of prostate cancer as determined with regard to human blood (Wetherill et al., 2005). Coating of BPA is present at the interior surface of food containers and beverage cans which has further importance for preventing corrosion and food contamination. BPA is present in the plastic bottles, and thus milk ingestion makes infants much exposed to it (Hoekstra and Simoneau, 2013).

2.5.1.2 BIOLOGICAL IMPLICATIONS DUE TO BISPHENOL-A (BPA)

2.5.1.2.1 Genotoxic Effects (GE) Occurring Because of Bisphenol-A (BPA)

No mutagenic effect was seen because of BPA in Ames test instead of metabolic activation (Schweikl et al., 1998). BPA may cause DNA damage in the eukaryotic cells which results in breakage of DNA strand into L5178Y lymphoma cells in mouse (Naik and Vijayalaxmi, 2009) whereas induced aneuploidy along with chromosomal aberrations present in the structures are known to occur in ER+ MCF-7 as well as CHO-K1 cells (Nakagawa and Suzuki, 2001; Shimizu et al., 2002; Tayama et al., 2008). DT40 cell lines are present in the mutant chicken, which are deficient in pathways of DNA repair and were studied by the researchers found chromosomal aberrations along with double-strand breaks (Lee et al., 2013). ROS generation is induced by the BPA which as a result cause induction in the DNA oxidative base production (Kovacic, 2010). No genotoxicity was observed on administered of BPA into the cell lines of humans namely LS174T, HepG2, ACHN (Audebert et al., 2011). Metabolites like BPA quinone are known to induce damage towards DNA through the formation of adducts having covalent nature along with deoxyguanosine (Lacroix et al., 2011; Kolšek et al., 2011). Alteration in gene and protein expression was also marked as an effect of BPA. According to a recent report, BPA showed an increase in the gene expression (BRACA1, BRCA2, BRCC3, and BCL2L11). It includes the genes which are involved in the repair of DNA and apoptosis in epithelial cells present in the human breast. BRCA1 or BRCA2 mutations are present in the women, and they are much more susceptible to the adverse effects which are caused by BPA (Fernandez et al., 2012).

2.5.1.2.2 Epigenetic Effects Caused Due to Bisphenol-A (BPA)

Low doses of BPA as utero or neonatal exposure may result in alterations caused in methylation of DNA, modifications in histone along with the modifications and changes in noncoding RNA expression patterns. Such changes may either up-regulate or down-regulate the various gene expressions causing permanent health effects like neural as well as immune disorders, infertility issues and various other diseases. Permanently occurring alterations in patterns of DNA methylation in different types of genes might be responsible for the cellular signaling process (Ahbab et al., 2014; Ferreira et al., 2015). Both folic acid and phytoestrogen genistein are sources of methyl group and are involved in preventing maternal supplementation (Dolinoy et al., 2007). In both animals and humans, BPA is known to induce prostate as well as breast cancer. The risk of cancer may be increased by BPA which affects various cell processes like methylation of DNA and remodeling of chromatin during the development stage (Keri et al., 2007). Prostate tissues show consistent methylation changes and neonatal exposure by rats towards the BPA result to high incidence of intraepithelial neoplasia with regard to prostate issue. Type 4 variant 4 (Pde4d4) of phosphodiester asegeneis is known to be hypomethylated along with regulatory CpG, which initially has an elevation in expression present in case of the adult rats (Ho et al., 2006; Prins et al., 2008). Neonatal BPA exposure causes alteration in the process of methylation as well as in nucleosome binding protein-1 (Nsbp1) expression along with genes which are similar to hippocalcin-1 (Hpcal1) present in case of the rodents like rat (Tang et al., 2012).

Various neonatal exposures towards BPA were known to induce the hypermethylation process in ER promoter areas present in the testis of rats. Various changes influencing epigenetic aspects were mediated by this phenomenon which as a result induces many harmful effects affecting the process of spermatogenesis as well as fertility (Doshi et al., 2011). miRNAs and BPA exposures in case of human placental cell lines are known for causing alteration in levels of miRNA expression. Mainly, miR-146a, which is strongly induced after BPA application. Slow cellular proliferation rates along with high sensitivity towards bleomycin leads to DNA induction and breaks (Avissar-Whiting et al., 2010; Cho et al., 2010).

2.5.1.2.3 Effect of Bisphenol-A (BPA) on Carcinogenicity

Evaluation of IARC for BPA is absent so far. Although, "bisphenol-A (BPA) diglycidyl ether" has been distributed under Group III carcinogens (IARC, 1999). BPA is shown to cause alteration in the development of mammary glands and an increase in tumor incidences at low doses in case of Sprague-Dawley rats (Soto et al., 2013). Instead, the comparable quantities of reference dosage, BPA is known to cause breast development along with chances of prostate cancers in the case of mice as well as rats (Keri et al., 2007). Recently potential of BPA in relation to malignant tumors was evaluated and then it was reported that the BPA act as carcinogen of mammary gland at different dosage which is detected in the urine or blood as demonstrated through epidemiological related studies. It is also known to cause hematopoietic cancers as well as testicular cancer in case of rodents (Huff, 2001). BPA promotes proliferation of OVCAR-3 cells, causing increase in the susceptibility with regard to cells of endometrium which affect estradiol and also causes inhibition of the induced genomic activity in 17-β-estradiol of ERβ inside DLD-1 colon cells (Ptak et al., 2011; Bolli et al., 2010; Cobellis et al., 2009).

2.5.1.2.4 Effect of Bisphenol-A (BPA) on the Reproductive Toxicity

BPA may lead to morphological as well as the alterations affecting functioning in case of both male as well as female genital tracts. Prenatal exposure to BPA dose can predispose the tissue resulting in earlier onset of disease, decreased fertility and cases of prostate cancers.

Many rodent studies have thrown light on the estrogenicity of BPA. Pregnant mice on the exposure to the natural BPA dosage, then opening of vagina along with the estrous cycle process were seen in the offspring. Prostate weight in mice is elevated by BPA and an advanced aging was also noted with reference to the reproductive processes (Markey et al., 2003). Development of the reproductive organs, secretion of testosterone along with production of sperms in mice is shown to be disrupted by the BPA (Richter et al., 2007) whereas some studies reveal that BPA is not responsible for the alterations in estrogenic function in case of rodents (Ryan et al., 2010).

2.5.1.2.5 Hepatotoxic Effects of Bisphenol-A (BPA)

DNA can be binding may occur due to the BPA metabolite bisphenol-o-quinone resulting into the process of hepatotoxicity induction (Atkinson and Roy, 1999). BPA is also known to cause damage to hepatocytes by the process of oxidative stress, and it resulted in a decrease in antioxidant enzymes along with glutathione (GSH) S-transferase activities reducing levels of GSH in rats. BPA (50 mg/kg) increase the biochemical levels of liver enzymes and causes a reduction in hepatic antioxidant gene expression. ROS generated by BPA reduce the antioxidant gene expression resulting in hepatotoxicity. It is also observed by that the low dosage of BPA leads to damage affecting hepatic cells which may happen because of mitochondrial dysfunction including production of ROS, hyperpolarization of mitochondrial trans-membrane, pro-inflammatory cytokine release and lipid peroxidation (Huc et al., 2012).

BPA exposure causes oxidative as well as pro-inflammatory damage in hepatocytes of rats. A decline in the GSH peroxidase activity was observed along with a decrease in peroxidation of lipids and pro-inflammatory cytokines, especially interleukin (IL)-6, along with tumor necrosis of factor-α. It was also observed at very minute doses. BPA causes a decrease in the rate of oxygen as well as in the production of ATP resulting in mitochondrial dysfunction (Moon et al., 2012).

2.6 POLYCHLORINATED BIPHENYLS (PCBS)

The synthetic as well as chlorinated compounds of hydrocarbon constituting of two rings of benzene linking with the single bond of carbon. Since 1929, PCBs are known to have been developed on a commercial scale with their applications in the manufacturing and production of inks, plasticizers, and paints. PCB congeners which are known to be highly chlorinated sometimes get adsorbed into the soil and sediments. Although, many congeners present inside the soil or sediment exhibit half-lives which may be extended from a few months to several years. Extent of congener chlorination including organic carbon (OC) and clay contents present in soil leads to increase in the PCBs adsorption. Biodegradation and volatilization are slow processes for the removal of PCB from water as well as soil. PCBs get accumulated inside the food chains and then they get rapidly absorbed by the gastrointestinal tract (GI) leading to their distribution and accumulation inside the liver as well as in the adipose tissue. They further pass-through placenta and are then secreted in milk, ultimately causing accumulation in case of the infants. Hydroxylation as well as subsequent conjugation cause metabolization of PCBs and metabolism rate vary with regard to the types of congeners.

2.6.1 EXPOSURE OF HUMANS

Inhalation of contaminated air and in gestation of contaminated food and water expose humans to PCBs. According to the report, estimated dietary uptake of PCBs is recorded to be 0.027 μg/kg body weight per day by adults at USA which was then decreased to 0.0005 μg/kg body weight per day during 1982–1984 and <0.001 μg/kg body weight per day between 1986 and 1991. Some neurological deficiencies present during early ages would get disappeared at the later stages of childhood. PCBs are perplexed by the exposures towards various halogenated contaminants present in the environment containing various adulterants. Chlorinated dibenzofurans (Concise International Chemical Assessment Document (CICAD)) deals with the contaminant toxicity which occurs due to the PCB heating.

2.6.2 BIOLOGICAL IMPLICATIONS

PCBs are carcinogens which may cause alteration in the functioning of the immune system leading to the adverse effect on the skin, thyroid, nervous system and sex steroid hormones, pancreas, kidney, liver, and cardiovascular system.

Humans exposed to PCBs are at major risk of getting infections, decreased cognitive function, and cancer along with adverse effects like hypothyroidism, infertility, diabetes, asthma, arthritis, heart disease, liver disease, hypertension, and other behavioral effects. Some effects like IQ deficits occur due to prenatal exposure and are irreversible, whereas other diseases like cancer can be treated if examined timely. Various PCB-associated diseases are as in subsections.

2.6.2.1 CANCER

PCBs can act as promoters for causing cancer, directly they are not carcinogens, but they may help in enhancing the effects caused due to various carcinogenic substances by ROS production and gene induction (Tharappel et al., 2002). Chromosomal aberrations are caused due to chronic exposures to PCBs (Silberhorn et al., 1990). A study shows that congeners similar to dioxin cause damaging of oxidative DNA (Oakley et al., 1996).

WHO classified Polychlorinated biphenyls (PCB) as 'probable human carcinogens, and dioxins are being entitled as 'known human carcinogens. Proof level of PCB carcinogenicity is missing in the humans and PCBs get accumulated along with many fat-soluble contaminants and that's why it cannot be stated that PCBs cause cancer while DDT and other sources like fat-soluble chemicals does not contribute to such diseases. The substance can be regarded as carcinogenic in the case of humans if it is carcinogenic in the case of animals having a maximum number of similar genes. Evaluation and consequences of PCB exposure with relation to human health, eliminating carcinogenic studies associated with animals do not have any merit as an occupational study as it is limited in efficiency because of various reasons like a smaller number of employees and thus less risk of incidence occurring in the small population. Many occupational studies have reported different types of cancers associated with PCB (Longnecker et al., 2003).

2.6.2.2 BRAIN CANCER

Brain cancer having an elevated incidence was reported among the people working at Indiana capacitor manufacturing plant (Sinks et al., 1992). A study was then published, and it was found that 1.91 (1.0–3.3). SMR along with desirable response towards dose ($p = 0.016$). It further concluded that the highest number of SMRs were found among the workers at the highest level of exposure towards PC (Ruder et al., 2005).

2.6.2.3 BREAST CANCER

A dose-dependent relation with total PCB exposure was reported based on different clinic trials based on the breast cancer-related studies. The largest and retrospective as well as epidemiological studies did not describe any such relation with regard to elevation of PCB levels (Moysich et al., 2002).

PCBs can cause potent induction of cytochrome enzyme CYP1A1, having several variant genotypes (Moysich et al., 1999).

2.6.2.4 RECURRENT INFECTIONS

Both antibody as well as immune response at cellular level and frequent occurring infections can occur because of PCB exposure. Immune suppression is considered as biological effect occurring at the lowest PCB concentration by the ATSDR, while the effects on neurobehavior even occur at low doses.

2.6.2.5 NEUROBEHAVIORAL EFFECTS

PCBs have been known to cause various harmful effects on the intellectual functions in case of the infants and small children. It was reported that cognitive children development occurred in cases where mothers were relied on dietary habits constituting of rice oil containing PCBs as well as dibenzofurans in countries like Taiwan in 1976 (Chen et al., 1985). For studying the neurological development for about the initial 48 hours of life span, the neonatal behavioral assessment scale was used. It was identified by the investigators that infants coming under the higher fish category show many abnormal reflexes and wide response towards the stress situation along with low habituation about repeated stimuli as compared to no fish as well as low fish babies. Thus, results are known to be consistent with general PCB effects affecting brain development prior to birth (Schantz et al., 2003).

2.6.2.6 HYPOTHYROIDISM

Structural similarity is present among the PCBs as well as thyroid hormones (THs). They have iodine instead of substitution of chlorine present on the two phenyl rings. Various animal-related studies reflected at multiple sites causes interference in PCBs with THs (Leemans et al., 2019). Thyroid functions in mother and infant were studied and it was marked that higher levels of organochlorine present in mother's milk had a correlation with low levels of T3 and T4 along with increased levels of thyroid-stimulating hormone (TSH). About 1.82 ppb is mean PCB level, and 4.75 ppb is the maximum level and statistically positive dosage dependent relationship among total PCBs as well as TSH levels along with negative relation with presence of free and total thyroxin (T4).

2.6.2.7 DISORDERS RELATED TO INFERTILITY AND REPRODUCTION

PCBs are known to be the inhibitors for male sex hormone synthesis (Troisi et al., 2020). Decreased levels of testosterone PCB-containing transformer fluids are seen in animals exposed to PCBs (Andric et al., 2000). PCBs also compete with testosterone for binding to testosterone receptor (Portigal et al., 2002). This causes a reduction in the masculinity, whereas in young humans, the serum levels are known to be inversely correlated with free levels of testosterone (Richthoff et al., 2003). Higher PCB levels are correlated with sperm mobility reduction (Richthoff et al., 2003; Bush et al., 1986).

2.6.2.8 CARDIOVASCULAR DISEASE AND SERUM LIPIDS

Hypertension as well as serum lipid levels account for risk factors leading to ischemic heart diseases (Wilsgaard et al., 2001). It was noted that blood levels containing PCBs are present among the people employed at sewage plant in various industries. They, along with their families were reported to have 17.4 and 75.1 ppb showing a dynamic relationship between triglycerides and concentrations of PCB serum. Hospitalization discharge and amount for diagnosis of coronary heart disease (CHD) as well as infection affecting myocardial tissues are some aspects that were reported in the residents of New York staying within a zip code containing waste site which is hazardous in nature constituting of various organic pollutants among which PCBs are known to be most popular and frequently persisting (Sergeev and Carpenter, 2005). The Hudson River and zip code subsets which are supposed to be 200 miles with reference to National Priority Site having contamination because of PCBs presence and are mainly present at areas with average income, which is quite high as compared to areas with low smoking rates. Residents are known to exercise rapidly leading to consumption of fruits and vegetables more often as compared to other parts of the state (Sergeev and Carpenter, 2005).

2.6.2.9 DIABETES

It is rarely accounted as an induced disease prevailing in an environment, with existing strong evidence that signifies the exposure of PCB in the environment leading to the chances of diabetes (Codru et al., 2007). It was concluded in a report that there is suggestive evidence for an association between exposure to dioxin and occurrence of the diabetes (Figure 2.2).

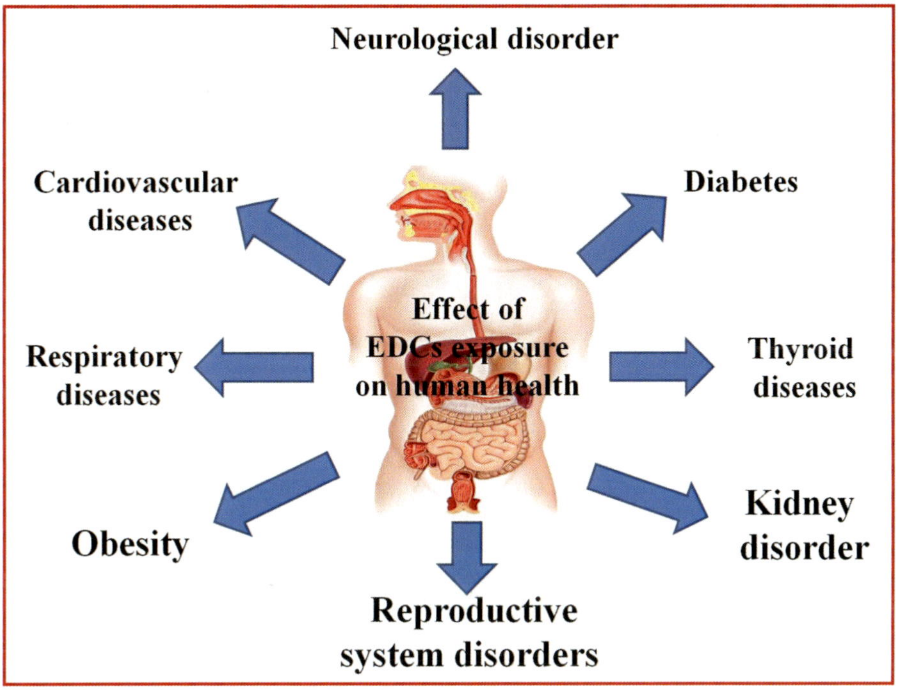

FIGURE 2.2 Effect of endocrine disrupting chemicals exposure on human body.

KEYWORDS

- **biological implication**
- **bisphenol-A**
- **endocrine disrupting chemical sources**
- **endocrine disruptors**
- **Environmental Protection Agency**
- **phthalates**
- **polychlorinated biphenyls**

REFERENCES

Ahbab, M. A., Ündeğer, Ü., Barlas, N. U. R. H. A. Y. A. T., & Başaran, N., (2014). In utero exposure to dicyclohexyl and di-n-hexyl phthalate possess genotoxic effects on testicular cells of male rats after birth in the comet and TUNEL assays. *Hum. Exp. Toxicol., 33*, 230–239.

Alonso-Magdalena, P., Ropero, A. B., Soriano, S., García-Arévalo, M., Ripoll, C., Fuentes, E., Quesada, I., & Nadal, Á., (2012). Bisphenol-A acts as a potent estrogen via non-classical estrogen triggered pathways. *Mol. Cell. Endocrinol., 355*, 201–207.

Andric, S. A., Kostic, T. S., Dragisic, S. M., Andric, N. L., Stojilkovic, S. S., & Kovacevic, R. Z., (2000). Acute effects of polychlorinated biphenyl-containing and-free transformer fluids on rat testicular steroidogenesis. *Environ. Health Perspect., 108*, 955–959.

Astill, B. D., Gingell, R., Guest, D., Hellwig, J., Hodgson, J. R., Kuettler, K., Mellert, W., et al., (1996). Oncogenicity testing of 2-ethyl hexanol in Fischer 344 rats and B6C3F1 mice. *Fundam. Appl. Toxicol., 31*, 29–41.

Atkinson, A., & Roy, D., (1995). *In vitro* conversion of environmental estrogenic chemical bisphenol A to DNA binding metabolite(s). *Biochem. Biophys. Res. Commun., 210*, 424–433.

Audebert, M., Dolo, L., Perdu, E., Cravedi, J. P., & Zalko, D., (2011). Use of the γH2AX assay for assessing the genotoxicity of bisphenol A and bisphenol F in human cell lines. *Arch. Toxicol., 85*, 1463–1473.

Avissar-Whiting, M., Veiga, K. R., Uhl, K. M., Maccani, M. A., Gagne, L. A., Moen, E. L., & Marsit, C. J., (2010). Bisphenol A exposure leads to specific microRNA alterations in placental cells. *Reprod. Toxicol., 29*, 401–406.

Benjamin, S., Pradeep, S., Josh, M. S., Kumar, S., & Masai, E., (2015). A monograph on the remediation of hazardous phthalates. *J. Hazard. Mater., 298*, 58–72.

Blom, A., Ekman, E., Johannisson, A., Norrgren, L., & Pesonen, M., (1998). Effects of xenoestrogenic environmental pollutants on the proliferation of a human breast cancer cell line (MCF-7). *Arch. Environ. Contam. Toxicol., 34*, 306–310.

Bolli, A., Bulzomi, P., Galluzzo, P., Acconcia, F., & Marino, M., (2010). Bisphenol A impairs estradiol-induced protective effects against DLD-1 colon cancer cell growth. *IUBMB, 62*, 684–687.

Bradley, E. L., Burden, R. A., Bentayeb, K., Driffield, M., Harmer, N., Mortimer, D. N., Speck, D. R., et al., (2013). Exposure to phthalic acid, phthalate diesters and phthalate monoesters from foodstuffs: UK total diet study results. *Food Addit. Contam. Part A., 30*, 735–742.

Bush, B., Bennett, A. H., & Snow, J. T., (1986). Polychlorobiphenyl congeners, p, p′-DDE, and sperm function in humans. *Arch. Environ. Contam. Toxicol., 15*, 333–341.

Chan, K. S., & Mu Lee, B., (2005). DNA methylation of estrogen receptor α gene by phthalates. *J. Toxicol. Environ. Health. Part A, 68*, 1995–2003.

Chen, P. H., Wong, C. K., Rappe, C., & Nygren, M., (1985). Polychlorinated biphenyls, dibenzofurans and quaterphenyls in toxic rice-bran oil and in the blood and tissues of patients with PCB poisoning (Yu-Cheng) in Taiwan. *Environ. Health Perspect., 59*, 59–65.

Cho, H., Kim, S. J., Park, H. W., Oh, M. J., Yu, S. Y., Lee, S. Y., Park, C., et al., (2010). A relationship between miRNA and gene expression in the mouse Sertoli cell line after exposure to bisphenol A. *BioChip J., 4*, 75–81.

Cobellis, L., Colacurci, N., Trabucco, E., Carpentiero, C., & Grumetto, L., (2009). Measurement of bisphenol A and bisphenol B levels in human blood sera from healthy and endometriotic women. *Biomed. Chromatogr., 23*, 1186–1190.

Codru, N., Schymura, M. J., Negoita, S., Rej, R., & Carpenter, D. O., (2007). Akwesasne Taskforce on the environment. Diabetes in relation to serum levels of polychlorinated biphenyls and chlorinated pesticides in adult Native Americans. *Environ. Health Perspect, 115*, 1442–1447.

Dalgaard, M., Østergaard, G., Lam, H. R., Hansen, E. V., & Ladefoged, O., (2000). Toxicity study of di (2-ethylhexyl) phthalate (DEHP) in combination with acetone in rats. *Pharmacol. Toxicol., 86*, 92–100.

Diamanti-Kandarakis, E., Bourguignon, J. P., Giudice, L. C., Hauser, R., Prins, G. S., Soto, A. M., Zoeller, R. T., & Gore, A. C., (2009). Endocrine-disrupting chemicals: An endocrine society scientific statement. *Endo. Rev., 30*, 293–342.

Dickerson, S. M., & Gore, A. C., (2007). Estrogenic environmental endocrine-disrupting chemical effects on reproductive neuroendocrine function and dysfunction across the life cycle. *Rev. Endocr. Metab. Disord., 8*, 143–159.

Dolinoy, D. C., Huang, D., & Jirtle, R. L., (2007). Maternal nutrient supplementation counteracts bisphenol A-induced DNA hypomethylation in early development. *Proc. Natl. Acad. Sci., 104*, 13056–13061.

Doshi, T., Mehta, S. S., Dighe, V., Balasinor, N., & Vanage, G., (2011). Hypermethylation of estrogen receptor promoter region in adult testis of rats exposed neonatally to bisphenol A. *Toxicol., 289*, 74–82.

Durmaz, E., Özmert, E. N., Erkekoğlu, P., Giray, B., Derman, O., Hıncal, F., & Yurdakök, K., (2010). Plasma phthalate levels in pubertal gynecomastia. *Pediat., 125*, 122–129.

Erickson, M. D., & Kaley, R. G., (2011). Applications of polychlorinated biphenyls. *Environ. Sci. Pollut. Res., 18,* 135–151.

Erkekoglu, P., & Kocer-Gumusel, B., (2016). Environmental effects of endocrine-disrupting chemicals: A special focus on phthalates and bisphenol A. *Environmental Health Risk-Hazardous Factors to Living Species* (pp. 155–190). London: Intech. Open.

Erkekoglu, P., Zeybek, N. D., Giray, B. K., Rachidi, W., Kızılgün, M., Hininger-Favier, I., Favier, A., et al., (2014). The effects of di (2-ethylhexyl) phthalate on rat liver in relation to selenium status. *Int. J. Experim. Pathol., 95*, 64–77.

Erkekoglu, P., Zeybek, N. D., Giray, B., Asan, E., & Hincal, F., (2012). The effects of di (2-ethylhexyl) phthalate exposure and selenium nutrition on Sertoli cell vimentin structure and germ-cell apoptosis in rat testis. *Arch. Environ. Contam. Toxicol., 62*, 539–547.

Erkekoglu, P., Zeybek, N. D., Giray, B., Asan, E., Arnaud, J., & Hincal, F., (2011). Reproductive toxicity of di (2-ethylhexyl) phthalate in selenium-supplemented and selenium-deficient rats. *Drug. Chem. Toxicol., 34*, 379–389.

Espandiari, P., Glauert, H. P., Lehmler, H. J., Lee, E. Y., Srinivasan, C., & Robertson, L. W., (2003). Polychlorinated biphenyls as initiators in liver carcinogenesis: Resistant hepatocyte model. *Toxicol. Appl. Pharmacol., 186*, 55–62.

Fernandez, S. V., Huang, Y., Snider, K. E., Zhou, Y., Pogash, T. J., & Russo, J., (2012). Expression and DNA methylation changes in human breast epithelial cells after bisphenol A exposure. *Int. J. Oncol., 41*, 369–377.

Ferreira, L. L., Couto, R., & Oliveira, P. J., (2015). Bisphenol A as epigenetic modulator: Setting the stage for carcinogenesis? *Eur. J. Clin. Investig., 45*, 32–36.

Fisher, J. S., (2004). Environmental anti-androgens and male reproductive health: Focus on phthalates and testicular dysgenesis syndrome. *Reprod., 127*, 305–315.

Flint, S., Markle, T., Thompson, S., & Wallace, E., (2012). Bisphenol A exposure, effects, and policy: A wildlife perspective. *J. Environ. Manage., 104*, 19–34.

Frederiksen, H., Jensen, T. K., Jørgensen, N., Kyhl, H. B., Husby, S., Skakkebæk, N. E., Main, K. M., et al., (2014). Human urinary excretion of non-persistent environmental chemicals: An overview of Danish data collected between 2006 and 2012. *Reprod., 147*, 555–565.

Heudorf, U., Mersch-Sundermann, V., & Angerer, J., (2007). Phthalates: Toxicology and exposure. *Int. J. Hyg. Environ. Health, 210*, 623–634.

Ho, S. M., Tang, W. Y., De Frausto, J. B., & Prins, G. S., (2006). Developmental exposure to estradiol and bisphenol A increases susceptibility to prostate carcinogenesis and epigenetically regulates phosphodiesterase type 4 variant 4. *Cancer Res., 66*, 5624–5632.

Hoekstra, E. J., & Simoneau, C., (2013). Release of bisphenol A from polycarbonate—a review. *Crit. Rev. Food Sci. Nutr., 53*, 386–402.

Huber, W. W., Grasl-Kraupp, B., & Schulte-Hermann, R., (1996). Hepatocarcinogenic potential of di (2-ethylhexyl) phthalate in rodents and its implications on human risk. *Crit. Rev. Toxicol., 26*, 365–481.

Huc, L., Lemarié, A., Guéraud, F., & Héliès-Toussaint, C., (2012). Low concentrations of bisphenol A induce lipid accumulation mediated by the production of reactive oxygen species in the mitochondria of HepG2 cells. *Toxicol. in Vitro, 26*, 709–717.

Huff, J., (2001). Carcinogenicity of bisphenol-A in Fischer rats and B6C3F1 mice. *Odontology., 89*, 0012–0020.

International Agency for Research on Cancer (IARC) Monographs, (2000). Di(2-ethylhexyl) phthalate. *IARC Monographs on the Evaluation of Carcinogenic Risks to Humans, 77*, 41–148.

International Agency for Research on Cancer (IARC), (1999). Agents classified by the IARC monographs (Vols. 1–112). Butyl benzyl phthalate (group 3). *Monograph, 101*(7), 73.

International Agency for Research on Cancer (IARC), (1999). *Monographs on the Evaluation of Carcinogenic Risks to Humans* (Vol. 71). Re-evaluation of some organic chemicals, hydrazine, and hydrogen peroxide, Lyon.

Iso, T., Watanabe, T., Iwamoto, T., Shimamoto, A., & Furuichi, Y., (2006). DNA damage caused by bisphenol A and estradiol through estrogenic activity. *Biol. Pharm. Bull., 29*, 206–210.

Johns, L. E., Ferguson, K. K., Soldin, O. P., Cantonwine, D. E., Rivera-González, L. O., Del, T. L. V. A., Calafat, A. M., et al., (2015). Urinary phthalate metabolites in relation to maternal serum thyroid and sex hormone levels during pregnancy: A longitudinal analysis. *Reprod. Bio. Endocrinol., 13*, 1–12.

Jones, H. B., Garside, D. A., Liu, R., & Roberts, J. C., (1993). The influence of phthalate esters on Leydig cell structure and function *in vitro* and *in vivo*. *Experim. Mol. Pathol., 58*, 179–193.

Kaufmann, W., Deckardt, K., McKee, R. H., Butala, J. H., & Bahnemann, R., (2002). Tumor induction in mouse liver: Di-isononyl phthalate acts via peroxisome proliferation. *Regul. Toxicol. Pharmacol., 36*, 175–183.

Keri, R. A., Ho, S. M., Hunt, P. A., Knudsen, K. E., Soto, A. M., & Prins, G. S., (2007). An evaluation of evidence for the carcinogenic activity of bisphenol A. *Reprod. Toxicol., 24*, 240–252.

Kolšek, K., Mavri, J., & Dolenc, M. S., (2012). Reactivity of bisphenol A-3, 4-quinone with DNA. A quantum chemical study. *Toxicol. in vitro, 26*, 102–106.

Kovacic, P., (2010). How safe is bisphenol A? Fundamentals of toxicity: Metabolism, electron transfer and oxidative stress. *Med. Hypotheses, 75*, 1–4.

Kuiper, G. G., Lemmen, J. G., Carlsson, B. O., Corton, J. C., Safe, S. H., Van, D. S. P. T., Van, D. B. B., & Gustafsson, J. A., (1998). Interaction of estrogenic chemicals and phytoestrogens with estrogen receptor β. *Endocrinol., 139*, 4252–4263.

Labow, R. S., Tocchi, M., & Rock, G., (1986). Contamination of platelet storage bags by phthalate esters. *J. Toxicol. Environ. Health. Part A Current Issues, 19*, 591–598.

Lacroix, M. Z., Puel, S., Collet, S. H., Corbel, T., Picard-Hagen, N., Toutain, P. L., Viguié, C., & Gayrard, V., (2011). Simultaneous quantification of bisphenol A and its glucuronide metabolite (BPA-G) in plasma and urine: Applicability to toxicokinetic investigations. *Talanta, 85*, 2053–2059.

Lee, H. R., Jeung, E. B., Cho, M. H., Kim, T. H., Leung, P. C., & Choi, K. C., (2013). Molecular mechanism(s) of endocrine-disrupting chemicals and their potent estrogenicity in diverse cells and tissues that express oestrogen receptors. *J. Cell. Mol. Med., 17*, 1–11.

Lee, S., Liu, X., Takeda, S., & Choi, K., (2013). Genotoxic potentials and related mechanisms of bisphenol A and other bisphenol compounds: A comparison study employing chicken DT40 cells. *Chemosphere., 93*, 434–440.

Leemans, M., Couderq, S., Demeneix, B., & Fini, J. B., (2019). Pesticides with potential thyroid hormone-disrupting effects: A review of recent data. *Front. Endocrinol., 10*, 743.

Longnecker, M. P., Wolff, M. S., Gladen, B. C., Brock, J. W., Grandjean, P., Jacobson, J. L., Korrick, S. A., et al., (2003). Comparison of polychlorinated biphenyl levels across studies of human neurodevelopment. *Environ. Health Perspect., 111*, 65–70.

Manikkam, M., Guerrero-Bosagna, C., Tracey, R., Haque, M. M., & Skinner, M. K., (2012). Transgenerational actions of environmental compounds on reproductive disease and identification of epigenetic biomarkers of ancestral exposures. *PloS One, 7*, 31901.

Markey, C. M., Coombs, M. A., Sonnenschein, C., & Soto, A. M., (2003). Mammalian development in a changing environment: Exposure to endocrine disruptors reveals the developmental plasticity of steroid-hormone target organs. *Evol. Dev., 5*, 67–75.

Meek, M. E., & Chan, P. K. L., (1994). Bis (2-ethylhexyl) phthalate: Evaluation of risks to health from environmental exposure in Canada. *J. Environ. Sci. Health Part C, 12*, 179–194.

Meeker, J. D., Singh, N. P., & Hauser, R., (2008). Serum concentrations of estradiol and free T4 are inversely correlated with sperm DNA damage in men from an infertility clinic. *J. Androl., 29*, 379–388.

Mettang, T., Alscher, D. M., Pauli-Magnus, C., Dunst, R., Kuhlmann, U., & Rettenmeier, A. W., (1999). Phthalic acid is the main metabolite of the plasticizer di (2-ethylhexyl) phthalate in peritoneal dialysis patients. *Adv. Perit. Dial., 15*, 229–233.

Mikołajewska, K., Stragierowicz, J., & Gromadzinska, J., (2015). Bisphenol A–Application, sources of exposure and potential risks in infants, children and pregnant women. *Int. J. Occup. Med. Environ. Health, 28,* 209–241.

Moon, M. K., Kim, M. J., Jung, I. K., Do Koo, Y., Ann, H. Y., Lee, K. J., Kim, S. H., et al., (2012). Bisphenol A impairs mitochondrial function in the liver at doses below the no observed adverse effect level. *J. Korean Med. Sci., 27*, 644–652.

Morgan, H. D., Sutherland, H. G., Martin, D. I., & Whitelaw, E., (1999). Epigenetic inheritance at the agouti locus in the mouse. *Nat. Genet., 23*, 314–318.

Moysich, K. B., Menezes, R. J., Baker, J. A., & Falkner, K. L., (2002). Environmental exposure to polychlorinated biphenyls and breast cancer risk. *Rev. Environ. Health., 17*, 263–278.

Naik, P., & Vijayalaxmi, K. K., (2009). Cytogenetic evaluation for genotoxicity of bisphenol-A in bone marrow cells of Swiss albino mice. *Mutation Research/Genetic Toxicology and Environmental Mutagenesis, 676*, 106–112.

Nakagawa, Y., & Suzuki, T., (2001). Metabolism of bisphenol A in isolated rat hepatocytes and oestrogenic activity of a hydroxylated metabolite in MCF-7 human breast cancer cells. *Xenobiotica., 31*, 113–123.

Nässberger, L., Arbin, A., & Östelius, J., (1987). Exposure of patients to phthalates from polyvinyl chloride tubes and bags during dialysis. *Nephron, 45*, 286–290.

Oakley, G. G., Roe, A. L., Blouin, R. A., Twaroski, T. P., Ganguly, T. C., Vore, M., Lehmler, H. J., & Robertson, L. W., (2001). *2, 4, 4'-Trichlorobiphenyl Increases STAT5 Transcriptional Activity* (Vol. 30, pp. 199–208). Molecular Carcinogenesis: Published in cooperation with the University of Texas MD Anderson Cancer Center.

Portigal, C. L., Cowell, S. P., Fedoruk, M. N., Butler, C. M., Rennie, P. S., & Nelson, C. C., (2002). Polychlorinated biphenyls interfere with androgen-induced transcriptional activation and hormone binding. *Toxicol. Appl. Pharmacol., 179*, 185–194.

Prince, M. M., Ruder, A. M., Hein, M. J., Waters, M. A., Whelan, E. A., Nilsen, N., Ward, E. M., et al., (2006). Mortality and exposure response among 14,458 electrical capacitor manufacturing workers exposed to polychlorinated biphenyls (PCBs). *Environ. Health Perspect., 114*, 1508–1514.

Prins, G. S., Tang, W. Y., Belmonte, J., & Ho, S. M., (2008). Developmental exposure to bisphenol A increases prostate cancer susceptibility in adult rats: Epigenetic mode of action is implicated. *Fertil. Steril., 89*, 41.

Ptak, A., Wróbel, A., & Gregoraszczuk, E. L., (2011). Effect of bisphenol-A on the expression of selected genes involved in cell cycle and apoptosis in the OVCAR-3 cell line. *Toxicol. Lett., 202*, 30–35.

Richter, C. A., Birnbaum, L. S., Farabollini, F., Newbold, R. R., Rubin, B. S., Talsness, C. E., Vandenbergh, J. G., et al., (2007). *In vivo* effects of bisphenol A in laboratory rodent studies. *Reprod. Toxicol., 24*, 199–224.

Ruder, A. M., Hein, M. J., Nilsen, N., Waters, M. A., Laber, P., Davis-King, K., Prince, M. M., & Whelan, E., (2006). Mortality among workers exposed to polychlorinated biphenyls (PCBs) in an electrical capacitor manufacturing plant in Indiana: An update. *Environ. Health Perspect., 114*, 18–23.

Rusyn, I., Kadiiska, M. B., Dikalova, A., Kono, H., Yin, M., Tsuchiya, K., Mason, R. P., et al., (2001). Phthalates rapidly increase production of reactive oxygen species *in vivo*: Role of Kupffer cells. *Mol. Pharmacol., 59*, 744–750.

Rusyn, I., Peters, J. M., & Cunningham, M. L., (2006). Modes of action and species-specific effects of di-(2-ethylhexyl) phthalate in the liver. *Crit. Rev. Toxicol., 36*, 459–479.

Ryan, B. C., Hotchkiss, A. K., Crofton, K. M., & Gray, Jr. L.E., (2010). In utero and lactational exposure to bisphenol A, in contrast to Ethinyl estradiol, does not alter sexually dimorphic behavior, puberty, fertility, and anatomy of female LE rats. *Toxicol. Sci., 114*, 133–148.

Schantz, S. L., Widholm, J. J., & Rice, D. C., (2003). Effects of PCB exposure on neuropsychological function in children. *Environ. Health Perspect., 111*, 357–576.

Schweikl, H., Schmalz, G., & Rackebrandt, K., (1998). The mutagenic activity of unpolymerized resin monomers in *Salmonella typhimurium* and V79 cells. *Mutat. Res. Genet. Toxicol. Environ. Mutagen., 415*, 119–130.

Sergeev, A. V., & Carpenter, D. O., (2005). Hospitalization rates for coronary heart disease in relation to residence near areas contaminated with persistent organic pollutants and other pollutants. *Environ. Health Perspect., 113*, 756–761.

Serrano, S. E., Karr, C. J., Seixas, N. S., Nguyen, R. H., Barrett, E. S., Janssen, S., Redmon, B., Swan, S. H., & Sathyanarayana, S., (2014). Dietary phthalate exposure in pregnant women and the impact of consumer practices. *Int. J. Environ. Res. Pub. Health., 11*, 6193–6215.

Sharpe, R. M., & Skakkebaek, N. E., (2008). Testicular dysgenesis syndrome: Mechanistic insights and potential new downstream effects. *Fertil. Steril., 89*, 33–38.

Shimizu, M., Ohta, K., Matsumoto, Y., Fukuoka, M., Ohno, Y., & Ozawa, S., (2002). Sulfation of bisphenol A abolished its estrogenicity based on proliferation and gene expression in human breast cancer MCF-7 cells. *Toxicol. in Vitro, 16*, 549–556.

Shintani, H., (1985). Determination of phthalic acid, mono-(2-ethylhexyl) phthalate and di-(2-ethylhexyl) phthalate in human plasma and in blood products. *J. Chromatogr. B: Biomed. Sci. Appl., 337*, 279–290.

Silberhorn, E. M., Glauert, H. P., & Robertson, L. W., (1990). Critical reviews in: Carcinogenicity of polyhalogenated biphenyls: PCBs and PBBs. *Crit. Rev. Toxicol., 20*, 440–496.

Sinks, T., Steele, G., Smith, A. B., Watkins, K., & Shults, R. A., (1992). Mortality among workers exposed to polychlorinated biphenyls. *Am. J. Epidemiol., 136*, 389–398.

Soto, A. M., Brisken, C., Schaeberle, C., & Sonnenschein, C., (2013). Does cancer start in the womb? Altered mammary gland development and predisposition to breast cancer due to in utero exposure to endocrine disruptors. *J. Mammary Gland Biol. Neoplasia, 18*, 199–208.

Tang, W. Y., Morey, L. M., Cheung, Y. Y., Birch, L., Prins, G. S., & Ho, S. M., (2012). Neonatal exposure to estradiol/bisphenol A alters promoter methylation and expression of Nsbp1 and Hpcal1 genes and transcriptional programs of Dnmt3a/b and Mbd2/4 in the rat prostate gland throughout life. *Endocrinol., 153*, 42–55.

Tavares, R. S., Escada-Rebelo, S., Correia, M., Mota, P. C., & Ramalho-Santos, J., (2016). The non-genomic effects of endocrine-disrupting chemicals on mammalian sperm. *Reprod., 151*, 1–13.

Tay, T. W., Andriana, B. B., Ishii, M., Tsunekawa, N., Kanai, Y., & Kurohmaru, M., (2007). Disappearance of vimentin in Sertoli cells: A mono (2-ethylhexyl) phthalate effect. *Int. J. Toxicol., 26*, 289–296.

Tayama, S., Nakagawa, Y., & Tayama, K., (2008). Genotoxic effects of environmental estrogen-like compounds in CHO-K1 cells. *Mutat. Res. Genet. Toxicol. Environ. Mutagen., 649*, 114–125.

Tharappel, J. C., Lee, E. Y., Robertson, L. W., Spear, B. T., & Glauert, H. P., (2002). Regulation of cell proliferation, apoptosis, and transcription factor activities during the promotion of liver carcinogenesis by polychlorinated biphenyls. *Toxicol. Appl. Pharmacol., 179*, 172–184.

Towae, F. K., Enke, W., Jäckh, R., & Bhargava, N., (1992). Phthalic acid and derivatives. *Ullmann's Encycl. Ind. Chem., 20*, 181–211.

Troisi, G. M., Barton, S. J., Liori, O., & Nyman, M., (2020). Polychlorinated biphenyls (PCBs) and sex hormone concentrations in ringed and grey seals: A possible link to endocrine disruption? *Arch. Environ. Contam. Toxicol., 78*, 513–524.

Vandenberg, L. N., Hauser, R., Marcus, M., Olea, N., & Welshons, W. V., (2007). Human exposure to bisphenol A (BPA). *Reprod. Toxicol., 24*, 139–177.

Waring, R. H., & Harris, R. M., (2005). Endocrine disruptors: A human risk? *Mol. Cell. Endocrinol., 244*, 2–9.

Wetherill, Y. B., Akingbemi, B. T., Kanno, J., McLachlan, J. A., Nadal, A., Sonnenschein, C., Watson, C. S., et al., (2007). *In vitro* molecular mechanisms of bisphenol A action. *Reprod. Toxicol., 24*, 178–198.

Wetherill, Y. B., Fisher, N. L., Staubach, A., Danielsen, M., De Vere, W. R. W., & Knudsen, K. E., (2005). Xenoestrogen action in prostate cancer: Pleiotropic effects dependent on androgen receptor status. *Cancer Res., 65*, 54–65.

Whaley, D. A., Keyes, D., & Khorrami, B., (2001). Incorporation of endocrine disruption into chemical hazard scoring for pollution prevention and current list of endocrine disrupting chemicals. *Drug. Chem. Toxicol., 24*, 359–420.

Wilsgaard, T., Jacobsen, B. K., Schirmer, H., Thune, I., Løchen, M. L., Njølstad, I., & Arnesen, E., (2001). Tracking of cardiovascular risk factors: The Tromsø study, 1979–1995. *Am. J. Epidemiol., 154*, 418–426.

World Health Organization, (2006). *The World Health Report 2006: Working Together for Health*. World Health Organization.

Wormuth, M., Scheringer, M., Vollenweider, M., & Hungerbühler, K., (2006). What are the sources of exposure to eight frequently used phthalic acid esters in Europeans? *Risk Anal., 26*, 803–824.

Wu, S., Zhu, J., Li, Y., Lin, T., Gan, L., Yuan, X., Xu, M., & Wei, G., (2010). Dynamic effect of di-2-(ethylhexyl) phthalate on testicular toxicity: Epigenetic changes and their impact on gene expression. *Int. J. Toxicol., 29*, 193–200.

Zawatski, W., & Lee, M. M., (2013). Male pubertal development: Are endocrine-disrupting compounds shifting the norms? *J. Endocrinol., 218*, 1–12.

CHAPTER 3

Impact of Endocrine Toxicants on a Sustainable Environment

RAMAN JASROTIA and SHAHNAZ TABASSUM

Department of Zoology, University of Jammu, Jammu–180006, Jammu & Kashmir, India, E-mail: ramanjasrotia18@gmail.com (R. Jasrotia)

ABSTRACT

The US Environmental Protection Agency (EPA) defined endocrine toxins as exogenous agents that alter the synthesis, secretion, transport, metabolism, and binding action of natural endogenous hormones present inside the body of an organism responsible for the regulation of various metabolic processes. Solvents from industries and their by-products like Polybrominated biphenyls (PBBs), pesticides (Dichlorobiphenyl trichloroethane), and fungicides (Vinclozolin) are the common synthetic endocrine disruptors (S-EDs) while the natural chemical includes genistein and coumestrol which are known to have a lower binding affinity. With the increase in population, rise in urbanization, and industrial development, everyone is focused on the sustainable environment, and anything that causes a direct threat to sustainability needs to be taken seriously, and efforts should be made to eliminate these threats. As of now, contaminants of these sustainable resources, like the anthropogenic chemicals and the secondary plant metabolites which are included in the naturally occurring contaminants, are found to have an alarming number of endocrine toxins. The endocrine toxins are slow to break down in the environment, which makes them potentially hazardous overtimes, so the effect of the endocrine toxins on the environment is gaining much attention in the scientific community because of the enormity of the dangerous threats they pose to various dimensions of

Environmental Endocrine Toxicants: Biology, Effects, and Management. Younis Ahmad Hajam, Rouf Ahmad Bhat, Khalid Rehman Hakeem, & Rajesh Kumar (Eds.)
© 2023 Apple Academic Press, Inc. Co-published with CRC Press (Taylor & Francis)

lives of living organisms and the environment. One of the best examples is the contaminants found in the reused water, which are used for agricultural purposes and even for drinking, thus leading to the severe difficulty faced by the water industry. Apart from this, these toxins play a major role in the reduced reproductive ability of many organisms. Some other harmful consequences of the endocrine toxins in wildlife are compromised immune systems (e.g., in seals) and deformed skeletal structures (e.g., in frogs). In humans, they are associated with decreased sperm count, cancer, malformation in newborn children, etc. The chapter thus is directed towards the effects of these endocrine toxins (both synthetic and natural) on humans and a sustainable environment so that it can act as key points to find out the solutions against threats posed by the EDs.

3.1 INTRODUCTION

The change in the function of the endocrine system leading to a hazardous effect on the health of the organism is caused by some chemicals, which are known as endocrine toxins. The endocrine system includes the organization of tissues interacting with each other and with the whole body through the secretions they produce. These chemical secretions are known to handle the various metabolic processes of the body, such as gamete formation, various developmental processes, etc. So, the disruption or any kind of alteration in their function leads to improper functioning of all the other mechanisms of the organisms resulting in various diseases or even fatality of the organism and also affecting the environment indirectly. A few of the health factors that occur due to endocrine toxicity are reproductive disorders, cancer, compromised immunity, asthma, diabetes, hyperactivity disorders, alteration in the biological clock, etc. Studies during the last decade focused on the interrelation between humans, wildlife, and the environment, hinting toward the adverse effect of these toxins on all of them, especially on human health. After getting the slightest idea of the level of their toxicity and their irreversible adverse effect on the life of organisms and its surrounding, this issue came to light and was discussed by the scientific community, and strategic plans on a global level were made to minimize their production in manufacturing places. Endocrine toxins were acknowledged as an emergency issue during the third session of the International Conference on Chemical Management held in 2012. Later in 2015, during the fourth session of ICCM, it was declared that all the required support would be provided for the research in the field of endocrine toxins.

The oldest organization researching the field of hormonal secretion, its effect, and other perspectives related to them is the Endocrine Society 1916, which comprises about 18,000 scientists, students, and workers from various other professions belonging to more than 100 countries. The definition for endocrine toxins provided by the Endocrine Society is "An endogenous chemical or mixture of chemicals that interferes with any aspect of hormone action. Hormones are natural chemicals produced in cells within the endocrine gland, which are located throughout the body." Exposure to these toxins can take place at home and in public places and can enter the human system through the food we consume, through fluid intake, or from the atmosphere. The effect of endocrine toxins varies in the individual at different life stages. For example, their effect is very strong during the fetal phase and can lead to an adverse effect on organs in the developmental stage, but the same chemical has a lesser effect on adults and sometimes doesn't even have any effect.

Their effect also depends on the level of concentration. In areas with a high level of contamination of endocrine toxins, like the industrial areas, all the potential chemicals produced by these manufacturing companies get frequently absorbed by the soil, which is later utilized by the inhabitants of the soil, such as plants, algae, and from this trophic level, they are passed onto the next level, i.e., animals through the food chain. The concentration of the endocrine toxins in the body of the organism gets accumulated and becomes highest as it reaches the top level of the food chain, which consists of humans. Thus, the tissue composition of humans has the highest concentration of these toxins. Among the innumerable chemicals produced in these industries, thousands of them are known to have potentially toxic effects on the human endocrine system. When humans were bio-monitored for the presence of the toxins, it was found that every person taken in for the assessment showed the presence of these chemicals in their urine, blood, tissue, and placenta.

3.2 COMMON ENDOCRINE TOXINS

Common examples of endocrine toxins are pesticides like DDT, fire retardants, organophosphates, mercury, arsenic, perchlorate, atrazine, glycol ethers, BPA, etc. All these compounds pose a great threat to the environment and the organisms living in it because they can be easily transported along with water and air from one location to another distant location, and this is proved by the presence of endocrine toxins in the organisms inhabiting the polar regions. Another factor that makes them even more dangerous is their ability to persist in the environment for years without undergoing degradation. The easiest way of getting exposed to these chemicals is when they are sprayed inside homes and over water bodies in the form of fungicides, pesticides, and algicides for removing the various infestations. Fungicides like vinclozolin are potential toxins of the endocrine system. As these chemicals are sprayed to exterminate the insects, fungi, or algae from our various places by alteration in the hormonal and reproductive function of the target animals, but due to the similarity between the physiological processes in most vertebrates and invertebrates, these chemicals also affect the humans in the same way as they affect the insects.

Few of these chemicals, like DDT, are banned in some countries. But most of the chemicals are still used all over the globe. Endocrine toxins are categorized into different classes, which are discussed in subsections.

3.2.1 PESTICIDES

3.2.1.1 DICHLORODIPHENYLTRICHLOROETHANE (DDT)

This is an insecticide used for the control of insect infestation in crops (both private and commercial purposes). It was used as a main insecticidal chemical during the 1940–1960s. But as its heavy toxic effect on organisms came to light, it was banned for use in agricultural practices in various countries across the world. In countries like Africa and India, it is used for lowering the population of the disease-transmitting insects which act as vectors for malaria, dengue, etc., as their use was restricted to vector control purposes only as per the Stockholm Convention. However, its illegal use continues in some countries. A large portion of DDT exposure all around the world is through the supply of food. Poultry, red meat, and dairy products comparatively have a higher proportion of DDT as they are stored in fats. DDE (which is a metabolite of DDT) has a longer half-life, so the fats storage detects DDE even if DDT is not detected.

The organism can be exposed to DDT even when they are in the fetal stage, where it is transferred through the placenta. A considerable amount of DDT was detected in the milk of breastfeeding mothers in countries like Sudan, Ukraine, India, the Philippines, etc. The level of DDT is found more in adults as they are exposed to DDT and have accumulated it in their bodies for a longer period as compared to the younger generation. Another reason for the higher level of DDT in them is because of the higher use of DDT in the previous years than now. It was found from a study conducted by Kreiss et al. (1981) that people living in the area of an old DDT manufacturing place had a different levels of DDT according to their age. People over the age of 70 had a DDT level of 350 ng/ml, while people younger than them had 159 ng/ml DDT. The higher level of the food chain is known to have highly accumulated DDT. The inhabitants of the Arctic region are predicted to be affected adversely by DDT accumulation in the future as they need the thyroid hormone system to adapt to the cold climatic condition, and the function of this system is considerably modified by this endocrine toxin. This insecticide was among the endocrine toxins that were recognized as the earliest toxins with several degrading effects on the endocrine and reproductive system of the body. Experiments conducted on laboratory animals showed a negative effect on the health of organisms. DDT was found to alter the important hormones of bodies like insulin, androgen, renin–angiotensinogen, and the neuroendocrine system, which led to the disruption of the normal metabolic pathways.

In a few cases, it also brings a change in the sex hormone ratio in males and females, thus modifying their reproductive capability. For example, decreased sperm quantity and cryptorchidism are seen in males living that are exposed to DDT (Aneck-Hahn et al., 2007). Their effect is most prominent in fetuses and infants. Exposure in young girls before the development of breast increases the chance of breast cancer later in life (Cohn et al., 2007). Other effects of endocrine toxicity are alteration in the menstrual cycle and early menopause. In females, shortening of the lactation phase due to exposure to DDT has been reported by Eskenazi et al. (2009). Studies done by Taylor et al. (2013) showed that countries that still use DDT have an increased rate of Type II diabetes patients as compared to countries that have stopped its use. DDT plays an important role in elevating the level of blood sugar, which is a key factor for diabetes. Insulin is known for reducing blood sugar levels secreted by the pancreas, but exposure to DDT alters the function of the pancreas by lowering its insulin-secreting ability.

3.2.1.2 CHLORPYRIFOS

These are pesticides formed of organ-phosphorus used to exterminate insects like lice, mosquitoes, termites, cockroaches, etc., from residential areas. They are also used as a pest control chemical in agriculture to lower the pest population on fruits, grains, and other crops. It is used in stored water to eradicate various larvae found in them. The chemical used in the formation of banana ripening bags on farms is made of organophosphates. Chlorpyrifos is extremely hazardous to organisms like honey bees and earthworms, which are extremely beneficial in agriculture.

The disintegration rate of chlorpyrifos is higher than DDT, but still, it remains in the soil for a long period. They get accumulated in the soil, which is frequently used for agricultural purposes. Inside the house, they are found on furniture, air, floor, and toys. Once applied, they can be seen on the surface for two weeks, even in areas that did not have any direct contact with them. During a study, chlorpyrifos was found to be present in a house even after it had been sprayed eight years ago for pest control (Wright et al., 1994). Among some common things, it is found in our day-to-day life commodities like fruits, vegetables, milk, rice, etc. This endocrine toxin has been detected in fishes inhabiting the Arctic region.

The endotoxin is removed from the adipose tissue in a relatively short period (shorted half-life). These chemicals do not show any accumulation property; instead, they are converted into some other metabolite that has harmful effects on the body. These metabolites are detected from the samples of blood, sperm, breast milk, urine, and even the hairs of the infants. Organophosphate pesticides have been detected in agricultural workers. A survey conducted in Egypt showed that the concentration of chlorpyrifos in the body of the workers depends upon the chemicals and their time span of exposure to that chemical (Lein et al., 2012). The primary way of exposure to them is through skin and inhalation. Their main effect is on the nervous system of the organism. Exposure to chlorpyrifos leads to neurotoxicity. They affect the endocannabinoid pathway of signaling, which plays a significant role in neural processes. Their toxicity results in reduced learning ability in rodents which is also related to the modification of the thyroid hormones (THs) (Cole et al., 2012). Reduced weight of the adrenal gland in rodents is also a consequence of this endocrine toxin. The toxicity of organophosphate in humans results in symptoms like vomiting, urination, and salivation, which happens due to modification in the neural system. These pesticides are found to be related to Parkinson's disease (Dhillon et al., 2008; Gatto et al., 2009), which was concluded after a study done in the United States. Some other common effects of organophosphate toxicity are muscle loss and memory problems.

One of the major risk factors of this endocrine toxin is susceptibility during developmental stages. Scientists strongly suggest that it should be banned due to its high toxicity, which results in neurological disorders (Saunders et al., 2012). Another study concluded that their toxicity in prenatal stages leads to compromised mental development in children. Earlier their adverse effect was found to be limited to the nervous system, but as new studies emerged, apart from neurological effects, they are also found to alter the lipid metabolism and functioning of the thyroid gland (TG).

3.2.1.3 PRODUCT TOXINS

Endocrine toxins are reported enormously from products we use in our day-to-day life that are frequently in our contact. Some examples are skincare, make-up products, electronics, toys, and cloth materials. If the room is not equipped with proper ventilation, chemicals from these substances are released and trapped inside the room, which are ultimately inhaled. Skincare products, washing soaps, and toothpaste consist of toxic chemicals that are absorbed in small quantities leading to hazardous effects.

Phthalates, belonging to a category of plasticizers, are substances generally found in aromatic essence candles, shampoo, air fresheners, toys, model clays, glue, ink, etc. They are known to cause a decrease in sperm count, obesity, and reproductive abnormalities in males. One of the most important toxins produced by daily-use products is inorganic lead. Lead is an element found in the earth's crust, and it occurs in the human surroundings through refining, mining, ornament industries, paints, cottage industries, etc. This process has drastically increased the lead pollution in the environment. Lead pollution is mainly reported in developing countries. The developed nations report that less use of lead in paints and petrol results in a reduced concentration of lead in the blood.

Lead is introduced into the body through contaminated eatables, inhalation of chemicals, tobacco, and utensils glazed with lead. The concentration of lead in the blood depicts its recent exposure, while its concentration or presence in the bone points towards its long exposure. The risk of lead toxicity is mainly seen in pregnant women and children. Lead accumulated in the bones of a pregnant woman can get transferred to the infant during lactation (Center for Disease Control and Prevention, 2010). Lead toxicity causes various cardiovascular, renal, and neurological disorders. In children, lead toxicity results in mental retardation, and in adults, one of the common symptoms of its toxicity is high blood pressure. It disrupts the functioning of the endocrine system and reproductive systems. They can change the level of estrogen in the body as they can activate estrogen receptors (ERs) (mimics estrogen).

3.2.1.4 ELECTRONICS

The persistent organic pollutants (POPs) used in textiles, electronics, computers, building materials, and foams are polybrominated diphenyl ethers (PBDEs). Earlier, three types of PBDEs were used, which are Penta

BDE, Octa BDE, and Deca BDE. The most common of them, the Penta BDE, is used in furniture; the other two are used in electronics. The Penta BDE is mainly responsible for the disruption of the TG and leads to the decreased secretion of the thyroid hormone. Penta BDE has a major resemblance with the thyroid hormone; thus, they can activate the thyroid hormone receptors and alter their secretion (Kitamura et al., 2008). During pregnancy, the demand for thyroid hormone is higher in the woman, and its insufficient availability results in life-long neurological disorders in the offspring and also have an adverse effect on the health of the mother. Neurological disorders are either caused by an alteration in the function of the brain or due to the modification in the thyroid hormone regulation (Eskenazi et al., 2013).

The brominated flame retardants (BFRs) enter the biotic system through the intake of food and dust containing the toxin. They remain indoors for a long period (longer half-life). Various flame retardants are found in recycled materials such as kitchen utensils. Their concentration is higher in children that feed on milk and are in direct contact with the dust. They are considered a strong endocrine toxin, whether it is in their real form or metabolite form.

3.3 TOXIC MATERIALS IN FOOD

3.3.1 BISPHENOL-A (BPA)

The utensils used for storing food, e.g., plastic containers, and the lining of food cans consists of bisphenol-A (BPA). Earlier plastic water bottles were made up of material consisting of BPA. But now, as the hazardous effect of BPA is known, they are replaced by BPA-free material. In some countries, the use of milk bottles for babies is banned; another source of BPA is the lining of processed food cans. The epoxy line, in general, helps the food to protect from degradation by any pathogen that leads to diseases like botulism, but some of the lined cans produce BPA. A few other sources of BPA are water pipelines and eyeglasses, which are all made of plastics.

By 2015, the production of BPA on a global level is thought to be more than 5.4 million metric tons. They have been detected in amniotic fluids, blood, urine blood, etc. As BPA mainly transfers from plastic, children are more prone to them due to their contact with toys made of plastic material. The lower concentration of BPA is found in individuals or communities who have lowered the use of plastic products in their daily lives. The exposure of BPA to human lives is through food and drinks stored in plastic containers, from where they are leached into the edibles. The use of microwave or re-heating of food stored in plastic containers increases its transfer to the food. Plastic bottles kept inside the car when it gets hot also transfer BPA to the liquid. The metabolism of BPA is faster as compared to other endocrine toxins. Their bioaccumulating properties are also very low. Their level in the body can easily be reduced by limiting the use of plastic products and canned food containing BPA and replacing them with other materials (Yang et al., 2011). On a global level, the oceans are filled with plastic bottles which don't undergo the recycling process. They ultimately increase the BPA level of the water, thus affecting the resources of oceans, ocean life, and the atmosphere.

BPA was synthesized for the first time in the year 1891, and in the year 1930, it was reported to mimic the estrogen hormone. They are known to alter the estrogen signaling pathway and modify the normal functions carried out by estrogen resulting in impaired reproductive functions and ovulation in females. They act through the mechanism of DNA methylation, thus carrying out small alterations in the genes. Their lower-level increases the risk of cancer, and their increased level is related to polycystic ovary syndrome (PCOS). As per the Stockholm convention, the use of BPA is banned because of their capability of transportation to distant areas and their high accumulation rate, due to which they are not easily degraded.

3.3.2 DIOXIN

In industries, dioxin is formed when chlorine or bromine reacts in the presence of oxygen or carbon. They have a long persisting ability and are potent carcinogens. They result in compromised immunity and harm reproductive health. Their toxication can be avoided by limiting or cutting down the use of animal products.

3.3.3 ATRAZINE

They are used as herbicides in agricultural fields and are known to be responsible for the conversion of sex in frogs. Male frogs get converted into female frogs, which are capable of producing viable eggs. In humans, they are associated with breast tumors, prostate cancer, and a delay in puberty. They can be avoided by using completely organic products and a water filter system that filters out atrazine.

3.3.4 PERCHLORATE

Perchlorates are the main component of rocket fuel. They affect the thyroid hormone, as they are good competitors of iodine. They can be filtered out from the water by a reverse osmosis filter.

3.3.5 ARSENIC

Arsenic enters our system through food and fluid. Their increased toxicity can lead to death, and minor toxicity results in lung cancer, skin cancer, etc. It alters the functioning of glucocorticoids.

3.4 MODE OF MECHANISM

The mode of action of endocrine toxins may be categorized into the following types:

- Oxidative stress;
- Modifying metabolism of steroid hormone;
- Targeting nuclear receptors;
- Endocrine toxins as hormone sensitizers.

3.4.1 OXIDATIVE STRESS

The endocrine toxin affects humans and wildlife through the production of reactive oxygen species (ROS), which leads to changes in the form of DNA, lipids, and protein components of the cell.

3.4.2 MODIFYING METABOLISM OF STEROID HORMONES

In this mechanism, the main target of the endocrine toxins is the enzymes that play a major role in the biosynthesis of the steroid hormones. In different organs of the body, enzymes like cytochrome P 400 play a key role in the synthesis of steroid hormones. Endocrine toxins like organotin target these enzymes and inhibit their function. Alteration of the aromatase enzyme in fish results in the development of the penis in females (Fent, 2003). Adreno-toxicity in dogs occurs due to mitotane and dichlorodiphenyldichloroethane

(DDD), which are derivatives of DDT. $MeSO_2$-DDE (methyl sulfonyl-DDE) inhibits P540c11 functioning and causes adrenodoxin effects in cell line Kin 8 and Y1 (Lund et al., 1995).

3.4.3 TARGETING NUCLEAR RECEPTORS

The expression of a particular gene is done after the binding of a particular ligand to the nuclear receptor. But some endocrine toxins mimic the hormones and bind to their nuclear receptors and cause the expression of its gene even if it is not needed, which brings about an undesirable change in the organism and even has lethal effects. Humans are exposed to organotin through contaminated seafood, polluted water, and fungicides. It activates the nuclear receptor for the differentiation of adipocytes, thus resulting in obesity. Other toxins such as phthalates and BPA act through this mechanism and cause various metabolic disorders.

3.4.4 ENDOCRINE TOXINS AS HORMONE SENSITIZERS

This mechanism affects the hormone conduction pathway and alters various stages of hormone production. Endocrine toxins like dioxins and polychlorinated biphenyls (PCBs) work through this mechanism. The higher level of hormonal pathway disruption in females is caused by oral contraceptives (Tabb et al., 2006). In rats, disrupted hormonal mechanism occurs with methoxy acetic acid that, hinders the beta estrogen receptor and acts as a risk factor for testicular cancer (Tirado et al., 2004).

In brief, the mechanism of action of endocrine toxins can be receptor-mediated or non-receptor mediated, and it may be genomic or non-genomic. They disrupt the hormonal pathway either through mimicking the original hormones or directly having an opposing action to that hormone.

3.5 EFFECT OF ENDOCRINE TOXINS ON HUMAN HEALTH AND THE ENVIRONMENT

One of the most alarming anthropogenic threats to the environment and biodiversity includes endocrine toxins. The amount and severity of their adverse effect on the health of an organism make it so important to discuss how to replace them with different substances and, if possible, totally cut down their utility. The aspects of human health affected by them are discussed in subsections.

3.5.1 REPRODUCTIVE HEALTH

In males, exposure to the endocrine toxins at any stage of life is known to have various irreversible adverse consequences starting with decreased production of sperm and poor semen quality. Toppari (2008) found that exposure to toxins that are anti-androgenic, e.g., procymidone and vinclozolin, results in cryptorchidism. According to Gray et al. (2006), fungicides act as an antagonist of androgens. Another condition, i.e., hypospadias, occurs when the male is exposed to toxins during the fetal stage. In females, endocrine toxins lead to PCOS, and endometriosis which further results in low fertility. If a pregnant woman is exposed to endocrine-disrupting chemicals, the offspring are born with various health problems. In certain cases, exposure to toxins in the prenatal stages leads to the development of an underweight fetus or even a miscarriage (Stillerman et al., 2008). Lead exposure is the main toxin responsible for the majority of miscarriage cases (Slama & Cordies, 2010).

3.5.2 THYROID EFFECTS

The THs regulate the normal physiology of humans, the development of the brain, and various other functions. Alteration in its function leads to delays in growth, mental retardation, etc. The functioning of the TG system is one of the most potent targets of endocrine toxins. Various disorders such as Graves's disease and hypothyroidism caused by the alteration in the level of THs are mainly reported in children (McGrogan et al., 2008). Chemicals like dioxins and flame retardants are structural mimics of thyroxine (T4) and triiodothyronine (T3), so these toxins easily bind with their receptors (WHO, 2012). Exposure to PCB and the functioning of THs are related to each other. The glucose balance in the body is disrupted by pesticides and BPA, which causes diseases such as diabetes and obesity.

3.5.3 CARCINOGENS

The Center for Disease Control (CDC) and prevention in the United States reported a list of 48 endocrine toxins that are known to have carcinogenic effects (Choi et al., 2004). EDCs have a mutagenic effect to a great extent. Increased incidence of leukemia has been reported in infants who are exposed to household pesticides. The main cause of lung cancer is tobacco smoke, which consists of nitrosamine. For controlling weeds, a chemical called atrazine is extensively used, which is related to mammary gland cancer. Xenoestrogen is predicted to be the major reason for breast cancer. Some DDT disrupts the functioning of the endocrine system and results in tumors of the breast.

3.5.4 NEUROLOGICAL DISORDERS

The pollutant exposure in children results in various neurological disorders depending on the time of exposure to the toxins (Weiss & Landrigan, 2000). Some disorders resulting from EDCs are autistic disorders, depressive disorders, and attention deficit hyperactivity disorder (ADHD). ADHD is prevalent in the population exposed to organophosphate pesticides. The fetus is the most vulnerable target of the chemicals like lead and mercury that pass into them through their mother and leads to neurological disorders (Needleman, 2009). Some other disorders are impaired hearing and vision and delay in growth. Schizophrenia is reported due to exposure to BPA. Methyl mercury causes oxidative stress in the brain of the hippopotamus.

3.5.5 EFFECT ON THE METABOLIC SYSTEM

Endocrine toxins may interfere with metabolic functioning and thus play a role in health conditions such as obesity and type-2 diabetes. Risk factors of type-2 diabetes include dioxin, hexachlorobenzene, and dichloro-diphenyl dichloroethylene (DDE).

3.5.6 CARDIOVASCULAR PROBLEMS

These problems are related to exposure to diethylstilbestrol (DES), a synthetic estrogen resulting in high blood pressure and heart stroke. Coronary heart disease (CHD) is also associated with a high level of organic polymers. The physiology of cardiac markers is disturbed by pesticides.

3.5.7 VULNERABLE POPULATION GROUP

Fetal development: Cell programming is affected if organisms are exposed to toxins in their developmental stages. This results in abnormal tissue and organ development. The effect of the same chemical differs in fetuses and adults. The pre-conception to the end phase of adolescence is the most vulnerable period for exposure to endocrine toxins. Puberty: About 200 years ago, the age of onset of the menstruation cycle was 17 years, but now it has been changed to 13 years (Aksglaede et al., 2008). Studies have reported that lead exposure causes a delay in the onset of puberty; no other EDC is found to be associated with menarche except for polybrominated biphenyl, which is related to early pubic hair development and early menarche.

3.6 EFFECT ON ENVIRONMENT

Endocrine toxins like mercury disrupt the atmospheric cycle; methyl mercury is the most commonly appearing form of mercury reported from the environment. Its property of being lipid-soluble and less solubility in water helps them to transfer easily to the aquatic environment. Mercury contamination in the soil leads to a decrease in the population of protozoans and other soil organisms except for the fungal community.

Another persistent endocrine toxin is DDT which is known to have adverse effects on soil and water as it is readily absorbed into the soil and gets accumulated in water through surface runoff. It is taken up by aquatic organisms and then accumulates in the food web. DDT is degraded very slowly in the environment; it takes approximately 50 years to disintegrate from the environment. Dichlorodiphenyltrichloroethane (DDT), an insecticide, breaks out into two metabolites, i.e., DDD and dichloro-diphenyl dichloroethylene (DDE). DDE results in thinning of eggshells in birds and hence the population decline of aquatic prey birds. DDT also has an adverse impact on the Arctic polar bear. DDT carried either by great continental water or by wind gets condensed in the cold or accumulated in the snow or hails. The polar bears feed on a large number of fishes (with absorbed DDT) before hibernation. The polar bear then starts to break down the fats during hibernation.

Exposure to arsenic inhibits plant growth and also lowers the rate of reproduction and photosynthesis in them. Arsenic cannot be eliminated once it enters the environment, so it keeps on adding up and is taken up by plants or gets collected on the surface of the water, and it increases the possibility of genetic material alteration in aquatic organisms. The pollution caused in various zones of the soil is greatly contributed by lead. In the places near farmlands where there is high exposure to lead, the quality of the soil is disturbed, which results in lead poisoning in organisms inhabiting that place. Vehicle fuels containing lead contribute to air pollution. By-products of the lead generated from industries result in soil and water pollution that causes disturbance of the ecosystem.

3.7 CONCLUSION

Endocrine toxins are found or released from various applications/things that are used in our daily lives, like pharmaceuticals, cosmetics, pesticides, toys, furniture, plastic bottles, etc. So, not getting exposed to them seems inevitable for humans and wildlife, but keeping in view the broad spectrum of their negative effects, their potential to cause various diseases, and adverse effects on the environment, demand for a global solution arises. It can be concluded in the present communication that endocrine-disrupting chemicals (EDCs) can enter the human system in various ways, which greatly affect the endocrine system that plays a vital role in reproduction and development. So, every possible thing must be done to tackle this situation. Some of the key points that are needed to be taken into consideration while discussing their management are:

- Scientists and workers should share the already available data and recent data on these toxins.
- People should be made aware of these toxins, so everyone should take possible precautions while purchasing or using such things at an individual level.
- Collaboration between developed and developing countries should be done to help the communities living in poor conditions and feeding on poor nutrition.
- Testing for endocrine toxins should be made better.
- Identification of EDCs that are yet to be known should be a priority so that their accumulation in the body can be prevented.
- Their exposure should be reduced by producing harmless alternatives for them.
- Policies should be made to limit their use.
- Some developmental stages are very vulnerable to toxins; thus, extra precautions should be taken in those phases.

KEYWORDS

- anthropogenic
- contaminations
- endocrine toxins
- hormones
- persistent organic pollutants
- polybrominated biphenyls
- polybrominated diphenyl ethers
- reproductive ability

REFERENCES

Aksglaede, L., Olsen, L. W., Sørensen, T. I., & Juul, A., (2008). Forty years trends in timing of pubertal growth spurt in 157,000 Danish school children. *PLoS One, 3*, e2728.

Aneck-Hahn, N. H., Schulenburg, G. W., Bornman, M. S., Farias, P., & De Jager, C., (2007). Impaired semen quality associated with environmental DDT exposure in young men living in a malaria area in the Limpopo province, South Africa. *J. Androl., 28*, 423–434.

Choi, S. M., Yoo, S. D., & Lee, B. M., (2004). Toxicological characteristics of endocrine-disrupting chemicals: Developmental toxicity carcinogenicity and mutagenicity, *J. Toxicol. Environ. Health., 7*, 1–23.

Cohn, B. A., Wolff, M. S., Cirillo, P. M., & Sholtz, R. I., (2007). DDT and breast cancer in young women: New data on the significance of age at exposure. *Environ. Health Perspect., 115*, 1406–1414.

Cole, T. B., Fisher, J. C., Burbacher, T. M., Costa, L. G., & Furlong, C. E., (2012). Neurobehavioral assessment of mice following repeated postnatal exposure to chlorpyrifos-Oxon. *Neurotoxicol. Teratol., 34*, 311–322.

Dhillon, A. S., Tarbutton, G. L., Levin, J. L., Plotkin, G. M., Lowry, L. K., Nalbone, J. T., & Shepherd, S., (2008). Pesticide/environmental exposures and Parkinson's disease in east Texas. *J. Agromedicine., 13*, 37–48.

Eskenazi, B., Chevrier, J., Rauch, S. A., Kogut, K., Harley, K. G., Johnson, C., Trujillo, C., et al., (2013). In utero and childhood polybrominated diphenyl ether (PBDE) exposures and neurodevelopment in the CHAMACOS study. *Environ. Health Perspect., 121*, 257–262.

Eskenazi, B., Chevrier, J., Rosas, L. G., Anderson, H. A., Bornman, M. S., Bouwman, H., Chen, A., et al., (2009). The Pine River statement: Human health consequences of DDT use. *Environ. Health Perspect., 117*, 1359–1367.

Fent, K., (2003). Ecotoxicological problems associated with contaminated sites. *Toxicol. Lett., 140*, 353–365.

Gatto, N. M., Cockburn, M., Bronstein, J., Manthripragada, A. D., & Ritz, B., (2009). Well-water consumption and Parkinson's disease in rural California. *Environ. Health Perspect., 117*, 1912–1918.

Gray, L. E., Wilson, V. S., Stoker, T., Lambright, C., Furr, J., Noriega, N., Howdeshell, K., Ankley, G. T., & Guillette, L., (2006). Adverse effects of environmental antiandrogens and androgens on reproductive development in mammals. *Int. J. Androl., 29*, 96–104.

Kitamura, S., Shinohara, S., Iwase, E., Sugihara, K., Uramaru, N., Shigematsu, H., Fujimoto, N., & Ohta, S., (2008). Affinity for thyroid hormone and estrogen receptors of hydroxylated polybrominated diphenyl ethers. *J. Health Sci., 54*, 607–614.

Kreiss, K., Zack, M. M., Kimbrough, R. D., Needham, L. L., Smrek, A. L., & Jones, B. T., (1981). Cross-sectional study of a community with exceptional exposure to DDT. *J. American Med. Assoc., 245,* 1926–1930.

Lein, P. J., Bonner, M. R., Farahat, F. M., Olson, J. R., Rohlman, D. S., Fenske, R. A., Lattal, K. M., et al., (2012). Experimental strategy for translational studies of organophosphorus pesticide neurotoxicity based on real-world occupational exposures to chlorpyrifos. *Neurotoxicol., 33*, 660–668.

Lund, B. O., & Lund, J., (1995). Novel involvement of a mitochondrial steroid hydroxylase (P450c11) in xenobiotic metabolism. *J. Biol. Chem., 270*, 20895–20897.

McGrogan, A., Seaman, H. E., Wright, J. W., & De Vries, C. S., (2008). The incidence of autoimmune thyroid disease: A systematic review of the literature. *Clin. Endocrinol., 69*, 687–696.

Needleman, H., (2009). Low level lead exposure: History and discovery. *Annals Epidemiol., 19*, 235–238.

Saunders, M., Magnanti, B. L., Carreira, S. C., Yang, A., Alamo-Hernández, U., Riojas-Rodriguez, H., Calamandrei, G., et al., (2012). Chlorpyrifos and neurodevelopmental effects: A literature review and expert elicitation on research and policy. *Environ. Health., 11*, S5.

Slama, R., & Cordier, S., (2010). Environmental contaminants and impacts on healthy and successful pregnancies. In: Woodruff, T. J., et al., (eds.), *Environmental Impacts on Reproductive Health and Fertility* (pp. 125–144). Cambridge, Cambridge University.

Stillerman, K. P., Mattison, D. R., Giudice, L. C., & Woodruff, T. J., (2008). Environmental exposures and adverse pregnancy outcomes: A review of the science. *Reprod. Scis., 15*, 631–650.

Tabb, M. M., & Blumberg, B., (2006). New modes of action for endocrine-disrupting chemicals. *Mol. Endocrinol., 20*, 475–482.

Taylor, K. W., Novak, R. F., Anderson, H. A., Birnbaum, L. S., Blystone, C., DeVito, M., Jacobs, D., et al., (2013). Evaluation of the association between persistent organic pollutants (POPs) and diabetes in epidemiological studies: A national toxicology program workshop review. *Environ. Health Persp., 121*, 774–783.

Tirado, O. M., Selva, D. M., Toràn, N., Suárez-Quian, C. A., Jansen, M., McDonnell, D. P., Reventós, J., & Munell, F., (2004). Increased expression of estrogen receptor beta in pachytene spermatocytes after short-term methoxyacetic acid administration. *J Androl.,* 25, 84–94.

Toppari, J., (2008). Environmental endocrine disruptors. *Sexual Develop., 2*, 260–267.

Weiss, B., & Landrigan, P. J., (2000). The developing brain and the environment: An introduction. *Environ. Health Persp., 108*, 373–374.

Wright, C. G., Leidy, R. B., & Dupree, H. E., (1994). Chlorpyrifos in the air and soil of houses eight years after its application for termite control. *Bull. Environ. Contam. Toxicol., 52*, 131–134.

Yang, C. Z., Yaniger, S. I., Jordan, V. C., Klein, D. J., & Bittner, G. D., (2011). Most plastic products release estrogenic chemicals: A potential health problem that can be solved. *Environ. Health Persp., 119*, 989–996.

CHAPTER 4

Endocrine Toxicants and Human Health

URUJ TAHIR, AYESHA HASSAN, and MUHAMMAD IBRAHIM

Department of Environmental Sciences, Faculty of Natural Sciences, University of Okara, Okara–56130, Pakistan,
E-mail: urujtahirjavaid@gmail.com (U. Tahir)

ABSTRACT

Human exposure to endocrine toxicants (ETs) and its related effects are of high concern and gained much attention. This chapter focuses on the basics of effect of endocrine intoxicants on endocrine glands, hormonal profile and the pathways disturbed due to intoxication. Endocrine toxicants are generated due to anthropogenic activities including manufacturing of different products. these products included insecticides, chemical generated during industrial processes, solvents etc. these endocrine disrupting products causes hormonal dysfunctions directly or indirectly viz. disruption of hormonal synthesis and their secretion, by modulating the function and metabolism). All these alteration finally disturbs the homeostasis of body. Therefore, there is need to pay attention in order to protect our body from intoxication. This chapter summarizes the effect of various endocrine disrupting chemicals and their action mechanism on humans.

4.1 INTRODUCTION

All multicellular organisms (including humans) require a communication system for coordination between tissues, organs, and various parts for proper functioning of their body, which also ensures maintenance of homeostasis (to meet physiological demands and environmental challenges like temperature,

Environmental Endocrine Toxicants: Biology, Effects, and Management. Younis Ahmad Hajam, Rouf Ahmad Bhat, Khalid Rehman Hakeem, & Rajesh Kumar (Eds.)
© 2023 Apple Academic Press, Inc. Co-published with CRC Press (Taylor & Francis)

exposure to toxicants and pathogens). Homeostasis refers to the maintenance and regulation of stable internal conditions of the living system that persists despite of fluctuations in the external environmental conditions. For instance, there must be no change in body temperature, pH, hormones, water balance, and concentration of glucose, various minerals, salts, and ions in the blood beyond the preset limits that allow survival (Hiller-Sturmhöfel and Bartke, 1998; Chou, 2005; Chou and Henderson, 2014; McEwen, 2016). Organisms respond to changes in internal and external environmental conditions via two communication systems: (a) the nervous system, and (b) the endocrine system (also known as the hormonal system). Nervous system comprising of complex network allows rapid transmission of information within different parts or regions of the body through electrical impulses and enables living organisms to respond appropriately to the conveyed information and/or interact with their surrounding environment. The nervous system plays a fundamental role in learning, thinking, speaking, memory, sensation, feelings, and functioning of all body parts. *Endocrine system*, on the other hand, is a communication system that produces and uses hormones from various glands as chemical signals, which travel through the bloodstream, and regulates the cell and organs functioning. Hence, both the communication systems complement and are interactive with each other, for instance, a stimulus originating from nervous system is involved in influencing the production of certain hormones or vice versa (Hiller-Sturmhöfel and Bartke, 1998; Lu et al., 2017). Endocrine toxicants (ETs) are the exogenous human synthesized chemicals and their mixtures, which are known to interfere with the hormonal secretions and actions. These toxicants are highly concerned because of their significant impacts on human health.

4.2 ENDOCRINE SYSTEM

The endocrine system consists of a network of ductless endocrine secretory organs and/or glands located throughout the body known as *endocrine glands*. The endocrine system regulates the coordination functions and communication throughout the human body by releasing chemical messengers known as *hormones*. Hormones are the chemical substances comprising of peptides, polypeptides, glycoproteins, steroids, and/or amino acid derivatives, and are utilized by the endocrine system for organ and tissue communication throughout the body (Hiller-Sturmhöfel & Bartke, 1998; Chou, 2005; Chou & Henderson, 2014). These hormones are involved in controlling of long-lasting responses, including metabolic rates, growth, developmental stages, and reproduction (Lu et al., 2017). Hormones are secreted by the endocrine glands into the circulatory (bloodstream) and lymphatic system, they are then transported to hormone secreting organs and/or other body parts, in response to internal physiological processes and external environmental changes to meet the demands of living organisms and facilitate them in adapting to fluctuating environmental conditions, via generating cellular responses in target organs (Atkinson et al., 2018). Basically, these hormones identify the target cells via specific receptors and get bound to these receptors, which triggers or regulates a series of cellular responses to increase or decrease certain types of biochemical reactions within the target cells, cellular activities and/or functions (Chou & Henderson, 2014). Some hormones are known to affect specific types of target cells, while other hormones influence different types of target cells throughout the human body (Hiller-Sturmhöfel & Bartke, 1998).

The endocrine glands which are responsible for hormonal secretions are located in the hypothalamus, pituitary, thyroid, parathyroid, pancreas, adrenals, gonads (i.e., ovaries in females and testicle in males), and adipose tissues. Other hormones secretory tissues/cells can be found in kidneys, liver, placenta, and gastrointestinal tract (GI) (Hiller-Sturmhöfel & Bartke, 1998; Chou, 2005; Chou & Henderson, 2014). Major targets of the endocrine modulation and regulation include muscles, bones, mammary glands, adipose tissues, and reproductive organs. Additionally, multiple hormones of the endocrine system are involved in the regulation of biochemical and cellular activities within the human body. For instance, sometimes a single stimulus might be triggering the synthesis and secretion of multiple hormones where each of the hormones will be involved in the generation of responses from various target cells, subjected to the existence of hormone specific receptors. The hormones with broad impacts on body metabolism and cellular activities possess receptors in almost all the body organs, tissues, and cells. These responses in turn regulate and exert impacts on the cellular growth, nutrients, and energy utilization/metabolism, release or retention of minerals, phosphorylation of proteins, expression of genes, enzymatic activities, and cell death (Chou, 2005; Chou & Henderson, 2014). Hormones are also known for central nervous system (CNS) and immune system modulation/ regulation. Hence, the basic function of the endocrine system is the maintenance of a steady or stable cellular environment despite of changes or fluctuations in internal physiological and external environmental challenges. Internal challenges encompass metabolic activities, growth, developmental stages, reproduction, and other stresses, while external challenges embrace nutrients availability, temperature, and exposure to different toxicants and pathogens (Chou & Henderson, 2014).

4.3 ENDOCRINE GLANDS AND HORMONES SECRETED BY THE ENDOCRINE GLANDS

The endocrine system is comprised of various types of specialized glands, which are discussed in subsections.

4.3.1 PINEAL GLAND

The pineal gland is a small gland in the brain that is involved in the secretion of melatonin. Melatonin regulates the body's rhythms including circadian rhythms and reproductive cycles (Lu et al., 2017).

4.3.2 HYPOTHALAMUS

Hypothalamus is the master endocrine gland that is located within the human brain and is responsible for the regulation and control of body functions (including hunger, thirst, sexual behaviors and functions, heartbeat, blood pressure, emotions, temperature, and wake and sleep cycles). Hypothalamus produces multiple hormones into the blood vessels that connect the hypothalamus and pituitary gland, hence controlling the hormones production from the pituitary gland. These hypothalamic hormones play very crucial roles in modulation and regulation of all the body functions either by releasing or inhibiting the hormones, which are transported to and act on the adjoining pituitary gland (Hiller-Sturmhöfel & Bartke, 1998; Lu et al., 2017).

4.3.2.1 MAJOR HYPOTHALAMIC HORMONES

4.3.2.1.1 Corticotropin Releasing Hormones (CRH)

Corticotropin-releasing hormones (CRH) is involved in the regulation of protein, carbohydrate, and fat metabolism in the body. It also regulates the water and sodium balance within the human body. This hormone is known to stimulate the secretion of adrenocorticotropic hormone (ACTH) from the anterior pituitary (AP).

4.3.2.1.2 Thyrotropin-Releasing Hormone (TRH)

Thyrotropin-releasing hormone (TRH) stimulates the production of thyroid-stimulating hormone (TSH) and prolactin (PRL) from AP gland for hormonal regulation of lactation. It is also involved in controlling the cellular metabolic processes.

4.3.2.1.3 Gonadotropin Releasing Hormone (GnRH)

Gonadotropin-releasing hormone (GnRH) in involved in regulation of sexual development (at the onset of puberty and maintains male and female physiology after puberty via regulating follicle stimulating and luteinizing hormones (LH)) and reproductive functions (pregnancy and lactation).

4.3.2.1.4 Growth-Hormone Releasing Hormone (GHRH)

Growth-hormone releasing hormone (GHRH) is involved in the release of growth hormone (GH) through AP gland for promoting the growth of organisms.

4.3.2.1.5 Somatostatin

Somatostatin is known to affect the muscle and bone growth by inhibiting the release of growth hormone (GHRH). It also inhibits the production of TSH and several intestinal hormones.

4.3.2.1.6 Dopamine

Dopamine functions primarily as a neurotransmitter but also partakes hormonal effects (i.e., repression/inhibition of PRL, modulation of motor control centers and activation of brain reward centers). PRL in addition to promote lactation also regulates the reproduction cycle, immune system, and metabolism (Lu et al., 2017; Shahid et al., 2021).

4.3.3 PITUITARY GLAND

Pituitary gland (also hypophysis) is attached at the base of the human brain directly below the hypothalamus. Production of hormones from pituitary glands are modulated and regulated by the hypothalamus via the hypothalamic-pituitary connection, hence both internal and external stimuli are influencing the secretion of pituitary hormones. The pituitary gland is known as the master gland of the body because it is involved in stimulating different endocrine glands for inducing synthesis, secretion, and production of their own hormones. It also plays a central role in the maintenance of whole-body homeostasis via receiving hormonal input from the neuroendocrine and secretion of regulatory hormones. The pituitary gland is comprised of two lobes (a) AP also called adenohypophysis or pars distalis; and (b) posterior pituitary also known as neurohypophysis. AP is comprised of 80% of the gland, thus, referred as true endocrine gland, while, posterior pituitary is an extension of the hypothalamus and formed from neural tissues (Hiller-Sturmhöfel & Bartke, 1998; Paez-Pereda et al., 2005; Perez-Castro et al., 2012; Wilkinson and Brown, 2015; Lu et al., 2017).

4.3.3.1 ANTERIOR PITUITARY (AP) HORMONES

Anterior pituitary (AP) is comprised of pars tuberalis and pars intermedia (intermediate lobe). Pars tuberalis consists of an extension of few cell layers surrounding the external region of lower hypophyseal stalk containing blood vessels, while, intermediate lobe s located in marginal area between anterior and posterior pituitary. AP secretes different hormones that either directly influence the target cells/organs or induce the production of hormones from the target glands via stimulation of target glands (for instance, thyroid glands (TGs), adrenal glands, or gonads). The hormones include TSH/thyrotropin, ACTH, GH, gonadotropins, and PRL is produced by lactotrophs. ACTH, TSH, and gonadotropins along with performing their specified functions on their target organs also act on other glands. On the other hand, GHs and PRL directly influence their target cells/organs.

4.3.3.1.1 Thyroid Stimulating Hormones (TSHs)

Thyroid stimulating hormones (TSHs)/thyrotropin are produced by thyrotrope cells. TSH is involved in the production of thyroid hormones (THs) (triiodothyronine [T3] and thyroxine [T4]) via stimulation of TG. These two hormones are involved in conversion food into energy. TG is also involved in the production of calcitonin which maintains calcium homeostasis.

4.3.3.1.2 Adrenocorticotropic Hormone (ACTH)

Adrenocorticotropic hormone (ACTH) is released by corticotroph cells. ACTH in involved in stimulation of adrenal cortex for inducing secretion of corticosteroid hormones or glucocorticoid (cortisol) along with production of male and female sex hormones in smaller quantities. Cortisol regulates protein, lipid, and carbohydrate metabolism.

4.3.3.1.3 Growth Hormone (GH)

Growth hormone (GH)/somatotropin a body stimulating hormone is secreted by somatotroph cells. It influences almost all the cells of body, and promotes the protein synthesis and carbohydrates metabolism and is involved in reduction of fats. It stimulates the bones growth via stimulation of somatomedin (IGF-1) released from the liver.

4.3.3.1.4 Gonadotropins

Gonadotropins are comprised of follicle stimulating hormones (FSH) and LHs, which modulate the secretion of sex hormones (in male testicles/female ovaries) and germ cells production (i.e., ova/egg cells or spermatozoa/sperm cells). FSH promotes the primary follicular growth and release of estradiol from female ovaries, whereas in males it is involved in the stimulation of sperm production and secretion of inhibin in male testicles. The secretion of FSH is in turn regulated by activin and follistatin. On the other hand, LH is involved in the stimulation of ovulation, corpus luteum formation and secretion of progesterone in females, while, in males it induces interstitial cells for the production of androgens (testosterone).

4.3.3.1.5 Prolactin (PRL)

Prolactin (PRL) is produced by lactotroph cells which plays vital role in development of females' breast, initiation of production of milk in mammary glands and maintenance of lactation after child birth.

4.3.3.2 INTERMEDIATE PITUITARY PRODUCES

4.3.3.2.1 α-Melanocyte Stimulating Hormone (α-MSH)

α-Melanocyte stimulating hormone (α-MSH) is involved in the simulation of melanophores which darken the skin color.

4.3.3.2.2 β-Endorphin

β-Endorphin severs as neuromodulator in brain, which regulates the release of neurotransmitters (Hiller-Sturmhöfel & Bartke, 1998; Paez-Pereda et al., 2005; Perez-Castro et al., 2012; Wilkinson & Brown, 2015; Lu et al., 2017).

4.3.3.3 POSTERIOR PITUITARY HORMONES

Posterior pituitary comprises of neural tissues and nerves terminals of axons with their cell bodies located in supraoptic nucleus (SON) and paraventricular nucleus (PVN) of hypothalamus. These axons are projected down from the hypothalamus via part of pituitary stalk (infundibulum) and end up in the posterior pituitary gland. The SON and PVN neurosecretory cells synthesize vasopressin (VP) and oxytocin hormones, which are transported to the axons and stored in posterior pituitary nerve terminals. Both these hormones are released into the blood stream on stimulation of hypothalamic SON and PVN neurons (composing posterior pituitary) and transported to their target cells. The hormones of the posterior pituitary.

4.3.3.3.1 Vasopressin (VP)

Vasopressin (VP) also known as anti-diuretic hormone (ADH) plays an important role in maintenance of electrolyte concentration in the body and promotion of water reabsorption from urine by the kidneys, thus conserves water and serves as ADH. VP is inhibited by an increase in the blood pressure/ high blood pressure.

4.3.3.3.2 Oxytocin

Oxytocin induces contractions of uterine muscles during childbirth and stimulates ejection of milk from mammary glands (Perez-Castro et al., 2012; Wilkinson & Brown, 2015; Lu et al., 2017; El-Sayed et al., 2021).

4.4 THYROID GLAND (TG)

Thyroid gland (TG) is located in front of the trachea (windpipe) just below the larynx (voice box). This gland produces triiodothyronine (T3) and thyroxine (T4). Both these hormones convert food into energy and regulate calcium homeostasis within the body (Hiller-Sturmhöfel & Bartke, 1998; Perez-Castro et al., 2012; Lu et al., 2017).

4.5 PARATHYROID GLANDS

Parathyroid glands are located behind the TG and secrets parathyroid hormones (PTHs) for regulation of calcium concentration in the blood by modulating bone metabolism. It also ensures an adequate supply of calcium for different functions throughout the body, for instance, transmission of signals within the cells and muscles movement. PTH also modulates calcium reabsorption from the urine and excretion of phosphates in urine. PHT stimulates the release of calcium stored in bones and bones resorption as well. Furthermore, PHT also modulates the absorption of calcium from food in the GI (Hiller-Sturmhöfel & Bartke, 1998; Kumar & Thompson, 2010; Lu et al., 2017).

4.6 ADRENAL GLANDS

Adrenal glands are located on the top of the kidneys and are comprised of the adrenal cortex (outer layer) and medulla (inner layer), both of these produce different hormones. The adrenal cortex secretes a wide range of hormones like androgens, aldosterone, and cortisol. On the other hand, the adrenal medulla is the main generator of noradrenaline and adrenaline hormones. These hormones are released in response to stress situations, e.g., fight or flight responses (Hiller-Sturmhöfel & Bartke, 1998; Lu et al., 2017).

4.7 PANCREAS

The pancreas is located behind the stomach in the abdominal cavity. It serves as an exocrine gland because several pancreatic cells produce a variety of digestive enzymes which are released in to gut for effective food digestion. Pancreas also functions as an endocrine gland as the islets of Langerhans synthesize insulin (lowers the blood glucose level) and glucagon (increases the blood glucose level) into the bloodstream. Insulin and glucagon work together and play a crucial role in the regulation of blood glucose levels (Hiller-Sturmhöfel & Bartke, 1998; Lu et al., 2017).

4.8 GONADS

Gonads (i.e., ovaries in females and testicle in males) release sex hormones. These hormones play vital roles in sexual development (onset of puberty) and reproduction (by producing gem cells, i.e., ova in ovaries and spermatozoa in testicles). The sexual hormones include testosterone or androgens which imply masculinization, estrogens or estradiol exerting feminization possessions, and progesterone which influences the uterus in preparation for and during pregnancy (Hiller-Sturmhöfel & Bartke, 1998; Lu et al., 2017).

4.9 THYMUS

The thymus is a lymphoid/immune organ and is a primary site for T cells development and/or lymphocytes. Thymus also secretes thymosins and thymopoietins, which play an important role in immune response regulation (Pearse, 2006; Choudhury et al., 2017; Lu et al., 2017).

4.10 CONCERNS ABOUT ENDOCRINE SYSTEM

Endocrine system constitutes one of the most sensitive coordination and communication system because a wide variety of complex chemical activities and functions (immediate responses or reactions and lifelong functions) within the human body are regulated by the endocrine organs and their respective hormones. All these hormones regulate and stabilize the normal and healthy functioning of the body and in turn, these hormones are regulated by the feedback systems of the body. Any kind of disturbance and disruption in their normal functioning and equilibrium may lead to abnormalities or may bring changes in the behavior, metabolism, growth, and development patterns, which ultimately will lead to adverse health consequences, thus affecting the human's health and their offsprings. Certain types of substances, whether naturally occurring or manmade, can impact the normal functioning of the endocrine system or their related organs and hormones. For instance, environmental contaminants/pollutants and certain types of drugs or plant chemicals (phytoestrogens exerting estrogen-like effects) can have the tendency to mimic, block, and interfere with the hormonal functions or activities of the endocrine system. The chemical messages are misinterpreted as a result of interference of such chemicals with the normal coordination and communication between messenger hormone and cell receptors, consequently lead to the generation of abnormal responses within the body (Bouchard, 2017; Encarnação et al., 2019; CCOHS, 2021).

4.11 TOXICANTS THAT CAUSE DISRUPTION OF ENDOCRINE SYSTEM

Endocrine toxicants (also known as endocrine disruptors (EDs)) refer to a diverse group of exogenous compounds or substances which have been found to interfere with the synthesis, production or secretion, transportation, binding, and elimination of the endocrine hormones involved in modulation and regulation of body homeostasis, behavior, development, and reproduction. These compounds can impact the endocrine system via inhibiting or stimulating (potentiating) the biosynthesis of hormones through direct or indirect mechanisms, thus leading to the production of hormones in decreased and increased amounts. The ETs can directly interact with the hormones and inhibit or stimulate their transcriptional activities, hormone binding proteins, alter the chemical signals sent by hormones and leads to degradation of hormones or their receptor cells. ETs by mimicking the natural hormones can bind to the receptor cells (thus preventing binding of correct hormones), which either generate the stronger signals compared to the naturally occurring hormones or may produce signals that express the responses at wrong time, consequently the body fails to respond appropriately. They can alter or block the ways through which natural hormones travel towards the target organs within the body and resultantly normal signals fails to occur. Moreover, these agents also act as genotoxicants because some of these toxic agents have the ability to alter the structure and function of DNA through their mechanism of action. The ETs even in very minute quantities can exert detectable health implications on exposed individuals and their offspring. Sometimes smaller amounts of these ETs can have cumulative impacts, and in certain circumstances, by-products of chemicals may be greatly harmful than their parent compounds. However, long-term consequences of ETs depend upon the genetic vulnerability of the intact individual, dose of endocrine toxicant, exposure duration and developmental window of exposure (Barouki et al., 2017; Li et al., 2017; Arambula & Patisaul, 2019; Encarnação et al., 2019).

ETs are ubiquitous in the environment, including air, drinking water and food as well. They can be tracked in our everyday life via household cleaning products, cosmetics, and personal care stuff, insect killer items, plastic products, preservatives, and furniture. Different class chemicals including pesticides, fungicides, industrial chemicals, heavy metals, plasticizers, pharmaceuticals, and naturally occurring compounds (like phytoestrogens) in plants and fungi have been classified as ETs. These chemicals in turn are related to numerous chronic ailments such as diabetes, thyroid eruptions, obesity, cardiovascular disease, ovarian diseases, reproductive aberrations/abnormalities, testicular carcinoma, early puberty, nerve or neurological disorders, breast cancers, endocrine related carcinoma, immune, metabolic, and various types of homeostatic disorders (Balaguer et al., 2017; Combarnous, 2017; Li et al., 2017; Slama et al., 2017; Arambula & Patisaul, 2019; Encarnação et al., 2019).

4.12 POTENTIAL ENDOCRINE TOXICANTS (ETS) AND HUMAN EXPOSURE

Major endocrine toxicants (ETs) are discussed in subsections.

4.12.1 PESTICIDES/HERBICIDES

Pesticides/herbicides are the synthetic chemicals or substances that are designed to control and eradicate pests, weeds/herbs, fungi, and animals.

These pesticides are comprised of insecticides, herbicides, fungicides, and rodenticides. Organophosphates, organochlorine, methoxychlor, chlorpyrifos, 2,4-dichlorophenoxyacetic acid (2,4-D), carbamates, DDT, atrazine, pyrethroids, dieldrin, vinclozolin, glyphosate, lindane are most the commonly found chemicals among the pesticides which are applied in various agricultural activities, household cleaning products and public places as well. Human are also exposed to these toxicants during chemicals packaging and handling, spraying of pesticides in agricultural or commercial settings, and consumption of these chemicals in the form of residues on vegetables, fruits, and in water, thus can find their way to the human body via skin, inhalation, or ingestion (Combarnous, 2017; Li et al., 2017; Arambula and Patisaul, 2019; Piazza and Urbanetz, 2019).

4.12.2 INDUSTRIAL CHEMICALS, SOLVENTS, AND BYPRODUCTS

The industrial chemicals like polyaromatic hydrocarbons (PHAs), polychlorinated biphenyls (PCBs), phthalates, phenolic compounds (bisphenol-A (BPA), alkylphenol), brominated flame retardants (BFRs), heavy metals, triclosan (TCS), parabens, perfluorochemicals, and dioxins find the applications in variety of industrial processes or are released as their byproducts. Their main sources include industrial effluents, municipal wastes/wastewater, landfills, incineration, customer care products, dairy, and canned food products, furniture products and carpet backing, etc. Human exposure to these toxicants mainly occurs through building materials, packaging, manufacturing, house dust inhalation, diet, drinking contaminated water, use of personal care and other consumer products (Li et al., 2017; Arambula & Patisaul, 2019; Piazza & Urbanetz, 2019).

4.12.3 NANOMATERIALS

Nanoparticles (comprising of 1 to 100 nm diameter) are being extensively utilized in consumer products, medicines, and industries. However, studies have reported that these nanomaterials exert detrimental health impacts. These materials enter the human body via lungs, intestinal tract, and skin from where it is translocated into various body parts or organs (Iavicoli et al., 2013; Larson et al., 2014; Lu et al., 2017; Priyam et al., 2018).

4.12.4 PLANT HORMONES

Plant hormones play a key role in plants growth and development. Plant hormones also called Phytohormones are synthesized in a specific part and gradually leached towards the target site (El-Maboud, 2019). Depending on the cell requirements, phytohormones have various functions in plants. Ripening of fruits, shedding of leaves, synthesis, and elongation of stem, leaves, and flowers are controlled by plant hormones (Simkin et al., 2020). Other functions include elongation of different plant parts, seed growth, and time of flowering. Different types of plants hormones are known, some of which are made up of various chemicals compounds while others are plant growth regulators that are produced by other organisms or synthesized by humans. Each type of plant hormones either inhibits or regulates certain physiological processes in plants. But certain plants hormones like phytoestrogens are suspected as ETs because they are very much similar to human estrogens. These phytoestrogens can be found in licorice, alfalfa, soy products, herbal supplements, medicinal herbs, legume seeds (beans and peas), celery, and fennel. Humans are exposed to such toxicants through food consumption (McLachlan et al., 2006; Saljoughian, 2007). Phytoestrogens along with having estrogenic activities also exert antiestrogenic effects via binding to estrogen receptors (ERs). They adversely affect the reproduction system and have also been reported to be involved in causing endocrine related breast cancers (Rice and Whitehead, 2006; Wocławek-Potocka et al., 2013; Rietjens et al., 2016).

4.12.5 PERSONAL CARE PRODUCTS (PCPS) AND COSMETICS

Personal care products (PCPs) and cosmetics are used on a routine basis. These products are used for skin care, removal of unpleasant body odor as well as personal hygiene. A previous study (Alani et al., 2013) revealed that average man uses six PCPs while an average woman uses 12 PCPs per day. Every year new PCPs and cosmetics products enter the market. The effects of PCPs and cosmetics on endocrine disruption are not fully understood. The current chapter focuses on chemicals used in PCPs and cosmetics. The possible sources of cosmetics are TCS, parabens, and phthalates. In addition, the present chapter also focuses on the impacts of PCPs and cosmetics on human health. Synthetic musks and ultraviolet filters adversely affect the reproductive system of both animals and humans. These chemicals are used in PCPs and are less studied as compared with phthalates, parabens, and TCS. Sources of synthetic musks are soap, perfume, and cosmetics. In these sources, it is used as fragrance. Common synthetic musks are 1,3,4,6,7,8-hexahydro-4,6,7,8,8,hexamethylcyclopentagamma-2-bezopyrane and 7-acetyl-1,1,3,4,4,6-hexamenthyl-tetraline. To absorb UVA as well as UVB, UV filters were prepared to protect human skin from sunburn. Common sources of UV filters are 3-(-4-methyl benzylidene)-camphor, octylmethoxycinnamate, octyl methoxycinnamate, and benzophenones (Krause et al., 2012). These UV filters are commonly used in sunscreens, creams, lipsticks, lotions, and hairspray (Witorsch and Thomas, 2010).

4.12.6 HEALTH CARE PRODUCTS

Humans are also exposed to ETs through health/medical care products including use of medicines (and some inactive ingredients in medicines formulation contain phthalates, parabens, etc., and medicines used as sex hormones or steroids), medical devices and equipment (microcapillary blood tubes, syringes, catheters, and blood bags, intravenous bags, cardiopulmonary bypass machines, infusion sets, etc., contain phthalates, parabens, TCS, BPA and perfluoroalkyl), exposures related to neonatal intensive care units and pharmaceutical residues via different sources. These toxicants not only provoke and promote the diseases but are also known to antagonize the efficacy of the disease treatments. Such ETs are known to cause immune and hormonal dysfunctions, carcinoma, and multiple health-related problems and disorders (Ingersle et al., 2003; Tijani et al., 2016; Genco et al., 2020).

4.13 PHTHALATES

Phthalates are chemicals widely used as an additive in PCPs, cosmetics, vinyl flooring as well as plastics as plasticizers (Xu et al., 2010). Phthalates are sub-divided into two groups called low molecular-weight phthalates and high-molecular phthalates. Low-molecular-weight phthalates such as emollients, humectants, and skin penetration enhancers are used as an additive in PCPs (Guo & Kannan, 2013). Banned phthalates were detected in some PCPs previously (Celeiro et al., 2015; Guo et al., 2014). New York City report showed the presence of Phthalates such as dibutyl phthalate (DBP) and diethyl phthalate (DEP) in nail polishes, skin tones and perfumes (Guo & Kannan, 2013). Similarly, DEP was detected in higher concentration in hand and body lotions in Tianjin, China (Guo et al., 2014). DEP and DBP were found in higher concentration in fragrances as well as in nail polishes in Canada PCPs and cosmetics products (Koniecki et al., 2011). Phthalate human exposure causes many diseases associated with serious human health risks (Hauser & Calafat, 2005). PCPs are frequently used on a daily basis therefore human exposure to phthalate occurs through many routes. Previous studies have confirmed the presence of phthalates such as DBP, DEP, DEHP, and DMP in PCPs (Parlett et al., 2013; Al-Saleh & Elkhatib, 2015). Phthalates severely effects ovaries development in females particularly follicle cells. In females, the primordial follicles regulate the reproductive cycle (Hirshfield, 1991).

4.14 PARABENS

In cosmetics and food products, parabens are largely used as antimicrobial as well as preservative agents. Human exposure to parabens occurs through inhalation, ingestion as well as dermal exposure (Liao et al., 2013). In females, Paraben's exposure is of great concern because it weakens the estrogen hormonal activity (Darbre & Harvey, 2008). In adults the detected concentration of Parabens was 5–50 mg/kg/day while in infants 15–230 mg/kg/day (Guo & Kannan, 2013). Parabens when enters the human body, can affect many biological systems. It can be found in urine, serum as well as seminal plasma (Frederiksen et al., 2011). Parabens were detected in amniotic fluids, suggesting that they can affect fetal growth and development. The effects of Parabens in human as well as animal ovarian development are scares (Taxyig et al., 2008).

4.15 TRICLOSAN (TCS)

Triclosan (TCS) was first developed in the 1960s. Its chemical formula is 5-chloro-2-(2,4-dichlorophenoxy) phenol) and it acts as an antibacterial agent. The main sources of TCS are PCPs and consumer products such as hand lotions, toothpastes, deodorant, shower gels and hand soap. In addition, it is also used in toothbrush, pizza cutters, mop handles, cutting boards, plastics, fabrics, and shoe inserts (Sabaliunas et al., 2003). Concentration of TCS ranges from 0.1% to 0.3% in PCPs and other antimicrobial products. TCS is detected in human blood and serum, breast milk and urine (Pycke et al., 2014). TCS daily human exposure varies range between from 0.047 to 0.073 $mg^{-1}kg^{-1}day^{-1}$ (Rodricks et al., 2010). TCS normally enters the human body through dermal contact or via ingestion. Common sources of increased

TCS in the human body are deodorants and air fresheners (Yang et al., 2015). The effects of TCS on human ovary are unknown. A previous study showed that mice paired ovary weights decreased after 13 weeks of exposure to TCS (125 $mg^{-1}kg^{-1}day^{-1}$ (Fang et al., 2015). Animal's study revealed that uterus structure is severely affected with TCS exposure. Relationship between men and TCS is unclear. Animals such as Wister rats showed no symptoms upon exposure to TCS 3–300 $mg^{-1}kg^{-2}day^{-1}$ (Zorrilla et al., 2009). Human study revealed that idiopathic male infertility is not associated with the presence of TCS in urine (Chen et al., 2013). In human males, the content did not change after 13 weeks of exposure to TCS (0–125 mg^{-1} kg^{-1} day^{-1} dermal exposure (Table 4.1) (Fang et al., 2015).

4.16 DISRUPTION OF NORMAL FUNCTIONING OF ENDOCRINE SYSTEM

EDs are linked with a wide range of health issues, mainly the metabolic variations resulted in long-term disease exposure worldwide comprising diabetes, obesity, and fatty liver disorder (Küblbeck et al., 2020). Endocrine disrupting chemicals (EDCs) alters the normal response of organs in particular situation. There are certain proteins in the human body known as "hormone receptors" which controls the effect of hormones on a particular cell. EDCs may directly inhibit the hormone action through interaction with the receptor or by normal delivery interference with the receptor (Diamanti et al., 2009; Zoeller et al., 2014). Major target receptors of EDCs are the nuclear receptors. The mechanism of action of EDCs and the range of health effects caused by them are under scientific debate from decades (Combarnous and Nguyen, 2019), but the recent findings through scientific investigations declare that disruption mechanisms are more extensive than previously reported (Diamanti-Kandarakis et al., 2009).

4.16.1 ENDOCRINE DISRUPTION MECHANISMS

The interference of EDCs with normal functioning of human physiology can occur by various means either through direct interaction with the hormone receptors or by indirect mechanism altering the endogenous hormonal levels and mess up the normal hormonal balance of human body ultimately resulting in metabolic, developmental, and reproductive disorders (Combarnous, 2017; Casals-Casas & Desvergne, 2011). EDCs can affect all levels of

TABLE 4.1 Endocrine Toxicants, Their Potential Sources and Routes of Human Exposure

Endocrine Toxicant	Category	Sources	Human Exposure	References
Organophosphates Carbamates Pyrethroids Chlorpyrifos Methoxychlor DDT	Insecticides	• Agricultural application and runoff • Industrial production and effluents • Household cleaning products	• Chemicals manufacturing and packaging • Spraying of pesticides in agricultural, public places or commercial settings • Consumption of residues on vegetables and fruits • Drinking contaminated water • Inhalation	Combarnous (2017); Li et al. (2017); Arambula and Patisaul (2019); Piazza and Urbanetz (2019)
Atrazine	Herbicide	• Agricultural application and runoff • Industrial production and effluents • Homegardens	• Chemicals manufacturing and packaging • Spraying agricultural commercial settings • Consumption of residues on vegetables and fruits • Drinking contaminated water • Inhalation	Piazza and Urbanetz (2019)
Vinclozolin	Fungicide	• Agricultural application and runoff • Industrial production and effluents	• Chemicals manufacturing and packaging • Spraying in agricultural or commercial settings • Consumption of residues on vegetables and fruits • Drinking contaminated water • Inhalation	Arambula and Patisaul (2019)
Lead	Heavy metals	• Paints • Smelting • Mining and refining • Lead acid batteries	• Dust inhalation • Drinking contaminated water • Children are exposed by licking of painted walls	Li et al. (2017); Arambula and Patisaul (2019); Piazza and Urbanetz (2019)

TABLE 4.1 *(Continued)*

Endocrine Toxicant	Category	Sources	Human Exposure	References
		• Lead based petrol/gasoline • Jewelry • Children toys and products	• Leads to abnormalities in brain development and genomic methylation	
Cadmium	Heavy metals	• Metallic pigments • Batteries • Plastics	• Dust inhalation • Drinking contaminated water • Eating food preserved in plastic containers • Exerts harmful effects to fetal umbilical cord and placental DNA • Bioaccumulates in kidneys and liver	Piazza and Urbanetz (2019)
Mercury	Heavy metals	Burning of charcoal	• Ingestion of contaminated fish • Affects the immune system and promotes inflammatory disorders and mitochondrial disruption (methylmercury)	Piazza and Urbanetz (2019)
Arsenic	Heavy metals	–	• Consumption of contaminated drinking water • Associated with immune disorders, cardiovascular diseases, neurotoxicity, carcinogenesis	Li et al. (2017); Arambula and Patisaul (2019)
Dioxin	–	• Industrial processes • Municipal waste incineration	Consumption of dairy products	Arambula and Patisaul (2019)
PCBs	Industrial chemical	• Industrial effluents and waste • Electrical transformers and capacitors • Hydraulic fluids	Drinking contaminated water	Li et al. (2017); Arambula and Patisaul (2019); Piazza and Urbanetz (2019)

TABLE 4.1 *(Continued)*

Endocrine Toxicant	Category	Sources	Human Exposure	References
		• Building materials • Additives, adhesives, resin, rubber, and paint industry		
Bisphenol-A (BPA)	• Industrial chemical • Artificial estrogen	• Medical devices • Food packaging • Epoxy resins and epoxy-based lining of food containers • Thermal paper • Polycarbonate plastic-based containers used in water and baby bottles	• Dietary ingestion • Exposure via air and drinking water • Alters endocrine system in multiple ways	Meeker (2010); Li et al. (2017); Arambula and Patisaul (2019)
Phthalates	Industrial chemical	• Plasticizers used in plastics manufacturing • Personal care products and cosmetics • Pharmaceuticals • Food packaging • Vinyl flooring and wall covering • Carpet backing	• Use of personal care and consumer products • Diet or consumption of contacted food	Meeker (2010); Arambula and Patisaul (2019)
Polybrominated diphenyl ethers	Brominated flame retardants	• Flame retardant • Textiles • Thermoplastics in	• Inhalation, dermal contact, and uptake of dust • Ingestion of contaminated food	Meeker (2010); Li et al. (2017); Arambula and Patisaul (2019)

TABLE 4.1 *(Continued)*

Endocrine Toxicant	Category	Sources	Human Exposure	References
		electronics (TV and computers) • Foam Furniture • Out- and in-door contaminant	Leaching of treated material into the environment	
Perfluorochemicals	–	• Nonstick cookware • Food packaging • Clothing and textiles • Waterproofing • Stain resistant • Lubricants	Exerts reproductive effects	Zlatnik (2016); Li et al. (2017)
Perchlorate	–	• Fertilizers • Rocket fuel	• Drinking contaminated water • Targets thyroid and causes hypothyroidism	Zlatnik (2016); Li et al. (2017)
Triclosan	–	Antibacterial	Use of personal care products, toothpastes	Li et al. (2017)
Phytoestrogens	Found in plant materials	Pulp mill effluents	Food consumption	Dodson et al. (2012)
Alkylphenols	–	• Used as surfactants • Personal care products	Use of personal care and consumer products	Helm et al. (2018)

cellular hormonal trails, but their interference with the hormone receptors linked with the group of nuclear receptors as well as extra-nuclear receptors are frequently reported (Swedenborg et al., 2009; Balaguer et al., 2019).

4.16.2 NUCLEAR RECEPTORS AS THE MAJOR TARGETS OF ENDOCRINE TOXICANTS (ETS)

All biological activities of human body, such as growth and development, reproduction, immunity, metabolism, and organ homeostasis and many other processes are control by a total number of 48 nuclear receptors. These nuclear receptors are the prime target of EDCs. Out of all NRs, endogenous ligands have been recognized for 24 family members. These receptors are categorized as ligand-dependent transcription factors due to their direct response to a wide range of hormonal and metabolic compounds which are water repellant, lipid soluble and very small in size (e.g., retinoic acid or estradiol). Another group of NRs is known as orphan receptors. In this class, regulatory ligands are yet unidentified or non-existent. Post-translational alterations and cell-specified expressions of co-regulatory proteins and many chemicals of medicinal and environmental importance are responsible for the regulation of transcriptional activity of orphan receptors (Balaguer et al., 2019). The duration and the quantitative limit of exposure are the main factors responsible for assessing risk associated with EDCs, but the exact acceptable threshold levels are still under debate. EDCs reveal various mechanisms of action comprising:

- Connecting with a hormone receptor resulting in activation of it signaling pathway;
- Connecting with a hormone receptor resulting in hindering of it signaling pathway;
- Interference with hormonal component indicator pathway downstream of a receptor;
- Stimulation of endogenous hormone biosynthesis;
- Inhibition of endogenous hormone biosynthesis;
- Linking with the flowing hormone-binding protein;
- Stimulation/inhibition of hormone-binding protein formation or degradation;
- Stimulation of hormone receptor expression;
- Inhibition of hormone receptor expression (Combarnous & Nguyen, 2019).

The first mechanism among all the above-mentioned mechanism is shared between all interactions related to hormones while the rest of them from 2 to 7 can cause imbalanced endocrine homeostasis which is not sequential to the direct hormonal-type mode of action by EDCs.

4.16.2.1 MECHANISM 1

The enormous interaction of manmade compounds of variable size and amount with relevant form and dimensions of hormones demonstrate that they may attract the receptors and get attached to them (Montes-Grajales and Olivero-Verbel, 2015). The receptors have not only modified enough to identify their respective cognate ligand but moreover inhibit the linking of additional endogenous molecules with similar assemblies. The receptors are secured against connecting with endogenous molecules approaching hormones. Yet they are not protected against interaction with pristine molecules, which may never come across before throughout the alteration process. Furthermore, the anthropogenic compound/molecules, besides of their lesser attraction towards receptors in contrast to hormones, can exist in relatively bulk amounts and interact with original hormones. Xenobiotic interaction with NRs of hormones is a good example of this mechanism. Undeniably, small-sized hormone-binding sites are present in these receptors, yet they can possibly lodge not only numerous industrial synthetic organic molecules but also the compounds of natural origin. Additionally, as a result of their transcriptional activity, target cell phenotype is intensely affected by the activation through xenobiotic substance. EDCs can also exhibit their interaction with membrane receptors. Numerous examples are available of this traditional mechanism, wherein EDCs perform actions like hormones through direct interface and activation of hormone receptors. However, the degradation of EDCs is probably less efficient as compared to the natural hormones hence show more activities *in vivo* due to the extended half-life in blood or in cells (Sheng et al., 2019).

4.16.2.2 MECHANISM 2

Some novel molecule-resembling hormones are capable of attaching to the receptor while others have the ability to freeze the receptor during their inactive condition resulting in abnormal endogenous hormonal action. Exogenous molecules can directly perform endocrine disruption by this mechanism of action. For instance, a group of compounds known as polychlorinated biphenyls (PCBs) can halt the transcription process by suppressing the attachment of T3 with the thyroid receptor site leading towards the dislocation of active TR complex from thyroid response element (TRE) (Boas et al., 2012). Treated wastewater samples were analyzed and numerous activities like anti-estrogenic, anti-androgenic, anti-progestronic, and anti-ER responses were observed (Combarnous & Nguyen, 2019; Rao et al., 2014). This mechanism involves binding of EDCs to the receptors as hormones through antagonistic approach rather than hormones. The mode of action can be verified *in vitro* while the cytotoxic responses of unidentified compounds can be misperceived *in vitro* analysis leading to the uncertain and biased interpretation of results (Marin-Kuan, 2017).

4.16.2.3 MECHANISM 3

Many EDCs interact with the endocrine system by direct contact with the pathways of hormonal signaling and associated components till receptor stimulation stream. There are certain molecules with generally variable structures as compared to hormones. E.g., In Prozac anti-depressant, fluoxetine (FLX) is used as a SSRI (selective serotonin reuptake inhibitor) active material responsible for modifications in many intracellular signaling pathways in several cellular forms (Liu et al., 2015; Stepulak, 2008).

4.16.2.4 MECHANISMS 4 AND 5

The biosynthesis (Mechanism 4) and degradation (Mechanism 5) of endogenous hormones can be affected either positively or negatively, which ultimately cause endocrine disruption without any direct interaction with hormone receptors. The physical structure of such molecules is diverse due to their indirect mode of action at the receptor level of hormones. Mechanism 4 can be explained well with the example of BPA which can cause inhibition of adiponectin discharge *in vitro* adipocytes in human body even at minor dosage (Qiu et al., 2013; Fenichel et al., 2013). Polybrominated diphenyl ethers (PBDEs) are frequently used flame retardants have been reported to perform the stimulation of liver enzymes responsible for glucuronidation resulting in an enhanced T4 abolition of T4 and blood concentration decline (Boas et al., 2012). Similarly, Parabens are well-known and commonly used ingredients in cosmetics cause inhibition of 17β-hydroxysteroid dehydrogenase and subsequently impede estrogen degradation resulting in an enhanced blood hormone level (Engeli et al., 2017).

4.16.2.5 MECHANISM 6

Several hormones, predominantly thyroid and steroid hormones which are hydrophobic in nature bind themselves with the proteins in the bloodstream. In addition to the transport proteins, they have the potential to bind with the minor water-repellent hormones due to the hydrophobic properties of EDCs. For instance, various compounds are reported to react with SHBG (steroid hormone-binding protein) or AFP (α-fetoprotein). They are capable of interacting with the transportation and concentration of steroid hormones in blood. This mechanism of action is adopted by the EDCs having structural similarities with the hormones in competition to bind with the hormone-binding transport proteins at circulating proteins' level. *In vitro* studies can be conducted for in-depth analysis of this mode of action.

4.16.2.6 MECHANISM 7

EDCs following this mode of action have dissimilar chemical structures from hormones. For example, the action of PBDEs comprises of down-regulation of carrier protein, i.e., transthyretin (TTR) leading to lesser level of T4 in blood. Several toxic compounds come under the category of EDCs when comparing this mechanism of action due to the fact that hormone-binding transport proteins are usually produced and/or demolished by the liver, which is a chief degrading organ and a core point of attack for toxic substances (Viluksela et al., 2004).

4.16.2.7 MECHANISM 8

Many EDCs interact with endocrine homeostasis through activation of endogenous hormone receptors. For instance, BPA is responsible for *in vitro* activation of the leptin receptor expression within cancerous cells of ovary (Ptak and Gregoraszczuk, 2012). Similarly, heavy metal specifically cadmium exposure of endothelial HUVEC cells can induce remarkable proliferation of two main enzymes including estradiol receptor β (ERβ) and Cyp19a1 at the level of mRNA and proteins leading towards a major drop down of androgen receptor (AR) when 24 hours of exposure were over (Fittipaldi et al., 2018).

4.16.2.8 MECHANISM 9

Another mechanism which can cause disruption of the endocrine system is the inhibition of receptor expression. It is reported that low oral intake of BPA by rats may lead to inhibition of estrogen receptors expression in hypothalamic cells in rats (Rebuli et al., 2014). Moreover, the *in vitro* and *in vivo* studies from the patients of breast and prostate cancer have also shown inhibition of androgen receptor expressions (Stossi et al., 2016).

4.17 HEALTH IMPACTS FROM EXPOSURE TO ENDOCRINE TOXICANTS (ETS)

Intensive scientific research on the effects of EDCs on the endocrine system and NRs have demonstrated their impacts on the performance of reproductive, immune, and nervous systems (Tabb & Blumberg, 2006).

4.18 EFFECTS OF EDCS ON METABOLISM

The possibility of association of EDCs for obesity and metabolic syndrome (MetS) development is obvious. The MetS, also known as syndrome 'X' or insulin resistance (IR) syndrome is the result of dysregulation of hormone metabolism. It is defined as a condition of multiple disorders involving disturbed glucose tolerance/lacking IR, dyslipidemia, hypertension, and rapid or unnecessary weight gain. *In vitro* exposure to EDCs can also lead to obesity and associated disorders (Phillips & Prins, 2008). Such kind of disorders once established are impossible to revert and may cause further health issues like diabetes and cardiovascular ailments. The genetic factors of developing obesity are also linked with the exposure to EDs (Newbold et al., 2007). The aryl hydrocarbon receptor, estrogen receptor and retinoid X receptor (RXR)/peroxisome proliferator-activated receptors are the three prime receptors in the human body which can be affected from the exposure to EDCs through any of the previously discussed mechanisms however, the exact quantity of exposure to EDCs is still need to be explored more (Swedenborg et al., 2009).

4.19 EFFECTS OF ENDOCRINE TOXICANTS (ETS) ON BONES

Studies have shown the impact of EDCs on bone degradation. Modification in bone paracrine hormonal formation can disturb the process of bone sculpting which may damage bone tissues. Alteration in the discharge of systemic hormones, cytokines, chemokines, and growth dynamics as well as bone marrow mesenchymal stem cell variation. The bone disrupting effects are evidenced due to the exposure of various assemblies of EDCs, including the perfluoroalkyl compounds, phthalate esters, organotin substances, BPA, alkyl-phenols, and dioxin (as well as similar to dioxin) compounds (Turan, 2021).

4.20 EFFECTS ON REPRODUCTIVE SYSTEM

The reproductive system is highly prone to the actions of EDCs due to the chemical structure of EDCs, which has the ability to bind with the sex gonadal hormones (Monneret, 2017). Animal models were used to study the effects of EDCs on ovarian development, which proved that prenatal exposure to these toxicants can be highly dangerous (Lauretta et al., 2019). Various epidemiological studies in different regions of the world came up with the conclusion that exposure to toxic substances is directly associated with reproductive disorders, including reduced sperm count in males besides mammary glands ailments and infertility in females (Diamanti-Kandarakis et al., 2009; Jagne et al., 2016). The exposure may cause mutilation to the developing ovary through interaction with germ cell nest destruction, meiosis, follicle growth and development. Genomic transcription of somatic cells resulting in disturbed physiological process of puberty in terms of delayed anticipated timings (Schug et al., 2011). Moreover, the EDCs are also reported to negatively affect male and female fertility (Zama & Uzumcu, 2010), polycystic ovarian syndrome (PCOS), testicular hypotrophy, and testicular dysgenesis syndrome (TDS) (Zoeller et al., 2012). The adversities associated with the exposure to phthalate, 17α-ethinylestradiol and BPA were also investigated in target animals and steroidogenesis as well as estrogen receptor pathway-induced meiotic disruption were reported (You & Song, 2021).

4.21 CARCINOGENIC EFFECTS OF ENDOCRINE TOXICANTS (ETS)

BPA is responsible for inducing changes in cell structure and prostate glands development which can magnify the risk of prostate cancer (Maffini et al.,

2006). The growth and development of prostate cancer is highly influenced by specific EDCs, which is evident from studies carried on animal models under the phenomena of prostate carcinogenesis. An intensification of prostate carcinogenesis in reported in target animals when exposed to PCBs, diethylstilbestrol (DES), cadmium, UV filters, BPA, and arsenic. Prostate glands, at the stage of fetal development in the uterus and neonatal stages, are at higher risk of abnormalities caused by EDCs. Newborns and younger children are highly vulnerable group of population fronting the exposure, whereas the risk gets enhanced with aging (Coster and Larebeke, 2012).

4.22 CONCLUSION

Human beings are presently exposed to a diversity of various toxic compounds, which can cause short- and long-term health issues. *In vitro* research and experiments on animal models verified the severity of risk associated with the exposure of anthropogenic toxicants, more specifically known as EDCs. Significant research studies evidenced that a range of EDCs, including PCDs, phthalates, halogenated compounds, BPA, etc., are adversely affecting human health by impairing normal functioning of the body. Disturbed growth and development, cancers, metabolic, reproductive, and hormonal disorders are all associated with the exposure of EDCs. Their mechanism of action in individuals may vary from person to person depending upon the duration and level of exposure hence determining the nature and severity of disorders. Prenatal exposure to EDCs may lead to abnormal fetal development, premature child birth. Newborns and children are highly prone to the effects on neurodevelopment, while adults may have an issue in growth and development.

KEYWORDS

- **communication strategies**
- **disruption mechanism**
- **endocrine hormones**
- **endocrine toxicants**
- **gonadotropin-releasing hormone**
- **growth hormone**

REFERENCES

Atkinson, S., Crocker, D. E., & Ortiz, R. M., (2018). Endocrine systems. In: *Encyclopedia of Marine Mammals* (3rd edn., pp. 318–328). Elsevier Inc. https://doi.org/10.1016/B978-0-12-804327-1.00118-7.

Balaguer, P., Delfosse, V., & Bourguet, W., (2019). Mechanisms of endocrine disruption through nuclear receptors and related pathways. *Curr. Opin. Endocr. Metab. Res., 7*, 1–8.

Balaguer, P., Delfosse, V., Grimaldi, M., & Bourguet, W., (2017). Structural and functional evidence for the interactions between nuclear hormone receptors and endocrine disruptors at low doses. *C. R. Biol., 340*, 414–420. doi: http://dx.doi.org/10.1016/j.crvi.2017.08.002.

Barouki, R., (2017). Endocrine disruptors: Revisiting concepts and dogma in toxicology. *C. R. Biol., 340*, 410–413. doi: http://dx.doi.org/10.1016/j.crvi.2017.07.005.

Boas, M., Feldt-Rasmussen, U., & Main, K. M., (2012). Thyroid effects of endocrine disrupting chemicals. *Mol. Cell Endocrinol., 355*(2), 240–248. doi: 10.1016/j.mce.2011.09.005.

Bouchard, P., (2017). Endocrine disruptors: Endocrine-disrupting chemicals, a multifaceted danger. *C. R. Biol., 340*, 401, 402.

Canadian Centre for Occupational Health and Safety (CCOHS), (2021). *Endocrine Disruptors*. Available from: https://www.ccohs.ca/oshanswers/chemicals/endocrine.html (accessed on 30 August 2022).

Casals-Casas, C., & Desvergne, B., (2011). Endocrine disruptors: From endocrine to metabolic disruption. *Annu. Rev. Physiol., 73*, 135–162.

Chaudhry, M. S., Velardi, E., Dudakov, J. A., & Marcel-Van-Den-Brink, R. M., (2016). Thymus: The next (Re) generation. *Immunol. Rev., 271*(1), 56–71. doi: 10.1111/imr.12418.

Chou, K., & Henderson, J., (2014). Endocrine system. *Encycl. Toxicol., 2*, 332–340.

Chou, K., (2005). Endocrine system. In: *Encyclopedia of Toxicology* (2nd edn., pp 170–174). © Elsevier Inc.

Combarnous, Y., & Nguyen, T. M. D., (2019). Comparative overview of the mechanism of action of hormones and endocrine disruptor compounds. *Toxics., 7*(5), 1–10.

Combarnous, Y., (2017). Endocrine disruptor compounds (EDCs) and agriculture: The case of pesticides. *C. R. Biolo., 340*, 406–409. doi: http://dx.doi.org/10.1016/j.crvi.2017.07.009.

Coster, S. M., & Larebeke, N. V., (2012). Endocrine disrupting chemicals: Associated disorders and mechanism of action. *J. Environ. Public Health,* 713696.

Diamanti-Kandarakis, E., Bourguignon, J. P., Giudice, L. C., Hauser, R., Prins, G. S., Soto, A. M., Zoeller, R. T., & Gore, A. C., (2009). Endocrine-disrupting chemicals: An endocrine society scientific statement. *Endocrine Rev., 30*, 293–342.

El Sayed, S. A., Fahmy, M. W., & Schwartz, J., (2021). Physiology, pituitary gland. In: *Stat Pearls*. Treasure Island (FL): Stat Pearls Publishing. Available from: https://www.ncbi.nlm.nih.gov/books/NBK459247/ (accessed on 30 August 2022).

Encarnação, T., Pais, A. A. C. C., Campos, M. G., & Burrows, H. D., (2019). Endocrine disrupting chemicals: Impact on human health, wildlife and the environment. *Sci. Prog., 102* (1), 3–42. doi: 10.1177/0036850419826802.

Fenichel, P., Chevalier, N., & Brucker-Davis, F., (2013). Bisphenol A: An endocrine and metabolic disruptor. *Ann. Endocrinol (Paris), 74*, 211–220. doi: 10.1016/j.ando.2013.04.002. 31.

Fittipaldi, S., Bimonte, V. M., Soricelli, A., Aversa, A., Lenzi, A., Greco, E. A., & Migliaccio, S., (2018). Cadmium exposure alters steroid receptors and proinflammatory cytokine levels in endothelial cells *in vitro*: A potential mechanism of endocrine disruptor atherogenic effect. *J. Endocrinol. Invest., 42*(6), 727–739. doi:10.1007/s40618-018-0982-1.

Genco, M., Anderson-Shaw, L., & Sargis, R. M., (2020). Unwitting accomplices: Endocrine disruptors confounding clinical care. *J. Clini. Endocrinol. Metabol., 105*(10), e3822-e3827. doi: 10.1210/clinem/dgaa358.

Heindel, J. J., Blumberg, B., Cave, M., Machtinger, R., Mantovani, A., Mendez, M. A., Nadal, A., et al., (2017). Metabolism disrupting chemicals and metabolic disorders. *Reprod. Toxicol., 68*, 3–33.

Helm, J. S., Nishioka, M., Brody, J. G., Rudel, R. A., & Dodson, R. E., (2018). Measurement of endocrine disrupting and asthma-associated chemicals in hair products used by black women. *Environ. Res., 165*, 448–458.

Hiller-Sturmhöfel, S., & Bartke, A., (1998). The endocrine system: An overview. *Alcohol. Health. Res. World, 22*(3), 153–164.

Iavicoli, I., Fontana, L., Leso, V., & Bergamaschi, A., (2013). The effects of nanomaterials as endocrine disruptors. *Intl. J. Mol. Sci., 14*(8), 16732–16801. doi: 10.3390/ijms140816732.

Ingerslev, F., Vaclavik, E., & Halling-Sørensen, B., (2003). Pharmaceuticals and personal care products: A source of endocrine disruption in the environment? *Pure Appl. Chem., 75*, 1881–1893. doi: 10.1351/pac200375111881.

Jagne, J., White, D., & Jefferson, F., (2016). Endocrine disrupting chemicals: Adverse effects of bisphenol A and parabens to women's health. *Water Air Soil Pollut., 227*, 182. doi: https://doi.org/10.1007/s11270-016-2785-3.

Küblbeck, J., Vuorio, T., Niskanen, J., Fortino, V., Braeuning, A., Abass, K., Rautio, A., et al., (2020). The EDCMET project: Metabolic effects of endocrine disruptors. *Int. J. Mol. Sci., 21*(8), 3021. doi: https://doi.org/10.3390/ijms21083021.

Kumar, R., & Thompson, J. R., (2011). The regulation of parathyroid hormone secretion and synthesis. *J. Am. Soc. Nephrol., 22*(2), 216–224. doi: 10.1681/ASN.2010020186.

Larson, J. K., Carvan, M. J., & Hutz, R. J., (2014). Engineered nanomaterials: An emerging class of novel endocrine disruptors. *Biol. Reprod., 91*(1), 1-8. doi: https://doi.org/10.1095/biolreprod.113.116244.

Lauretta, R., Sansone, A., Sansone, M., Romanelli, F., & Appetecchia, M., (2019). Endocrine disrupting chemicals: Effects on endocrine glands. *Front. Endocrinol., 10*, 1–7.

Li, X., Gao, Y., Wang, J., Ji, G., Lu, Y., Yang, D., Shen, H., et al., (2017). Exposure to environmental endocrine disruptors and human health. *J. Publ. Health Emerg.,* 1, 8–21. doi: 10.21037/jphe.2016.12.09.

Liu, X. L., Luo, L., Mu, R. H., Liu, B. B., Geng, D., Liu, Q., & Yi, L. T., (2105). Fluoxetine regulates mTOR signaling in a region-dependent manner in depression-like mice. *Sci. Rep., 5*, 16024. doi: 10.1038/srep16024.

Lu, X., Zhu, Y., & Zhu, T., (2017). Endocrine system. In: *Adverse Effects of Engineered Nano Materials* (339–354). Elsevier Inc. http://dx.doi.org/10.1016/B978-0-12-809199-9.00014-8.

Maffini, M. V., Rubin, B. S., Sonnenschein, C., & Soto, A. M., (2006). Endocrine disruptors and reproductive health: The case of bisphenol-A. *Mol. Cell. Endocrinol.,* 179–186.

Marin-Kuan, M., Fussell, K. C., Riederer, N., Latado, H., Serrant, P., Mollergues, J., Coulet, M., & Schilter, B., (2017). Differentiating true androgen receptor inhibition from cytotoxicity-mediated reduction of reporter-gene transactivation *in-vitro*. *Toxicol. In-Vitro, 45*, 359–365. doi: 10.1016/j.tiv.2017.03.014.

McEwen, B. S., (2016). Central role of the brain in stress and adaptation: Allostasis, biological embedding, and cumulative change. In: *Stress: Concepts, Cognition, Emotion, and Behavior* (pp. 39–55). Elsevier Inc. doi: http://dx.doi.org/10.1016/B978-0-12-800951-2.00005-4.

McLachlan, J. A., Simpson, E., & Martin, M., (2006). Endocrine disruptors and female reproductive health. *Best Pract. Res. Clin. Endocrinol. Metabol., 20*(1), 63–75. doi: 10.1016/j.beem.2005.09.009.

Meeker, J. D., (2010). Exposure to environmental endocrine disrupting compounds and men's health. *Maturitas., 66*, 236–241.

Monneret, C., (2017). What is an endocrine disruptor? *C. R. Biol.,* 340(9, 10), 403–405.

Montes-Grajales, D., & Olivero-Verbel, J., (2015). EDCs data bank: 3D-structure database of endocrine disrupting chemicals. *Toxicol., 327*, 87–94. doi: 10.1016/j.tox.2014.11.006.10.

Newbold, R. R., Padilla-Banks, E., Snyder, R. J., Phillips, T. M., & Jefferson, W. N., (2007). Developmental exposure to endocrine disruptors and the obesity epidemic. *Reprod. Toxicol.,* 23, 290–296.

Paez-Pereda, M., Kuchenbauer, F., Arzt, E., & Stalla, G. K., (2005). Regulation of pituitary hormones and cell proliferation by components of the extracellular matrix. *Braz. J. Med. Biolo. Res., 38*, 1487–1494.

Pearse, G., (2006). Normal structure, function, and histology of the thymus. *Toxicol. Pathol., 34*, 504–514. doi: 10.1080/01926230600865549.

Perez-Castro, C., Renner, U., Haedo, M. R., Stalla, G. K., & Arzt, E., (2012). Cellular and molecular specificity of pituitary gland physiology. *Physiol. Rev., 92*, 1-38. doi: 10.1152/physrev.00003.2011.

Phillips, L. K., & Prins, J. B., (2008). The link between abdominal obesity and the metabolic syndrome. *Curr. Hypertens. Rep., 10*, 156–164.

Piazza, M. J., & Urbanetz, A. A., (2019). Environmental toxins and the impact of other endocrine disrupting chemicals in women's reproductive health. *JBRA Assisted Reprod., 23*(2), 154–164. doi: 10.5935/1518-0557.20190016.

Priyam, A., Singh, P. P., & Gehlout, S., (2018). Role of endocrine-disrupting engineered nanomaterials in the pathogenesis of type 2 diabetes mellitus. *Front. Endocrinol., 9*, 704. doi: https://doi.org/10.3389/fendo.2018.00704.

Ptak, A., & Gregoraszczuk, E. L., (2012). Bisphenol A induces leptin receptor expression, creating more binding sites for leptin, and activates the JAK/Stat, MAPK/ERK and PI3K/Akt signaling pathways in human ovarian cancer cells. *Toxicol. Lett., 210*, 332–337.

Qiu, L. L., Wang, X., Zhang, X. H., Zhang, Z., Gu, J., Liu, L., Wang, Y., et al., (2013). Decreased androgen receptor expression may contribute to spermatogenesis failure in rats exposed to low concentration of bisphenol A. *Toxicol. Lett., 219*(2), 116–124.

Rao, K., Li, N., Ma, M., & Wang, Z., (2014). *In vitro* agonistic and antagonistic endocrine-disrupting effects of organic extracts from wastewater of different treatment processes. *Front. Environ. Sci. Eng. Chin., 8*, 69–78. doi: 10.1007/s11783-013-0502-7.

Rebuli, M. E., Cao, J., Sluzas, E., Delclos, K. B., Camacho, L., Lewis, S. M., Vanlandingham, M. M., & Patisaul, H. B., (2014). Investigation of the effects of subchronic low dose oral exposure to bisphenol A (BPA) and Ethinyl estradiol (EE) on estrogen receptor expression in the juvenile and adult female rat hypothalamus. *Toxicol. Sci., 140*(1), 190–203. doi:10.1093/toxsci/kfu074.

Rice, S., & Whitehead, S. A., (2006). Phytoestrogens and breast cancer promoters or protectors? *Endocrine-Related Cancer, 13*, 995–1015. doi:10.1677/erc.1.01159.

Rietjens, I. M. C. M., Louisse, J., & Beekmann, K., (2016). The potential health effects of dietary phytoestrogens. *British J. Pharmacol., 174*, 1263-1280. doi: https://doi.org/10.1111/bph.13622.

Robin, E., Nishioka, D. M., Laurel, J., Laura, L. J., Perovich, B. J. G., & Rudel, R. A., (2012). Endocrine disruptors and asthma-associated chemicals in consumer products. *Environ. Health. Perspect., 120*(7), 935–943. doi: 10.1289/ehp.1104052.

Saljoughian, M., (2007). Focus on phytoestrogens. *U.S. Pharma., 32*(12), HS-27–HS-32.

Schug, T. T., Janesick, A., Blumberg, B., & Heindel, J. J., (2011). Endocrine disrupting chemicals and disease susceptibility. *J. Steroid Biochem. Mol. Biol., 127*(3, 5), 204–215.

Shahid, Z., Edinen, A., & Singh, G., (2021). Physiology, hypothalamus. In: *Stat Pearls*. Treasure Island (FL): Stat Pearls Publishing. Available from: https://www.ncbi.nlm.nih.gov/books/NBK535380/ (accessed on 30 August 2022).

Sheng, Z., Wang, C., Ren, F., Liu, Y., & Zhu, B., (2019). Molecular mechanism of endocrine-disruptive effects induced by bisphenol A: The role of transmembrane G-protein estrogen receptor 1 and integrin alphavbeta3. *J. Environ. Sci. (China)., 75*, 1–13. doi: 10.1016/j.jes.2018.05.002.

Slama, R., Vernet, C., Nassan, F. L., Hauser, R., & Philippat, C., (2017). Characterizing the effect of endocrine disruptors on human health: The role of epidemiological cohorts. *C. R. Biol., 340*, 421–431. doi: http://dx.doi.org/10.1016/j.crvi.2017.07.008.

Stepulak, A., Rzeski, W., Sifringer, M., Brocke, K., Gratopp, A., Kupisz, K., Turski, L., & Ikonomidou, C., (2008). Fluoxetine inhibits the extracellular signal regulated kinase pathway and suppresses the growth of cancer cells. *Cancer Biol. Therap., 7*, 1685–1693.

Stossi, F., Dandekar, R. D., Bolt, M. J., Newberg, J. Y., Mancini, M. G., Kaushik, A. K., Putluri, V., et al., (2016). High throughput microscopy identifies bisphenol AP, a bisphenol A analog, as a novel AR down-regulator. *Oncotarget., 7*(13), 16962–16974. doi:10.18632/oncotarget.7655.

Swedenborg, E., Ruegg, J., Makela, S., & Pongratz, I., (2009). Endocrine disruptive chemicals: Mechanisms of action and involvement in metabolic disorders. *J. Mol. Endocrinol., 43*(1), 1–10.

Tabb, M. M., & Blumberg, B., (2006). New modes of action for endocrine disrupting chemicals. *Mol. Endocrinol., 20*, 475–482.

Tijani, J. O., Fatoba, O. O., Babajide, O. O., & Petrik, L. F., (2016). Pharmaceuticals, endocrine disruptors, personal care products, nanomaterials and perfluorinated pollutants: A review. *Environ. Chem. Lett., 14*, 27–49. doi: https://doi.org/10.1007/s10311-015-0537-z.

Turan, M. D. S., (2021). Endocrine disrupting chemicals and bone. *Best Pract. Res. Clin. Endocrinol. Metabol., 2021*, 101495. doi: https://doi.org/10.1016/j.beem.2021.101495.

Viluksela, M., Raasmaja, A., Lebofsky, M., Stahl, B. U., & Rozman, K. K., (2004). Tissue specific effects of 2,3,7,8-tetrachlorodibenzo-p-dioxin (TCDD) on the activity of 50-deiodinases I and II in rats. *Toxicol. Lett., 147*, 133–142.

Wilkinson, M., & Brown, R., (2015). The pituitary gland and its hormones. In: *An Introduction to Neuroendocrinology* (pp. 45–56). Cambridge: Cambridge University Press. doi: 10.1017/CBO9781139045803.004.

Wocławek-Potocka, I., Mannelli, C., Boruszewsk, D., Kowalczyk-Zieba, I., Waśniewski, T., & Skarżyński, D. J., (2013). Diverse effects of phytoestrogens on the reproductive performance: Cow as a model. *Intl. J. Endocrinol., 2013*, 650984. doi: https://doi.org/10.1155/2013/650984.

You, H. H., & Song, G., (2021). Review of endocrine disruptors on male and female reproductive systems. *Comparat. Biochem. Physiol. Part C: Toxicol. Pharmacol., 244*, 109002. doi: https://doi.org/10.1016/j.cbpc.2021.109002.

Zama, A. M., & Uzumcu, M., (2010). Epigenetic effects of endocrine-disrupting chemicals on female reproduction: An ovarian perspective. *Front. Neuroendocrinol., 31*(4), 420–439. doi: 10.1016/j.yfrne.2010.06.003.

Zlatnik, M. G., (2016). Endocrine-disrupting chemicals and reproductive health. *J. Midwifery Women's Health, 61*(4), 442–455. doi: 10.1111/jmwh.12500.

Zoeller, R. T., Bergman, A., Becher, G., Bjjeregaard, P., Bornman, R., Brandt, I., Iguchi, T., et al., (2014). A path forward in the debate over health impacts of endocrine disrupting chemicals. *Environ Health., 13*, 118. doi: https://doi.org/10.1186/1476-069X-13-118.

Zoeller, R. T., Brown, T. R., Doan, L. L., Gore, A. C., Skakkebaek, N. E., Soto, A. M., Woodruff, T. J., & Saal, F. S. V., (2010). Endocrine-disrupting chemicals and public health protection: A statement of principles from the endocrine society. *Endocrinol., 153*(9), 4097–4110. doi: 10.1210/en.2012-1422.

CHAPTER 5

Toxicological Manifestations of Endocrine Toxicants

YOUNIS AHMAD HAJAM,[1] SEEMA RAI,[2] and MOHD. SALIM RESHI[3]

[1]*Department of Life Sciences and Allied Health Sciences, Sant Baba Bhag Singh University, Padhiana, Jalandhar, Punjab, India*

[2]*Department of Zoology, Guru Ghasidas Vishwavidayalaya, Bilaspur, Chhattisgarh–495001, India*

[3]*Department of Zoology, School of Biosciences and Biotechnology, Baba Ghulam Shah Badshah University, Rajouri, Jammu and Kashmir–185234, India, E-mail: reshisalim@gmail.com*

ABSTRACT

Endocrine toxicity results when a disrupting chemical inhibits the synthesis, secretion, transportation, or action/elimination of hormones, and finally, it leads to tissue dysfunction, growth, and development. Endocrine-disrupting chemicals (EDCs) lead to endocrinal hormonal imbalance and cause endocrine toxicity. EDCs can be grouped into industrial, agricultural, pharmaceutical, phthalates, bisphenol-A (BPA), and polybrominated biphenyls (PBBs). Industrial chemicals are further divided into PCBs (polychlorinated biphenyls, dioxins, and alkyl-phenols. Also, agricultural toxic chemicals are included as herbicides, fungicides, pesticides, insecticides, and phytoestrogens. Humans and other mammals are the top predators in the food chain. So that's why they store larger doses of EDCs. In humans, it exposes lifelong to many compounds may bring about cumulative, preservative, and synergic properties. Many EDCs causing wandering effects at altered concentrations depending upon doses. Furthermore, due to different actions of hormone receptors, different

Environmental Endocrine Toxicants: Biology, Effects, and Management. Younis Ahmad Hajam, Rouf Ahmad Bhat, Khalid Rehman Hakeem, & Rajesh Kumar (Eds.)
© 2023 Apple Academic Press, Inc. Co-published with CRC Press (Taylor & Francis)

EDCs apply non-traditional responses to the dose. Endocrine toxicants (ETs) can affect the central nervous system (CNS), and recent evidence shows that some EDCs may affect CNS and damages hypothalamic and pituitary gland functions therefore, disturbing the actions of peripheral glands. Therefore, the present chapter aimed to discuss about the toxicological manifestation of the EDCs.

5.1 INTRODUCTION

Endocrine toxicity results when a disrupting chemical inhibits the synthesis, secretion, transportation, or action/elimination of hormones and finally it leads to tissue dysfunction, growth, and development. Endocrine disrupting chemicals (EDCs) leads to endocrinal hormonal imbalance, and causes endocrine toxicity. It inhibits the hormonal systems and cause adverse effects such as developmental, immunological, and reproductive effects, etc. It can be found in different types of products such as detergents, food, toys, cosmetics, plastic bottles, and pesticides, etc. Though, endocrine toxicants (ETs)/disruptors effects on human health and the environment. Research studied showed that these toxicants effects on human health. It increases the progression of diseases included as diabetes, obesity, infertility, and cancers. These chemical substances act as endocrine modulators, active compounds, and hormones, etc. (Yang et al., 2015). EDCs can be grouped into: industrial, agricultural, pharmaceutical, phthalates, bisphenol-A (BPA), and polybrominated biphenyls (PBBs). Industrial chemicals are further divided into PCBs (polychlorinated biphenyls, dioxins, and alkylphenols. Also, agricultural toxic chemicals are included as herbicides, fungicides, pesticides, insecticides, and phytoestrogens (De Coster and van Larebeke, 2012; Kabir et al., 2015; Monneret, 2017). Heavy metals are also included as EDCs like mercury, lead, and mercury (De Coster and van Larebeke, 2012; Lubrano et al., 2013). EDC affects/contaminated human body through different pathways. Most common is direct contact or food intake characterize most common routs/pathways (Gore et al., 2015; Kabir et al., 2015; Balaguer et al., 2017). Moreover, EDCs might enter the food chain through these routes and their accumulation in animal tissues, mainly adipose tissue takes place. Most of the ETs are highly lipophilic (Gore et al., 2015; Heindel et al., 2015; Sargis, 2015). These features understand the accumulation of EDCs in adipose tissue for many years (Barouki, 2017). The superior predators of food chain are human beings and other mammalian species. When a person is exposed towards any endocrine disrupting chemical, it may causes serious health issues according to the concentration of doses. These adverse health issues may occur due to structural changes in cell surface receptors on which hormones bind (Kabir et al., 2015; Barouki, 2017). The complication of the harmful effects of EDCs and showing the evidence of timing of exposure in the endocrine system (Zoeller et al., 2012; Lubrano et al., 2013; Barouki, 2017; Unuvar and Buyukgebiz, 2012). The key sources of ETs / disruptors are pesticides and fungicides. Meanwhile, in the year 1996, the EPA (Environmental Protection Agency) has been emerging and endorsing the EDSP (endocrine disruption screening program), which purposes to generate etiquettes to research accuracy, effective, and replicable recognition approaches, create an importance about the constituents/elements to be investigated, evaluate the useful strategies in order to differentiate the accurate analysis for any exposure condition (Zoeller et al., 2012; Gore et al., 2015; Marty, 2018).

ETs can imitate the natural hormones action in the body, and bind or stimulate several hormone receptors, i.e., glucocorticoid receptor, thyroid hormone receptor, androgen receptor (AR) and estrogen receptor, etc. (Mnif et al., 2011; Schug et al., 2011; Monneret, 2017; Balaguer et al., 2017). In particular, the same constituents may activate certain receptor agonist isoforms and blocking antagonist other isoforms. Similarly, these components are selective to such estrogen receptor modulators behavior (Lubrano et al., 2013). ETs have the potential to inhibit the hormone synthesis, and hormone exclusion. Thus, the synthesis, transportation, and metabolism of hormones decrease the concentration of endogenous hormones (Zoeller et al., 2012; De Coster and van Larebeke, 2012). Beside this, EDCs should be considered that affect endocrine system (additive or synergistic effects) (Barouki, 2017). The risk of process is tremendously complex and causes toxicity associated complications (Barouki, 2017; Campion et al., 2012). ETs mainly targeted the hypothalamic-pituitary-thyroid axis, hypothalamic-pituitary-adrenal axis, and hypothalamic-pituitary-gonadal axis. It may target on most of organ system such as liver, kidney, and reproductive system (Gore et al., 2015; Kabir et al., 2015). However, there are various studies which showed that ETs can affect the central nervous system (CNS) and some recent studies revealed that some endocrine disruptive chemicals may affect and damages the functions of hypothalamic and pituitary gland consequently, disturbing the activities of peripheral glands as well (De Coster and van Larebeke, 2012; Kabir et al., 2015). EDCs may disturb homeostasis and vary in size and number of adipose tissues. It impaired the regulation of appetite, metabolic rate, adipose tissue, and the generation of adipocytokine. These chemicals or toxicants are also known as obesogens because of physical inactivity and low-calorie diet (Heindel, 2015). Diethylstilbestrol (DES) is a chemical having estrogens, and it has the ability to bind with great empathy. Estrogen receptor-α and estrogen receptor-β have a significant function in regulating fattiness and central or peripheral energy stability. It induces adipogenesis and causes obesity or overweight (Gore et al., 2015; Heindel, 2015). Moreover, it is the beginning of various prolonged metabolic ailments, including diabetes, dyslipidemia, and obesity. It can be characterizing human contact with environmental toxicants and enhanced chronic diseases (Figure 5.1) (Gore et al., 2015; Heindel, 2015).

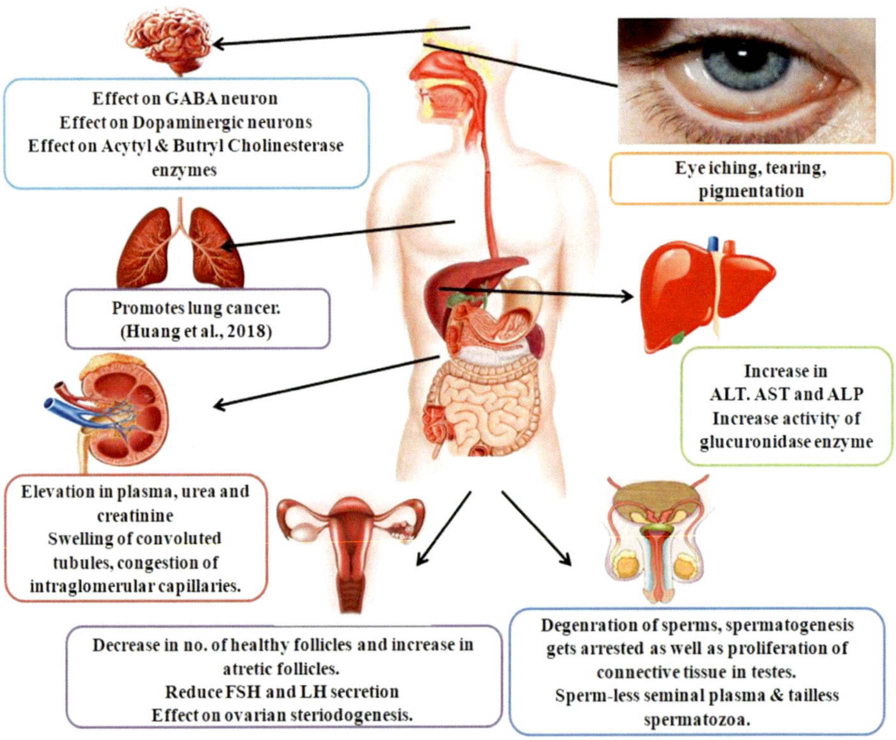

FIGURE 5.1 The effect of endocrine toxicants on various parts of the human body.

5.2 EXPOSURE AND REGULATION OF ENDOCRINE DISRUPTORS (EDS)

The exposure to ETs is mainly from the major route, i.e., food. BPA occurs or contaminated through food and water (Cichna-Markl, 2012). Mostly, household products contain pollutants and become an important source of endocrine disruptor (ED) exposure (Charles, 2009). EDs accumulate in the different organs of the body (Sharpe and Irvine, 2004). Phthalates can easily release into the environment and produced industrially and naturally (Bertelsen et al., 2013). But non-phthalates are produced industrially and degraded alkylphenol ethoxylates by environmental activities. It initiates primarily from the degradation of non-phthalates ethoxylates which are commonly used as industrialized surfactants (Soares et al., 2008). There are a number of environmental affluences, such as heavy metals, which are reported as ETs. In the human body, various tissues like liver, kidney, and testis are mainly exposed to toxicity of heavy metals (Siu et al., 2009). From the various industries and agricultural practices, toxic chemicals are discharged into the natural environment (Järup and Akesson, 2009). Tobaccos is thought to be the key source of human exposure to toxic chemicals. Mostly, disclosure by way of tobaccos is key provenance to human disclosure towards heavy metals (Pappas et al., 2006). With classifying indication of the destructiveness of ETs, their use is heavily controlled and the human body weight of the ETs' deterioration. The first step is to reduce body burden and also its production. The second step, due to the high number of EDs lowers the body weight (Darbre, 2017).

5.3 ENDOCRINE TOXICANTS (ETS)

5.3.1 BISPHENOL-A (BPA)

Bisphenol-A (BPA) is a carbon-based synthetic compound, and the chemical formula is $(CH_3)_2C\ (C_6H_4OH)_2$. BPA belongs to diphenylmethane derivatives and isophenols. Since 1957, it is commercially use and poorly soluble in water. In 2010, US Food and Drug Administration (USFDA) reported that the identified risks in infants, mother womb and children. Human consumes food with the intake of EDs (BPA) and direct contamination of food or through water. BPA is leaked from the food lining and beverages, and it is commonly used as plastics forms. This ingredient is used to protect from direct food and beverages (Goodson et al., 2002). Chemicals, plastics, paper, heavy metal waste and staining can add/enter directly into the environment in the form of BPA. BPA is an endocrine toxicant that mimics estrogen hormone and negative effect on human health. It increases the risk of obesity, hormonal imbalance, and reproductive dysfunction. BPA disclosure with human foreskin fibroblast chambers from hypospadias. It is some of the supreme familiar hereditary deformities with worldwide incidences 0.2% to 1.0%. They documented 29-upregulated genes and 42-down-controlled genes after the disclosure of bisphenols-A, signifying interrelated purposes in relation to bisphenols-A (Yin et al., 2016; Qin et al., 2012).

5.3.2 DI-2-ETHYLHEXYL PHTHALATE (PHTHALATE CATEGORY)

Di-2-Ethylhexyl phthalate (DEHP) is an organic compound and its chemical formula is [C_6H_4 ($C_8H_{17}COO$)] (Meeker et al., 2009). It is constituent of 2-ethyl hexanol (branched chain) and phthalic acid (diester bond). Mostly it is used as plasticizer and manufacturing PVC (polyvinyl chloride) (Jamarani et al., 2018). DEHP are produced 3 billion (per kgs) annually all over the world and causes environmental pollution. It can be absorbed from water and food and known as ED. It affects development and causes cardiotoxicity (Sharman et al., 1994; Swan, 2008). Generally, children who are exposed to phthalates showing more complications as compared to adults (Sathyanarayana et al., 2008; López-Carrillo et al., 2010). The current issue is the connection between endocrine disruption and obesity. Phthalate (Di-2-ethylhexyle phthalate) metabolites revealed that the significant relationships between obesity and insulin resistance (IR) (Desvergne et al., 2009). Phthalates inhibits the action of testosterone in the testis, it follows PPAR-α and p53 signaling pathway. These signaling pathways involved in disruption and can be signal of phthalate disclosure. PPAR-α is a collective genetic factor that deviates in phthalate exposure in cardiac muscle functioning (Posnack et al., 2012). It is a significant controller of lipid metabolic breakdown and growth of peroxisome (You et al., 2014; Zhou and Flaws, 2017). There are three main genes, GJA1, CLDN6, and MMP2 showing defect in testicular carcinoma cells (Yao et al., 2012). GJA1 considerably downcast controlled and CLDN6, MMP2 are considerably up controlled. It performs cell-cell interaction and causes breast cancer cells as well as other cancers (Hsieh et al., 2012). In cancer cells, (LEF1 lymphoid enhancer-binding factor 1) is the most important gene, and contribution in the Wnt signaling pathway. It plays an important role in differentiation and follicle morphogenesis (Rishikaysh et al., 2014).

5.3.3 NONYLPHENOL (NP)

Nonylphenol (NP) is commonly known as alkylphenols and used in lubricating, additives, detergents, and emulsifiers, etc. (Soares et al., 2008).

"Alkylphenol ethoxylates" and "NP ethoxylates" are the commercially important non-ionic surfactants originate from the NPs. These originated compounds are further used in plastics, paints, detergents, etc. NPs can also act as xenoestrogen like ETs because these bind to the receptors of estrogen, might show estrogen-like activity and can inhibit natural estrogens. NPs have been shown to mimic the natural hormone 17β-estradiol and to compete with the endogenous hormone for binding with estrogen receptors ERα and ER (Soares et al., 2008). NPs can cause toxicity to the testis and immune system simultaneously. Ajj et al. (2013) detected the changes in the expression of DNA cytosine-5-methyltransferase gene (DNMT3) caused by the ETs (Ajj et al., 2013). During development, DNA cytosine-5-methyltransferase gene (DNMT3) is obligatory for genome-wide de novo methylation and is known to be DNA methylation patterns (Koh and Rao, 2013). An illustrative disease associated with DNMT3 is "immunodeficiency-centromeric instability-facial anomalies syndrome 1" (ICF3) where various research studies show limited hypomethylation of DNA present in small parts in the patient's genome (Wijmenga et al., 2000; Jiang et al., 2005). Another study, NPs is showing alteration in miRNAs expression (Choi et al., 2011). The analysis of the study showed that the mouse exposed to NP having changes in their miRNAs level, tracking the micro-RNA targets. CCNE1 and "peroxisome proliferator-activated receptor-α (PPAR-α)" are the two genes and altered in miRNAs. CCNE1 is a crucial factor of the cell cycle at the G1/S (start) modification, and PPAR-α gene is called as ligand-activated transcription factor and lipid metabolism (key regulator) (Gorla-Bajszczak et al., 1999). Cell to cell interaction and cellular association were activated by 20 genes, and these genes are responsible for cell cycle and cell death.

5.4 ENDOCRINE ASPECTS OF EDCS

According to the endocrine system aspects, ETs have the ability to bind endocrine nuclear receptors, or reversed agonists or as antagonists (Gore et al., 2015). These toxicants activate hormone receptors and mimic the natural hormonal action (Mnif et al., 2011; Schug et al., 2011; Monneret, 2017). But in some cases; it activates/blocks the certain isoforms. ETs may delay hormone metabolism and selective for estrogen receptor modulators (Gore et al., 2015; Heindel et al., 2015). Certainly, these toxicants are causing various chronic metabolic diseases, i.e., obesity, diabetes, hypertension, and dyslipidemia.

5.4.1 PITUITARY GLAND

In diencephalic system, it represents a special target of ETs, which may alter the functioning of CNS. EDCs imitating the neurotransmitter action and binds to the endocrine receptors. It acts on the pituitary gland and influences the various endocrine axes. As a result, the clinical manifestations of endocrine system have been related to the exposure to pollutants and causes circadian disruption or delayed puberty (Soriano-Guillén et al., 2019; Sen and Sellix, 2016).

5.4.2 REPRODUCTIVE SYSTEM

In the reproductive system, most of ETs mimic sex hormones and most susceptible endocrine axis the action of EDCs. It is able to bind the endocrine receptors and interfere with hormonal signals (Monneret, 2017). Main five categories of ETs described by USEPA with anti-androgenic properties, first is synthetic estrogens or drugs, second is phytoestrogens, third is pesticides (such as organophosphates, carbamates, and organochlorines, etc.), fourth is plasticizers (such as PVC) and last is industrial substances or its by-products such as dioxins, flame retardants, dioxins, etc. (Harding et al., 2006). When assessing their initial developmental stages effects on organism/human during the timing of ETs exposure should be considered when assessing their effects on the organism (Sweeney et al., 2015). Various ETs damage the developing ovary by restricting germ cell damage. During meiosis process, it causes destruction of follicles (Sweeney et al., 2015). During the alterations in the genetic transcription of somatic cells, which may be delays the timing of physiological process (Schug et al., 2011). Additionally, ETs look to have a reliable role in declining both female and male fertility. ETs correlated with ovarian dysfunction (polycystic ovarian syndrome and testicular hypotrophy (Schug et al., 2011). The signs of developmental disorders might be deliberated by damaged spermatogenesis, testis cancer, and hypospadias. Nowadays, testicular hypotrophy is increasing because of the harmful environmental effects. It is due to the disruption of physiological embryonic programming and gonadal development during fetal life (Virtanen et al., 2005; Schug et al., 2011; Skakkebaek et al., 2003). Usually, "phthalates, acetaminophen, and polybrominated diphenyl ethers (PBDE)" plays an important role in the beginning of TDS. In females, recent studies suggested that chronic exposure to ETs can change the fertility of females, and may cause endometriosis (Minguez-Alarcon and Gaskins, 2017). Phthalates and

BPA may be implicated in the beginning of endometriosis happening in 10% of fertile women. It causes infertility in 50% affected people/animals (Caserta et al., 2013). DDT (dichlorodiphenyltrichloroethane), phthalates, BPA, parabens, and TCS are exposed to ovarian pathology (Marques-Pinto and Carvalho, 2013). Pollutants and ETs may affect the male reproductive system. Due to sedentary lifestyle and obesity plays a critical role and decrease male fertility. EDCs such as BPA, vinclozolin, and phthalates are well-known therapeutic drugs (Handelsman and Cooper, 2013).

5.4.3 EFFECT ON ADRENAL GLAND

Merely some investigations have evaluated the consequences of the chemical ETs on the adrenal gland, particularly about the hazards associated with disclosure to chemical elements. This scarcity of proof is not certainly understandable, seeing that the appropriate functionality of the HPA (hypothalamic-pituitary-adrenal) alliance is essential for the life of humans and is a general objective for numerous drugs and chemicals. Without a doubt, adrenal gland shows some organizational and bio-chemical characteristics which construct ultimate marks for ETs, for example, lipophilic structure, a raised blood flow because of the raised level of polyunsaturated fatty acids on the cell membrane, and occurrence of Cytochrome P450 enzymes generating free radicals and deadly metabolic products (Harvey, 2016). The maximum examined consequence of ETs on this axis is characterized by their capability to affect the metabolism and bio-synthesis of steroid hormones via the intonation or the suppression of enzymes convoluted in the synthesis of steroids. In metabolic pathways of adrenal steroidogenesis 5-α reductase, aromatase, along with "3-β, 11-β, and 17-β hydroxysteroid dehydrogenases" have a vital role and their accurate role is hindered by ETs, in specific xenoestrogens damage adrenal role preventing these enzymes (Kabir et al., 2015). Similarly, StAR (steroidogenic acute regulatory protein), which controls the initial stage of synthesis of steroids inside adrenal, is an objective of ETs (Hampl et al., 2016). Hundreds of drugs and chemicals may co-operate along with the HPA axis, and each phase of steroid synthesis may be exaggerated by endocrine disrupting compounds along with every chemical disruptor may turn changing dissimilar phase of steroidogenesis. In assessing the disease-causing impacts of EDCs on HPA axis, it must be reflected that yet an inadequate damage of appropriate role of adrenal can have major consequence on well-being of humans. In this perception, bio-accumulation of these chemical compounds in adipose soft tissue could produce a cocktail, along with proven results that may be witnessed merely later numerous years of persistent, low-dose disclosure. Among the chemicals, Hexachlorobenzene is capable of disturbing role of corticoid hormone in Wistar rats (Braverman et al., 2005).

5.4.4 INHIBITION OF ENDOCRINE RECEPTORS

Endocrine system is greatly influenced by EDs at every level. Initially, they can interrupt the enzymes action convoluted in the synthesis of steroids. These enzymes might be repressed, as can the enzymes have tangled in estrogens metabolic breakdown. Some PCBs metabolic products prevent the action of sulfotransferase enzyme, which results in a rise of circulating estradiol (Kester et al., 2000). The passage of hormones is moreover subjected by assured chemical compounds able to interrelate along with the obligatory locations of sex hormone binding globulin, therefore contending with endogenous estrogens. The maximum investigated way of act of ETs is their capability to impasse and stimulate ERs (estrogen receptors) in objective cell (Celiket al., 2008). On the other hand, it is well-known that the two ERs (estrogen receptors) facilitate different biological properties in many cells, including the brain, vascular system, bone, and mammary glands in both the males and female's body. As a result, since ERα and ERβ express dissimilar distribution of cells and different physiological roles, ETs could exhibit agonist or antagonist action during development. In view of the substantial changes in organizational characteristics and comparative empathy of ligand binding of the estrogen receptors, ETs may prompt dissimilar configurational variations in the postsecondary arrangement of the ERs (estrogen receptors), influencing the employment of co-factors in a different way. These communications between ERs (endocrine receptors) and co-activators or co-repressors are perilous phases in ER-facilitated transcriptional regulation and subsequently the intonation of the communication of ER-target genes. Additionally, the effect of genistein dependent upon various aspects containing the expression of particular co-factors, the ERα and ERβ ratio, and the degree of communication of definite intracellular kinases, comprising tyrosine kinases of cytoplasm and it is often tissue specific. In cancer cells, genistein has been evaluated to reveal both propagative and non-propagative properties (Jagadeesh et al., 2006). In general, ETs perform in hundreds to thousands folds larger levels as compared to estradiol but may have chemical or synergic properties with endogenous estradiol or after once they are found in mixture (Kortenkamp and Altenburger, 1998).

Moreover, the capability of some ETs to perform as agonists in particular cells and as antagonists in the other cells give rise to the occurrence and practice of particular ERs (estrogen receptors) regulators, in specific for anti-hormonal dealings, for example, raloxifene and tamoxifen. A few EDs can also have an impact on the estrogen receptor non-proteomics passageways and prompt an endocrine interruption. According to research accomplished on organizationally dissimilar EDs exhibited that at their raised levels, DES, and BPA are capable of triggering ERs (endocrine receptors) through the stimulation of mitogen-stimulated phosphotidyl inositol 3-kinase and protein kinase inside breast cancer cells. Moreover, a few EDs have been observed to activate the protein kinase C (PKC) (Li et al., 2006). Remarkably, PKC has been stated to control the transcriptional activity of ERα. Consequently, additive or synergic consequences among the passageways to syndicate the stimulation of estrogen receptor indicating could be possible. Cd (Cadmium) is recognized as a chief endocrine toxicant which influences the hormonal production and its control (Henson and Chedrese, 2004). Certainly, cadmium simulated synthesis of progesterone in JC-410 porcine granulosa cells and stimulated the ERα and mimic estrogen in various cells, for example uterus and mammary gland and breast cancer cell lines (Smida et al., 2004; Johnson et al., 2003; Brama et al., 2007). Cd regulates the gene expression and activity in LN cap cells of androgen receptors (Martin et al., 2002). In male rodents, cadmium expressively ups and downs the numerous hormones circulating, e.g., "testosterone, follicle-stimulating hormone (FSH), and luteinizing hormone (LH)" (Lafuente et al., 2004). Furthermore, its compact LH receptor, cyclic adenosine monophosphate (cAMP), and STAR protein levels in the testis (Gunnarsson et al., 2007). Cadmium exerts influence on the circadian array discharge of noradrenaline, a controller of hormone secretion from the hypothalamus, which give rise to variations in the regular pattern of LH and plasma testosterone levels. Additionally, plasma concentrations in hormones released from the pituitary gland, e.g., FSH, LH, adrenocorticotropic, and prolactin (PRL) hormone were changed after exposure of cadmium (Lafuente et al., 2003).

5.4.5 EDCS AND ENDOCRINE GLANDS CANCER

According to the research conducted by a number of scientific associations, including "European Commission; European Environmental Agency; The Endocrine Society; WHO; IARC" concluded the connection in the middle of EDCs and cancer influencing prostate, testis, breast, and thyroid signifying that the disclosure to a few EDCs may be a threat aspect in the growth of the tumors (Gore et al., 2015; Lamb et al., 2014; Trasande et al., 2016). PDBE (Penta and decabromodiphenyl ether), organochlorides, PCBs, DDT (dichloro diphenyl trichloroethane), DDE (dichloro diphenyl dichloroethylene), fungicides, pesticides, Ar, and Cd have been projected to show a part in testicular cancer. Apart from that, TCDD (2,3,7,8-Tetrachlorodibenzo-p-dioxin), PCBs, pesticides, and solvents seem to be convoluted in cancer of the thyroid. A number of other compounds are capable of damage thyroid functions. Zeng et al. (2017) employed records retrieved from a study of Connecticut and observed the connotation with work-related disclosure to pesticides and biocides. PCBs, dioxins, cadmium, ethylene oxide, phytoestrogens, and furans may subsidize in the progress of breast cancer. In conclusion, PCBs, arsenic, pesticides, and cadmium look like to subsidize prominently to prostate cancer. According to the current analysis undertaken by Benedetti et al. (2017) stated that the occurrence of these tumors in particular contaminated zones in Italy, i.e., NPCSs (National Priority Contaminated Sites). A considerably greater frequency of cancer, including prostate and breast was detected in these zones; however, the occurrence of breast cancer appears to be amplified in respective zone, though the occurrence of prostate cancer seems to proliferate in specific zones and to decline in others (Benedetti et al., 2017). This information recommends that an extensive variety of chemical substances may subsidize to growth of breast cancer where certain chemicals may disturb prostate carcinogenesis. As above-mentioned, EDCs may attach to and stimulate or obstruct numerous endocrine receptors (Kim et al., 2018), performing as whole, limited, or reversed agonist or antagonist. The capability to bind AhR (Aryl hydrocarbon receptor) through its subsequent signifies a recognized act of ETs; AhR is known as a cell nuclear receptor and performs as ligand stimulated transcript element that intellects the occurrence of external complexes for example "persistent organic pollutants (POPs)" and generate cytochrome P450 enzymes stimulation which is required for excrete out them from the body, being convoluted in controlling the reaction to XREs (xenobiotic responsive elements) (Fujii-Kuriyama and Kawajiri, 2010; Mackowiak and Wang, 2016). In addition to the purification and bio-activation of XREs (xenobiotic responsive elements), AhR has been exposed to normalize necessary enzyme expression towards breakdown of endogenous elements such as hormones and to control a considerably varied variety of cellular cycles concerning cell propagation, cell differentiation and cell death (De Coster and van Larebeke, 2012). Tetrachlorodibenzo P-dioxin (TCDD) is recognized as an AhR ligand, and has significant carcinogenic action, signifying that AhR stimulation may perform a function in carcinogenesis (De Coster and van Larebeke, 2012; Kim and Lee, 2017). Additionally, AhR may be stimulated by a number of EDCs, including PCBs, furans, dioxins, and polycyclic aromatic hydrocarbons (PAHs), though with dissimilar effectiveness (Fujii-Kuriyama and Kawajiri, 2010). Without of its ligands, AhR stays in the cytoplasm supported via a cluster of proteins like co-chaperone (Swedenborg et al., 2009). AhR interacting protein performs an essential function in adapting the feedback to XREs (xenobiotic responsive elements); by way of a material of detail, its transmutation carries almost the damage of its role and accordingly a variation of organic passageways organized via AhR triggering a damage in the reaction to XREs (xenobiotic responsive elements) (Fujii-Kuriyama and Kawajiri, 2010). Apart from that, due to the incidence of ligands, AhR arrives into the nucleus and attached to translocator inside nucleus, recognized as ARNT (AhR nuclear translocator), and its co-activators generating a molecular compound whose role is to attach xenobiotic responsive elements triggering certain target-genes (Mackowiak and Wang, 2016; Cannavo et al., 2017). In addition, the detoxification and bio-activation of XREs (xenobiotic responsive elements), AhR has been exposed to control enzymes expression which are necessary to the metabolic breakdown of endogenous substances such as hormones and to control an ample extensive variety of cellular passageways comprising cell propagation, cell division and cell death (De Coster and van Larebeke, 2012). The damage of role of AIP (Acute intermittent porphyria) brings about through its transmutation displayed a noticeable connotation with FIPAs (familiar isolated pituitary adenomas); especially it has been witnessed that FIPAs demonstrate unfavorable feedback to remedial healings, enhanced sternness, and prior age of beginning (Jennings et al., 2009; Beckers et al., 2013). A few indications displayed the connection amid endocrine disruptive compounds and the growth of FIPAs, underlining that AhR may be attacked by ETs with mutagenic effects (Daly et al., 2006); in specific, the linkage amid a substantial disclosure to carcinogenic, capable to bind AhR, and an amplified prevalence of FIPAs has been examined (Lloyd and Grossman, 2014).

5.5 DEVELOPMENTAL TOXICITY

A number of studies concluded that numerous hormones perform serious function in all phases of the developing practice, it is to be expected that exogenous ecological elements which are capable to parodist, obstruct, or control messengers of endogenous compound be able to create negative progressive results (Birnbaum, 1994). DDT (Dichloro diphenyl trichloroethane) signifies a standard pattern of EDCs. Disclosure of humans to DDT (Dichloro diphenyl trichloroethane) throughout pregnancy effects in small head circumferences and low birth weight in children (Fein et al., 1984). In animals like bald eagles, DDT (Dichloro diphenyl trichloroethane) turned out to be exposed to influence developing being because of eggshell retreating and furious (Faber and Hickey, 1973). "TCDD" is also one of the examples of ETs, and has been revealed to be a developing and propagative chemical in various classes, to modify the system of hormones, comprising the thyroid, and to bring about the variations in equilibrium (Jones and Hajek, 1995). Embryonic estradiol disclosure in birds influenced behavioral and endocrine reactions in males and compressed production in females (Ottinger et al., 2001).

5.5.1 CARCINOGENICITY

The assumption that endocrine interference is able to generate cancer in humans is established as the contributing association amid DES disclosure of gravid women and cancer in glandular cells of the cervix and vagina into their female children (Herbst et al., 1971). Hormone-associated danger aspects for uterine and breast cancer and restricted proof of a connotation amid body organization affliction stages of "1,1-dichloro 2,2-bis (p-chlorophenyl) ethylene" or "PCBs" and threat of breast cancer has also been predictable (Laden et al., 1999). Exposure of women towards particular determined pesticides, including organochlorines, e.g., 1,1-dichloro 2,2-bis(p-chlorophenyl) ethylene, "PCBs," or "PAHs," have an augmented collective cancer threat, i.e., breast cancer in their lifespan (Davis et al., 1993). Along with these medical outcomes of endocrine-disturbing chemicals mechanisms, a few of the other harmful results detected in the common social population including augmented amounts of breast cancer on the order of 1% per year in people above the age of 50 years (Feuer and Wun, 1992), raised frequency of "endometriosis" and declined the age of arrival, and amplified amounts of cancer of testis and prostate on the imperative of 2–3 periods (Adami et al., 1994). All these interpretations have been gathered a significant body of confirmation demonstrating the occurrence of cancer of testis in men has increased considerably. There is an increasing awareness in the probable amount of endocrine disruptive compounds, but not any procedure has been projected for their achievement, which might describe their improvement of the progress of tumors of testicular (Ohlson and Hardell, 2000). Though, slight is identified relating to the reasons behind testicular cancer, although genetics, diet, endocrine prominence, age, and aspects of environmentally friendly threat have been recommended (Nam et al., 1999).

5.5.2 MUTAGENICITY

The mutagenic properties of 48 ETs have been reported in various Mutagenicity test systems including the *Salmonella* mutagenicity assay (SAL), the "sister chromatid exchange (SCE)," genotoxic effects (GE), the chromosome aberration assay (ABS), and the "micronucleus test (MN)." As ETs are in maximum suitcases neither mutagens nor severe toxicants at ambient deliberations, they may be validly unconfined into the surroundings (Colborn et al., 1993). Immuno-toxicity due to the retaliation of the immune system to the lesser concentration of chemicals and its significance for sustaining resistance of respective host counter to illness, significant prominence has newly been employed on the growth of immune endpoints that are able to expecting disclosure or reaction from environmental chemicals (Zelikoff et al., 2000). In the case of human beings, purposeful changes of the immune system were newly detected in cord blood lymphocytes in the newborn from an isolated population of maritime uncovered to methylmercury and organochlorines via eating upon seafood (Belles-Isles, 2002). It has been also established that a possible connotation occurs amid auto-immune disorders (Noller et al., 1988). In individuals exposed to the PCBs and DDT (dichloro diphenyl trichloroethane), a declined immune functionality and an augmented occurrence of infections were noted (Svensson et al., 1994). In bottlenose dolphins damage in function of immune system has been found due to the exposure towards the PCBs and dichloro diphenyl trichloroethane (Lahvis et al., 1995). It has been reported that estrogen promotes selective B and T cells hyperactivities by dysregulating B and T cells equilibrium, and this could become the essential source for the stimulation of auto-immunity by estrogen (Ahmed et al., 1999). Consequently, alarms turn out to be upraised the disclosure of fetuses to estrogenic chemical elements in time of the greatly serious phase of development of immune may change the immune system. Chemical compounds were categorized as immune-toxicants by the collected works investigation if they created destruction in one or more of the assessments, including the natural killer (NK) cell activity, hindered-type antipathy reaction, antibody-forming cell assay, cytotoxic T-lymphocyte, etc.

5.5.3 NEUROTOXICITY

It is usually believed that continuous disclosure to chemical compounds can negatively impact the nervous system. Comparing with adults, the growing nervous system is more susceptible to chemicals because of various factors, including lack of a protective blood brain barrier (BBB), rates of excretion, differences in metabolizing enzymes, and different rates of binding affinity towards various target proteins (Spyker, 1975; Tilson, 1998). Neuro-developmental variations because of inquisitive with neuro-endocrine function could be prompted by numerous chemical elements or categories of chemicals, i.e., PCBs, pesticides, dioxins, metals, phytoestrogens, and synthetic steroids (Kavlock et al., 1996; Tilson and Kavlock, 1997). Neuroendocrine dysregulation may take place due to disruption in the modulation of neuroendocrine response systems. Perhaps, a disruption in the reliability of the "hypothalamic-pituitary system" influences many types of behaviors, containing sexually dimorphic behaviors and reproductive behaviors in animals (Ottinger et al., 2002; Tilson, 1998). Numerous dissimilar endpoints comprising valuation of motor activity, an assessment of sensual meanings, and an assessment of variations in culture routine and reasoning behaviors can demonstrate neurotoxic consequences of chemical compounds (Figure 5.2) (Kulig, 1996).

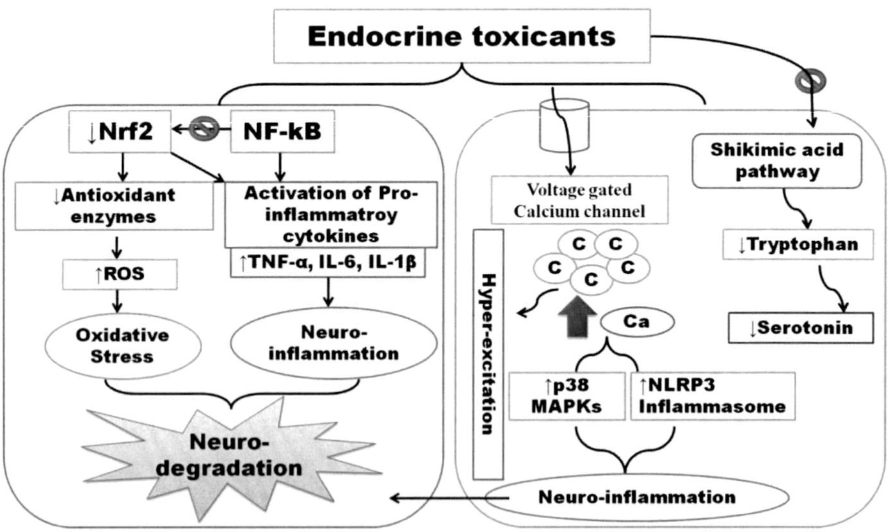

FIGURE 5.2 The neurotoxicity caused by the endocrine disrupting chemicals.

5.5.4 HORMONAL EFFECTS

Due to estradiol or other estrogens, numerous ETs seem like to perform as estrogen or anti-estrogen by interrelating with estrogen receptors (White et al., 1994; Cheek et al., 1998). ETs are passively diffused across the cell membrane to interrelate with hormonal receptors inside the cell (McLachlan and Arnold, 1996). The stimulated estrogen receptor ligand compound attached with high empathy and certainty to small DNA arrangements named hormone retaliation components, and prompts any gene stimulation or conquest (Landel et al., 1995). A few ETs including dioxin, which is the array of PHAHs (polyhalogenated aromatic hydrocarbons), seem to employ an anti-estrogenic action by interrelating with receptors of aryl hydrocarbon and ultimately reducing the action of the estrogen receptor (Routledge and Sumpter, 1997). It has been established that other EDCs containing p,p′-DDE (1,1-Dichloro-2,2-bis(*p*-chlorophenyl) ethylene) behave like an anti-androgen by attaching to the receptor of androgen and preventing its action (Kelce et al., 1995). There is a diminutive proof that ETs attach to TR (thyroid receptors). The thyroid hormone nuclear receptors are facilitated the activities of thyroid hormones (THs). This occurs because of xenobiotic as they have the potential to change the signaling of thyroid hormone via the thyroid receptor by both attaching straight to the thyroid receptor or incidentally by changing the thyroid receptor phosphorylation or by other auxiliary proteins communications (De-Vito et al., 1999). Review on functioning property of pituitary and thyroid gland (TG) exposed the properties of pesticides comprising mancozeb and amitrole that changed "triiodothyronine (T3), thyroxin (T4), or thyroid-stimulating hormone (TSH)" concentrations (Wilson et al., 1996; Hurley, 1998). In addition to this, a procedure was projected via the "EDSTAC (Endocrine Disruptor Screening and Testing Advisory Committee)" to identify the thyroid function and "hypothalamic–pituitary–gonadal (HPG)" system progress during pre-pubertal disclosure to endocrine disruptive chemicals (Stoker et al., 2000). The numerous hormonal-modifying things of endocrine disruptive chemicals, including thyroid hormone effects, anti-estrogenic activity, estrogenic activity, and anti-androgenic activity. ETs with estrogenic properties had a developed compassion than those with non-estrogenic properties, which comprised hormonal controlling properties excluding the estrogenic type. Generally, ETs with estrogenic properties were exposed to have a higher carcinogenic prospective than non-estrogenic mediators. Even though endocrine disruptive chemicals agonize and antagonize the influence of endogenous hormones and are capable to disturb the production and breakdown of endogenous hormones, for example, "1,2-dibromo-3-chloropropane (DBCP) (Schug et al., 2013), trifluralin (Rawlings, 1998), heptachlor (Wango et al., 1997), and 2,4,5-T (Périan et al., 2020)" or to disturb the production and breakdown of the hormonal receptor, for example, hexachlorobenzene (Alvarez et al., 2000). Metals have numerous endocrine-controlling actions. For instance, cadmium was exposed to restrict the steroidogenesis inside rat's ovary (Piasek and Laskey, 1999), modifies the quantity and empathy of estradiol receptors in rats and formed motherly hormonal inequity subsequent in a small number of gestations, and mercury prompts an elevated rate of synthesis of progesterone, convoyed by a small amount of translation to 17-beta-estradiol inside fish oocytes (Villeneuve et al., 2008).

5.5.5 OBESITY

EDs perform additional function in obesity and the metabolic encoding of risk of obesity. Even though, up until today, statistics have been insufficient; a few epidemiological and *in-vitro* analysis proposed a connection amid environmental chemical disclosure and obesity (Grün and Blumberg, 2009). It has been studied that obesogens are the EDs responsible for obesity. Obesogens have been revealed to attack transcription controllers instituted in networks of genes that are utilized to regulate homeostasis of intracellular lipid along with adipocytes proliferation and differentiation. The main assembly of controllers that is attacked is an assembly of nuclear hormone receptors recognized as "peroxisome proliferator-activated receptors (PPAR)" γ and δ. These hormonal receptors can sense a diversity of metabolic ligands comprising lipophilic hormones, dietary fatty acids, and their metabolic products and accountable upon the ligand's concentration, control in transcription of genes convoluted in matching the alterations in fatty acid equilibrium inside the body. The PPAR must hetero-dimerize with additional receptor recognized as the 9-cis RXR (retinoic acid receptor) in direction to develop proper and active function as both metabolic devices and transcription supervisors. The RXR receptor is the second chief mark of obesogens afterward to the PPAR receptors (Grün and Blumberg, 2009). The significant controller in this procedure is the PPARγ (peroxisome proliferator-activated receptors-γ), which links with the RXR receptors and attached DNA marks as a hetero-dimer to openly control the expression at the transcriptional phase (Tyagi et al., 2011). PPAR-γ is deliberated to be the principal controller of adipogenesis and perform major parts in just about all phases of adipocyte biology (Peter and Bruce, 2008). When the "ligand-binding domain" of PPAR-γ was altered such that the receptor was indifferent to recognized agonists, the capability of pre-adipocytes to distinguish into adipocytes in cell culture was natural (Walkey and Spiegelman, 2008). Numerous EDs are recognized to influence PPAR-γ action and prompt adipogenesis process. Distinguished amid these are organotins, for example, triphenyltin and tributyltin, and some phthalates (Hiromori et al., 2009). Phthalates and triorganotins also have the capability to prompt differentiation of adipocyte in a diversity of cell culture prototypes (Kanayama et al., 2005). Other ETs are recognized to stimulate adipogenesis, but possibly do not perform through "PPAR-γ." These comprise organophosphate pesticides, monosodium glutamate, PBDEs, and BPA (Rubin and Soto, 2009). Though numerous ETs are linked with the formation of adipocytes and obesity in different models of animal. Also, tributyltin is the only ED identified to cause embryonic impacts on fatty cells through stimulating PPAR-γ. Prenatal disclosure to tributyltin in mice bring about a significant growth in the quantity of triglycerides in tissues of newborn which usually have slight to not any fat at all, even though, the testing did not differentiate, in case extra lipid was deposited in prevailing cells, further newly formed cells or in both. Former ETs are expected to stimulate the formation of adipocytes, in the embryonic stage, even if it is not able to cause any major metabolic imbalances. For example, assured PBDEs and PCBs decrease thyroid functions by way of the antibacterial chemical compound TCS. The mechanism of action is not fully assured, but promising approaches contain interfering with the production, passage, breakdown, or allowance of thyroid hormone (Hallgren et al., 2001; Paul et al., 2010).

KEYWORDS

- **bisphenol-A**
- **endocrine disrupting chemicals**
- **endocrine toxicants**
- **hormones**
- **nonylphenol**
- **toxicity**

REFERENCES

Adami, H. O., Bergström, R., Möhner, M., Zatoôski, W., Storm, H., Ekbom, A., Tretli, S., et al., (1994). Testicular cancer in nine northern European countries. *Int. J. Cancer., 59*, 33–38.

Ahmed, S. A., Hissong, B. D., Verthelyi, D., Donner, K., Becker, K., & Karpuzoglu-Sahin, E., (1999). Gender and risk of autoimmune diseases: Possible role of estrogenic compounds. *Environ. Health Perspect., 107*, 681–686.

Ajj, H., Chesnel, A., Pinel, S., Plenat, F., Flament, S., & Dumond, H., (2013). An alkylphenol mix promotes seminoma derived cell proliferation through an ER alpha36-mediated mechanism. *PloS One., 8*, 61758.

Alvarez, L., Randi, A., Alvarez, P., Piroli, G., Chamson–Reig, A., Lux–Lantos, V., & De Pisarev, D. K., (2000). Reproductive effects of hexachlorobenzene in female rats. *J. Appl. Toxicol., 20*, 81–87.

Balaguer, P., Delfosse, V., Grimaldi, M., & Bourguet, W., (2017). Structural and functional evidence for the interactions between nuclear hormone receptors and endocrine disruptors at low doses. *C. R. Biol., 340*, 414–420.

Barouki, R., (2017). Endocrine disruptors: Revisiting concepts and dogma in toxicology. *C. R. Biol., 340*, 410–413.

Beckers, A., Aaltonen, L. A., Daly, A. F., & Karhu, A., (2013). Familial isolated pituitary adenomas (FIPA) and the pituitary adenoma predisposition due to mutations in the aryl hydrocarbon receptor-interacting protein (AIP) gene. *Endocr. Rev., 34*, 239–277.

Belles-Isles, M., Ayotte, P., Dewailly, E., Weber, J. P., & Roy, R., (2002). Cord blood lymphocyte functions in newborns from a remote maritime population exposed to organochlorines and methylmercury. *J. Toxicol. Environ. Health Part A, 65*, 165–182.

Benedetti, M., Zona, A., Beccaloni, E., Carere, M., & Comba, P., (2017). Incidence of breast, prostate, testicular, and thyroid cancer in Italian contaminated sites with presence of substances with endocrine disrupting properties. *Int. J. Environ. Res. Pub. Health., 14*, 355.

Bertelsen, R. J., Carlsen, K. C. L., Calafat, A. M., Hoppin, J. A., Håland, G., Mowinckel, P., Carlsen, K. H., & Løvik, M., (2013). Urinary biomarkers for phthalates associated with asthma in Norwegian children. *Environ. Health Perspect., 121*, 251–256.

Birnbaum, L. S., (1994). The mechanism of dioxin toxicity: Relationship to risk assessment. *Environ. Health Perspect, 102*, 157–167.

Brama, M., Gnessi, L., Basciani, S., Cerulli, N., Politi, L., Spera, G., Mariani, S., et al., (2007). Cadmium induces mitogenic signaling in breast cancer cell by an ERα-dependent mechanism. *Mol. Cell. Endocrinol., 264*, 102–108.

Braverman, L. E., He, X., Pino, S., Cross, M., Magnani, B., Lamm, S. H., Kruse, M. Bet al., (2005). The effect of perchlorate, thiocyanate, and nitrate on thyroid function in workers exposed to perchlorate long-term. *J. Clin. Endocrinol. Metab., 90*, 700–706.

Campion, S., Catlin, N., Heger, N., McDonnell, E. V., Pacheco, S. E., Saffarini, C., Sandrof, M. A., & Boekelheide, K., (2012). Male reprotoxicity and endocrine disruption. *Mol. Clin. Environ. Toxicol.*, 315–360.

Cannavo, S., Trimarchi, F., & Ferrau, F., (2017). Acromegaly, genetic variants of the aryl hydrocarbon receptor pathway and environmental burden. *Mol. Cell. Endocrinol., 457*, 81–88.

Caserta, D., Bordi, G., Ciardo, F., Marci, R., La Rocca, C., Tait, S., Bergamasco, B., et al., (2013). The influence of endocrine disruptors in a selected population of infertile women. *Gynecol. Endocrinol., 29*, 444–447.

Celik, L., Lund, J. D. D., & Schiøtt, B., (2008). Exploring interactions of endocrine-disrupting compounds with different conformations of the human estrogen receptor α ligand binding domain: A molecular docking study. *Chem. Res. Toxicol., 21*, 2195–2206.

Charles, G., Corbel, S., Carré, M. C., Roques-Carmes, T., & Zahraa, O., (2009). Design of a micro-channel reactor for decomposition of organic pollutants in wastewater treatment. In *Interact. Des. Manuf.*, 449–452.

Cheek, A. O., Vonier, P. M., Oberdörster, E., Burow, B. C., & McLachlan, J. A., (1998). Environmental signaling: A biological context for endocrine disruption. *Environ. Health Perspect., 106*, 5–10.

Choi, J. S., Oh, J. H., Park, H. J., Choi, M. S., Park, S. M., Kang, S. J., Oh, M. J., et al., (2011). miRNA regulation of cytotoxic effects in mouse Sertoli cells exposed to nonylphenol. *Reprod. Biol. Endocrinol., 9*, 1–11.

Cichna-Markl, M., (2012). Sample clean-up by sol-gel immunoaffinity chromatography for the determination of bisphenol A in food and urine. *Methods, 56*, 186–191.

Colborn, T., Vom, S. F. S., & Soto, A. M., (1993). Developmental effects of endocrine-disrupting chemicals in wildlife and humans. *Environ. Health Perspect., 101*, 378–384.

Daly, A. F., Jaffrain-Rea, M. L., Ciccarelli, A., Valdes-Socin, H., Rohmer, V., Tamburrano, G., Borson-Chazot, C., et al., (2006). Clinical characterization of familial isolated pituitary adenomas. *J. Clin. Endocrinol. Metabol., 91*, 3316–3323.

Darbre, P. D., (2017). Endocrine disruptors and obesity. *Curr. Obes. Rep., 6*, 18–27.

Davis, D. L., Bradlow, H. L., Wolff, M., Woodruff, T., Hoel, D. G., & Anton-Culver, H., (1993). Medical hypothesis: Xenoestrogens as preventable causes of breast cancer. *Environ. Health Perspect., 101*, 372–377.

De Coster, S., & Van, L. N., (2012). Endocrine-disrupting chemicals: Associated disorders and mechanisms of action. *J. Environ. Pub. Health.*

Desvergne, B., Feige, J. N., & Casals-Casas, C., (2009). PPAR-mediated activity of phthalates: A link to the obesity epidemic? *Mol. Cell. Endocrinol., 304*, 43–48.

DeVito, M., Biegel, L., Brouwer, A., Brown, S., Brucker-Davis, F., Cheek, A. O., Christensen, R., et al., (1999). Screening methods for thyroid hormone disruptors. *Environ. Health Perspect., 107*, 407–415.

Faber, R. A., & Hickey, J. J., (1973). Eggshell thinning, chlorinated hydrocarbons, and mercury in inland aquatic bird eggs, 1969 and 1970. *Pestic Monit J., 7*, 27–36.

Fein, G. G., Jacobson, J. L., Jacobson, S. W., Schwartz, P. M., & Dowler, J. K., (1984). Prenatal exposure to polychlorinated biphenyls: Effects on birth size and gestational age. *J. Pediat., 105*, 315–320.

Feuer, E. J., & Wun, L. M., (1992). How much of the recent rise in breast cancer incidence can be explained by increases in mammography utilization? A dynamic population model approach. *Am. J. Epidemiol., 136*, 1423–1436.

Fujii-Kuriyama, Y., & Kawajiri, K., (2010). Molecular mechanisms of the physiological functions of the aryl hydrocarbon (dioxin) receptor, a multifunctional regulator that senses and responds to environmental stimuli. *Proc. Jpn. Acad. Series B, 86*, 40–53.

Goodson, A., Summerfield, W., & Cooper, I., (2002). Survey of bisphenol A and bisphenol F in canned foods. *Food Add. Contam., 19*, 796–802.

Gore, A. C., Chappell, V. A., Fenton, S. E., Flaws, J. A., Nadal, A., Prins, G. S., Toppari, J., & Zoeller, R. T., (2015). EDC-2: The endocrine society's second scientific statement on endocrine-disrupting chemicals. *Endocr. Rev., 36*, 1–150.

Gorla-Bajszczak, A., Juge-Aubry, C., Pernin, A., Burger, A. G., & Meier, C. A., (1999). Conserved amino acids in the ligand-binding and τi domains of the peroxisome proliferator-activated receptor α are necessary for heterodimerization with RXR. *Mol. Cell. Endocrinol., 147*, 37–47.

Grün, F., & Blumberg, B., (2009). Endocrine disruptors as obesogens. *Mol. Cell. Endocrinol., 304*, 19–29.

Gunnarsson, D., Nordberg, G., & Selstam, G., (2007). Differential effects of cadmium on the gene expression of seven-transmembrane-spanning receptors and GAPDH in the rat testis. *Toxicol. Lett., 168*, 51–57.

Hallgren, S., Sinjari, T., Håkansson, H., & Darnerud, P., (2001). Effects of polybrominated diphenyl ethers (PBDEs) and polychlorinated biphenyls (PCBs) on thyroid hormone and vitamin A levels in rats and mice. *Arch. Toxicol., 75*, 200–208.

Hampl, R., Kubátová, J., & Stárka, L., (2016). Steroids and endocrine disruptors—History, recent state of art and open questions. *J. Steroid Biochem. Mol. Bio., 155*, 217–223.

Handelsman, D. J., & Cooper, T. G., (2013). Falling sperm counts and global estrogenic pollution: What have we learned over 20 years? *Asian. J. Androl., 15*, 159.

Harding, A. K., Daston, G. P., Boyd, G. R., Lucier, G. W., Safe, S. H., Stewart, J., Tillitt, D. E., & Van, D. K. G., (2006). Endocrine disrupting chemicals research program of the US environmental protection agency: Summary of a peer-review report. *Environ. Health Perspect., 114*, 1276–1282.

Harvey, A., Modak, A., Déry, U., Roy, M., Rinfret, S., Bertrand, O. F., Larose, É., et al., (2016). Changes in CYP2C19 enzyme activity evaluated by the [13C]-pantoprazole breath test after co-administration of clopidogrel and proton pump inhibitors following percutaneous coronary intervention and correlation to platelet reactivity. *J. Breath Res., 10*, 017104.

Heindel, J. J., Newbold, R., & Schug, T. T., (2015). Endocrine disruptors and obesity. *Nature Reviews Endocrinol., 11*, 653–661.

Henson, M. C., & Chedrese, P. J., (2004). Endocrine disruption by cadmium, a common environmental toxicant with paradoxical effects on reproduction. *Exp. Biol. Med., 229*, 383–392.

Herbst, A. L., Ulfelder, H., & Poskanzer, D. G., (1971). Adenocarcinoma of the vagina. In: *Problems of Birth Defects* (pp. 217–220). Springer, Dordrecht.

Hiromori, Y., Nishikawa, J. I., Yoshida, I., Nagase, H., & Nakanishi, T., (2009). Structure-dependent activation of peroxisome proliferator-activated receptor (PPAR) γ by organotin compounds. *Chem-Biol. Interact., 180*, 238–244.

Hsieh, T. H., Tsai, C. F., Hsu, C. Y., Kuo, P. L., Hsi, E., Suen, J. L., Hung, C. H., et al., (2012). n-butyl benzyl phthalate promotes breast cancer progression by inducing expression of lymphoid enhancer factor 1. *PloS One.*

Hurley, P. M., (1998). Mode of carcinogenic action of pesticides inducing thyroid follicular cell tumors in rodents. *Environ. Health Perspect., 106*, 437–445.

Hyun, K. D., Min, C. S., Soo, L. D., Roh, T., Jun, K. S., Yoon, S., Kook, K. M., et al., (2018). Risk assessment of endocrine disrupting phthalates and hormonal alterations in children and adolescents. *J. Toxicol. Environ. Health, Part A, 81*, 1150–1164.

Jagadeesh, S., Kyo, S., & Banerjee, P. P., (2006). Genistein represses telomerase activity via both transcriptional and posttranslational mechanisms in human prostate cancer cells. *Cancer Res., 66*, 2107–2115.

Jamarani, R., Erythropel, H. C., Nicell, J. A., Leask, R. L., & Marić, M., (2018). How green is your plasticizer? *Polymers, 10*, 834.

Järup, L., & Åkesson, A., (2009). Current status of cadmium as an environmental health problem. *Toxicol. Appl. Pharmacol., 238*, 201–208.

Jennings, J. E., Georgitsi, M., Holdaway, I., Daly, A., Tichomirowa, M., Beckers, A., Aaltonen, L. A., et al., (2009). Aggressive pituitary adenomas occurring in young patients in a large Polynesian kindred with a germline R271W mutation in the AIP gene. *European J. Endocrinol., 161*, 799–804.

Jiang, Y. L., Rigolet, M., Bourc'his, D., Nigon, F., Bokesoy, I., Fryns, J. P., Hulten, M., et al., (2005). DNMT3B mutations and DNA methylation defect define two types of ICF syndrome. *Human Mutat., 25*, 560–563.

Johnson, M. D., Kenney, N., Stoica, A., Hilakivi-Clarke, L., Singh, B., Chepko, G., Clarke, R., et al., (2003). Cadmium mimics the *in vivo* effects of estrogen in the uterus and mammary gland. *Nat. Med., 9*, 1081–1084.

Jones, L. A., & Hajek, R. A., (1995). Effects of estrogenic chemicals on development. *Environ. Health Perspect., 103*, 63–67.

Kabir, E. R., Rahman, M. S., & Rahman, I., (2015). A review on endocrine disruptors and their possible impacts on human health. *Environ. Toxicol. Pharmacol., 40*, 241–258.

Kanayama, T., Kobayashi, N., Mamiya, S., Nakanishi, T., & Nishikawa, J. I., (2005). Organotin compounds promote adipocyte differentiation as agonists of the peroxisome proliferator-activated receptor γ/retinoid X receptor pathway. *Mol. Pharmacol., 67*, 766–774.

Kavlock, R. J., Daston, G. P., DeRosa, C., Fenner-Crisp, P., Gray, L. E., Kaattari, S., Lucier, G., et al., (1996). Research needs for the risk assessment of health and environmental effects of endocrine disruptors: A report of the US EPA-sponsored workshop. *Environ. Health Perspect., 104*, 715–740.

Kelce, W. R., Stone, C. R., Laws, S. C., Gray, L. E., Kemppainen, J. A., & Wilson, E. M., (1995). Persistent DDT metabolite p, p'–DDE is a potent androgen receptor antagonist. *Nature, 375*, 581–585.

Kester, M. H., Bulduk, S., Tibboel, D., Meinl, W., Glatt, H., Falany, C. N., Coughtrie, M. W., et al., (2000). Potent inhibition of estrogen sulfotransferase by hydroxylated PCB metabolites: A novel pathway explaining the estrogenic activity of PCBs. *Endocrinol., 141*, 1897–1900.

Kim, J. T., & Lee, H. K., (2017). Childhood obesity and endocrine disrupting chemicals. *Ann. Pediat. Endocrinol. Metab., 22*, 219.

Koh, K. P., & Rao, A., (2013). DNA methylation and methylcytosine oxidation in cell fate decisions. *Curr. Opin. Cell Bio., 25*, 152–161.

Kortenkamp, A., & Altenburger, R., (1998). Synergisms with mixtures of xenoestrogens: A reevaluation using the method of isoboles. *Sci. Total Environ., 221*, 59–73.

Kulig, B., Alleva, E., Bignami, G., Cohn, J., Cory-Slechta, D., Landa, V., O'Donoghue, J., & Peakall, D., (1996). Animal behavioral methods in neurotoxicity assessment: SGOMSEC joint report. *Environ. Health Perspect., 104*, 193–204.

Laden, F., Neas, L. M., Spiegelman, D., Hankinson, S. E., Willett, W. C., Ireland, K., Wolff, M. S., & Hunter, D. J., (1999). Predictors of plasma concentrations of DDE and PCBs in a group of US women. *Environ. Health Perspect., 107*, 75–81.

Lafuente, A., Cano, P., & Esquifino, A. I., (2003). Are cadmium effects on plasma gonadotropins, prolactin, ACTH, GH and TSH levels, dose-dependent? *Biometals, 16*, 243–250.

Lafuente, A., González-Carracedo, A., Romero, A., Cano, P., & Esquifino, A. I., (2004). Cadmium exposure differentially modifies the circadian patterns of norepinephrine at the median eminence and plasma LH, FSH and testosterone levels. *Toxicol. Lett., 146*, 175–182.

Lahvis, G. P., Wells, R. S., Kuehl, D. W., Stewart, J. L., Rhinehart, H. L., & Via, C. S., (1995). Decreased lymphocyte responses in free-ranging bottlenose dolphins (*Tursiops truncatus*) are associated with increased concentrations of PCBs and DDT in peripheral blood. *Environ. Health Perspect., 103*, 67–72.

Lamb, IV. J. C., Boffetta, P., Foster, W. G., Goodman, J. E., Hentz, K. L., Rhomberg, L. R., Staveley, J., et al., (2015). Comments on the opinions published by Bergman et al. (2015) on critical comments on the WHO-UNEP State of the science of endocrine disrupting chemicals (Lamb et al., 2014). *Regul. Toxicol. Pharmacol., 73*, 754–757.

Landel, C. C., Kushner, P. J., & Greene, G. L., (1995). Estrogen receptor accessory proteins: Effects on receptor-DNA interactions. *Environ. Health Perspect., 103*, 23–28.

Li, X., Zhang, S., & Safe, S., (2006). Activation of kinase pathways in MCF-7 cells by 17β-estradiol and structurally diverse estrogenic compounds. *J. Steroid Biochem. Mol. Bio., 98*, 122–132.

Lloyd, C., & Grossman, A., (2014). The AIP (aryl hydrocarbon receptor-interacting protein) gene and its relation to the pathogenesis of pituitary adenomas. *Endocrine., 46*, 387–396.

López-Carrillo, L., Hernández-Ramírez, R. U., Calafat, A. M., Torres-Sánchez, L., Galván-Portillo, M., Needham, L. L., Ruiz-Ramos, R., & Cebrián, M. E., (2010). Exposure to phthalates and breast cancer risk in northern Mexico. *Environ. Health Perspectives, 118*, 539–544.

Lubrano, C., Genovesi, G., Specchia, P., Costantini, D., Mariani, S., Petrangeli, E., Lenzi, A., & Gnessi, L., (2013). Obesity and metabolic comorbidities: Environmental diseases? *Oxid. Med. Cell. Longev.*

Mackowiak, B., & Wang, H., (2016). Mechanisms of xenobiotic receptor activation: Direct vs. indirect. *Biochim. Biophys. Acta Gene Regul. Mech., 1859*, 1130–1140.

Marques-Pinto, A., & Carvalho, D., (2013). Human infertility: Are endocrine disruptors to blame? *Endocr. Connect., 2*, 15–29.

Martin, M. B., Voeller, H. J., Gelmann, E. P., Lu, J., Stoica, E. G., Hebert, E. J., Reiter, R., et al., (2002). Role of cadmium in the regulation of AR gene expression and activity. *Endocrinol., 143*, 263–275.

Marty, M. S., Borgert, C., Coady, K., Green, R., Levine, S. L., Mihaich, E., Ortego, L., et al., (2018). Distinguishing between endocrine disruption and non-specific effects on endocrine systems. *Regul. Toxicol. Pharmacol., 99*, 142–158.

McLachlan, J. A., & Arnold, S. F., (1996). Environmental estrogens. *Am. Sci., 84*, 452–461.

Meeker, J. D., Sathyanarayana, S., & Swan, S. H., (2009). Phthalates and other additives in plastics: Human exposure and associated health outcomes. *Phil. Trans. R. Soc. B Sci., 364*, 2097–2113.

Mínguez-Alarcón, L., & Gaskins, A. J., (2017). Female exposure to endocrine disrupting chemicals and fecundity: A review. *Curr. Opin. Obstet. Gynecol., 29*, 202.

Mnif, W., Hassine, A. I. H., Bouaziz, A., Bartegi, A., Thomas, O., & Roig, B., (2011). Effect of endocrine disruptor pesticides: A review. *Int. J. Environ. Res. Pub. Health, 8*, 2265–2303.

Monneret, C., (2017). What is an endocrine disruptor? *C. R. Biol., 340*, 403–405.

Nam, R. K., Fleshner, N., Rakovitch, E., Klotz, L., Trachtenberg, J., Choo, R., Morton, G., & Danjoux, C., (1999). Prevalence and patterns of the use of complementary therapies among prostate cancer patients: An epidemiological analysis. *J. Urol., 161*, 1521–1524.

Noller, K. L., Blair, P. B., O'Brien, P. C., Melton, L. J., Offord, J. R., Kaufman, R. H., & Colton, T., (1988). Increased occurrence of autoimmune disease among women exposed in utero to diethylstilbestrol. *Fertil. Steril., 49*, 1080–1082.

Ohlson, C. G., & Hardell, L., (2000). Testicular cancer and occupational exposures with a focus on xenoestrogens in polyvinyl chloride plastics. *Chemosphere., 40*, 1277–1282.

Ottinger, M. A., Abdelnabi, M. A., Henry, P., McGary, S., Thompson, N., & Wu, J. M., (2001). Neuroendocrine and behavioral implications of endocrine disrupting chemicals in quail. *Horm. Behav., 40*, 234–247.

Ottinger, M. A., Abdelnabi, M., Quinn, M., Golden, N., Wu, J., & Thompson, N., (2002). Reproductive consequences of EDCs in birds: What do laboratory effects mean in field species? *Neurotoxicol. Teratol., 24*, 17–28.

Pappas, R. S., Polzin, G. M., Zhang, L., Watson, C. H., Paschal, D. C., & Ashley, D. L., (2006). Cadmium, lead, and thallium in mainstream tobacco smoke particulate. *Food Chem. Toxicol., 44*, 714–723.

Paul, D. A., Mackley, A., & Yencha, E. M., (2010). Thyroid function in term and late preterm infants with respiratory distress in relation to severity of illness. *Thyroid, 20*, 189–194.

Périan, S., Cerutti, C., Forcet, C., Tribollet, V., & Vanacker, J. M., (2020). A cell-based method to detect agonist and antagonist activities of endocrine-disrupting chemicals on GPER. *Front. Endocrinol., 11*, 547.

Peter, T., & Bruce, M. S., (2008). Fat and beyond: The diverse biology of PPAR-gamma. *Annu. Rev. Biochem., 77*, 289–312.

Piasek, M., & Laskey, J. W., (1999). Effects of *in vitro* cadmium exposure on ovarian steroidogenesis in rats. *J. Appl. Toxicol., 19*, 211–217.

Posnack, N. G., Swift, L. M., Kay, M. W., Lee, N. H., & Sarvazyan, N., (2012). Phthalate exposure changes the metabolic profile of cardiac muscle cells. *Environ. Health Perspect., 120*, 1243–1251.

Qin, X. Y., Sone, H., Kojima, Y., Mizuno, K., Ueoka, K., Muroya, K., Miyado, M., et al., (2012). Individual variation of the genetic response to bisphenol a in human foreskin fibroblast cells derived from cryptorchidism and hypospadias patients. *PLoS One, 7*, 52756.

Rishikaysh, P., Dev, K., Diaz, D., Qureshi, W. M. S., Filip, S., & Mokry, J., (2014). Signaling involved in hair follicle morphogenesis and development. *Int. J. Mol. Sci., 15*, 1647–1670.

Routledge, E. J., & Sumpter, J. P., (1997). Structural features of alkylphenolic chemicals associated with estrogenic activity. *J. Biol. Chem., 272*, 3280–3288.

Rubin, B. S., & Soto, A. M., (2009). Bisphenol A: Perinatal exposure and body weight. *Mol. Cell. Endocrinol., 304*, 55–62.

Sargis, R. M., & Simmons, R. A., (2019). Environmental neglect: Endocrine disruptors as underappreciated but potentially modifiable diabetes risk factors. *Diabetologia, 62*, 1811–1822.

Sathyanarayana, S., Karr, C. J., Lozano, P., Brown, E., Calafat, A. M., Liu, F., & Swan, S. H., (2008). Baby care products: Possible sources of infant phthalate exposure. *Pediat., 121*, 260–268.

Schug, T. T., Abagyan, R., Blumberg, B., Collins, T. J., Crews, D., DeFur, P. L., Dickerson, S. M., et al., (2013). Designing endocrine disruption out of the next generation of chemicals. *Green Chem., 15*, 181–198.

Schug, T. T., Janesick, A., Blumberg, B., & Heindel, J. J., (2011). Endocrine disrupting chemicals and disease susceptibility. *J. Steroid Biochem. Mol. Biol., 127*, 204–215.

Sen, A., & Sellix, M. T., (2016). The circadian timing system and environmental circadian disruption: From follicles to fertility. *Endocrinol., 157*, 3366–3373.

Sharman, M., Read, W. A., Castle, L., & Gilbert, J., (1994). Levels of di-(2-ethylhexyl) phthalate and total phthalate esters in milk, cream, butter and cheese. *Food Add. Contam., 11*, 375–385.

Siu, E. R., Mruk, D. D., Porto, C. S., & Cheng, C. Y., (2009). Cadmium-induced testicular injury. *Toxicol. Appl. Pharmacol., 238*, 240–249.

Skakkebæk, N. E., Holm, M., Hoei-Hansen, Christina, Jørgensen, N., & Rajpert-De Meyts, E. W. A., (2003). Association between testicular dysgenesis syndrome (TDS) and testicular neoplasia: Evidence from 20 adult patients with signs of maldevelopment of the testis. *Apmis., 111*, 1–11.

Smida, A. D., Valderrama, X. P., Agostini, M. C., Furlan, M. A., & Chedrese, J., (2004). Cadmium stimulates transcription of the cytochrome p450 side chain cleavage gene in genetically modified stable porcine granulosa cells. *Biol. Reprod., 70*, 25–31.

Soares, A., Guieysse, B., Jefferson, B., Cartmell, E., & Lester, J. N., (2008). Nonylphenol in the environment: A critical review on occurrence, fate, toxicity and treatment in wastewaters. *Environ. Int., 34*, 1033–1049.

Soriano-Guillen, L., & Argente, J., (2019). Central precocious puberty, functional and tumor-related. *Best Pract. Res. Clin. Endocrinol. Metab., 33*, 101262.

Spyker, J. M., (1975). Assessing the impact of low level chemicals on development: Behavioral and latent effects. *Behav. Pharmacol.,* 161–180.

Stoker, T. E., Laws, S. C., Guidici, D. L., & Cooper, R. L., (2000). The effect of atrazine on puberty in male Wistar rats: An evaluation in the protocol for the assessment of pubertal development and thyroid function. *Toxicol. Sci., 58*, 50–59.

Svensson, B. G., Hallberg, T., Nilsson, A., Schütz, A., & Hagmar, L., (1994). Parameters of immunological competence in subjects with high consumption of fish contaminated with persistent organochlorine compounds. *Int. Arch. Occup. Environ. Health, 65*, 351–358.

Swan, S. H., (2008). Environmental phthalate exposure in relation to reproductive outcomes and other health endpoints in humans. *Environ. Res., 108*, 177–184.

Swedenborg, E., Rüegg, J., Mäkelä, S., & Pongratz, I., (2009). Endocrine disruptive chemicals: Mechanisms of action and involvement in metabolic disorders. *J. Mol. Endocrinol., 43*, 1–10.

Sweeney, M. F., Hasan, N., Soto, A. M., & Sonnenschein, C., (2015). Environmental endocrine disruptors: Effects on the human male reproductive system. *Rev. Endocr. Metab. Disord., 16*, 341–357.

Tilson, H. A., & Kavlock, R. J., (1997). The workshop on endocrine disruptor research needs: A report. *Neurotoxicol., 18*, 389–392.

Tilson, H. A., & Kodavanti, P. R., (1998). The neurotoxicity of polychlorinated biphenyls. *Neurotoxicol., 19*, 517–525.

Tilson, H. A., (1998). Developmental neurotoxicology of endocrine disruptors and pesticides: Identification of information gaps and research needs. *Environ. Health Perspect., 106*, 807–811.

Trasande, L., Zoeller, R. T., Hass, U., Kortenkamp, A., Grandjean, P., Myers, J. P., DiGangi, J., et al., (2016). Burden of disease and costs of exposure to endocrine disrupting chemicals in the European Union: An updated analysis. *Androl., 4*, 565–572.

Tyagi, S., Gupta, P., Saini, A. S., Kaushal, C., & Sharma, S., (2011). The peroxisome proliferator-activated receptor: A family of nuclear receptors role in various diseases. *J. Adv. Pharmaceut. Technol. Res., 2*, 236.

Ünüvar, T., & Büyükgebiz, A., (2012). Fetal and neonatal endocrine disruptors. *J. Clin. Res. Pediatr. Endocrinol., 4*, 51.

Villeneuve, D. L., Blake, L. S., Brodin, J. D., Cavallin, J. E., Durhan, E. J., Jensen, K. M., Kahl, M. D., et al., (2008). Effects of a 3β-hydroxysteroid dehydrogenase inhibitor, trilostane, on the fathead minnow reproductive axis. *Toxicol. Sci., 104*, 113–123.

Virtanen, M., Kivimäki, M., Joensuu, M., Virtanen, P., Elovainio, M., & Vahtera, J., (2005). Temporary employment and health: A review. *Int. J. Epidemiol., 34*, 610–622.

Waldbillig, N. R. S. C. D., (1998). Effects of the pesticides carbofuran, chlorpyrifos, dimethoate, lindane, triallate, trifluralin, 2, 4-D, and pentachlorophenol on the metabolic endocrine and reproductive endocrine system in ewes. *J. Toxicol. Environ. Health. Part A, 54*, 21–36.

Walkey, C. J., & Spiegelman, B. M., (2008). A functional peroxisome proliferator-activated receptor-γ ligand-binding domain is not required for adipogenesis. *J. Biol. Chem., 283*, 24290–24294.

Wango, E. O., Onyango, D. W., Odongo, H., Okindo, E., & Mugweru, J., (1997). *In vitro* production of testosterone and plasma levels of luteinizing hormone, testosterone and cortisol in male rats treated with heptachlor. *Comp. Biochem. Physiol. C. Pharmacol. Toxicol. Endocrinol, 118*, 381–386.

White, R., Jobling, S., Hoare, S. A., Sumpter, J. P., & Parker, M. G., (1994). Environmentally persistent alkylphenolic compounds are estrogenic. *Endocrinol., 135*, 175–182.

Wijmenga, C., Hansen, R. S., Gimelli, G., Björck, E. J., Davies, E. G., Valentine, D., Belohradsky, B. H., et al., (2000). Genetic variation in ICF syndrome: Evidence for genetic heterogeneity. *Human Mutat., 16*, 509–517.

Wilson, K., Grenfell, B. T., & Shaw, D. J., (1996). Analysis of aggregated parasite distributions: A comparison of methods. *Funct. Ecol.*, 92–601.

Yang, O., Kim, H. L., Weon, J. I., & Seo, Y. R., (2015). Endocrine-disrupting chemicals: Review of toxicological mechanisms using molecular pathway analysis. *J. Cancer Prev., 20*, 12.

Yao, P. L., Lin, Y. C., & Richburg, J. H., (2012). Mono-(2-ethylhexyl) phthalate (MEHP) promotes invasion and migration of human testicular embryonal carcinoma cells. *Biol. Reprod., 86*, 160, 161.

Yin, L., Dai, Y., Jiang, X., Liu, Y., Chen, H., Han, F., Cao, J., & Liu, J., (2016). Role of DNA methylation in bisphenol A exposed mouse spermatocyte. *Environ. Toxicol. Pharmacol., 48*, 265–271.

You, H., Chen, S., Mao, L., Li, B., Yuan, Y., Li, R., & Yang, X., (2014). The adjuvant effect induced by di-(2-ethylhexyl) phthalate (DEHP) is mediated through oxidative stress in a mouse model of asthma. *Food. Chem. Toxicol., 71*, 272–281.

Zelikoff, J. T., Raymond, A., Carlson, E., Li, Y., Beaman, J. R., & Anderson, M., (2000). Biomarkers of immunotoxicity in fish: From the lab to the ocean. *Toxicol. Lett., 112*, 325–331.

Zeng, F., Lerro, C., Lavoué, J., Huang, H., Siemiatycki, J., Zhao, N., Ma, S., et al., (2017). Occupational exposure to pesticides and other biocides and risk of thyroid cancer. *Occup. Environ. Med., 74*, 502–510.

Zhou, C., & Flaws, J. A., (2017). Effects of an environmentally relevant phthalate mixture on cultured mouse antral follicles. *Toxicol. Sci., 156*, 217–229.

Zoeller, R. T., Brown, T. R., Doan, L. L., Gore, A. C., Skakkebaek, N. E., Soto, A. M., Woodruff, T. J., & Vom, S. F. S., (2012). Endocrine-disrupting chemicals and public health protection: A statement of principles from the Endocrine Society. *Endocrinol., 153*, 4097–4110.

CHAPTER 6

Mechanistic Approach of Various Endocrine Toxicants

YOUNIS AHMAD HAJAM,[1] PREM LATA,[2] DIKSHA,[2] and ROUF AHMAD BHAT[3]

[1]*Department of Life Sciences and Allied Health Sciences, Sant Baba Bhag Singh University, Padhiana, Jalandhar, Punjab, India*

[2]*Division Zoology, Department of Biosciences, Career Point University, Hamirpur, Himachal Pradesh–176041, India, E-mail: premlata1617@gmail.com (P. Lata)*

[3]*Sher-e-Kashmir University of Agricultural Sciences and Technology, Kashmir, Jammu and Kashmir, India*

ABSTRACT

Endocrine toxicants (ETs) are an exogenous mediator that hinders the secretion or release of hormones present in the human body. It has been thought that ETs exert their effect mainly through the receptors of various hormones such as receptors of estrogen, androgens, progesterone, thyroid, and retinoid. ETs show their effects through the receptors of the nucleus, non-nuclear receptors of steroid hormones, receptors of non-steroid like serotonin or noradrenaline, orphan receptors like aryl hydrocarbon, various pathways involved in the synthesis of steroids, and different mechanisms that join upon the endocrine or reproductive system. These multifaceted compounds go through the food chain and gather in the higher animals, such as humans, eagles, bears, and other voracious animals. Animals can also be exposed to these chemicals through drinking of polluted water, inhaling in the polluted air, taking contaminated food, or contact with the polluted soil.

Environmental Endocrine Toxicants: Biology, Effects, and Management. Younis Ahmad Hajam, Rouf Ahmad Bhat, Khalid Rehman Hakeem, & Rajesh Kumar (Eds.)
© 2023 Apple Academic Press, Inc. Co-published with CRC Press (Taylor & Francis)

Those people are mainly at higher risk for exposure who work in chemical industries, pesticides, fungicides, and therefore for developing a reproductive or endocrine system-related deformities. A number of ETs were considered to have long half-lives. Due to this characteristic, these were important for their utilization in industries, but these are relatively harmful to animals and human life. This chapter summarizes about the mechanism of various ETs.

6.1 INTRODUCTION

The United States Environmental Protection Agency (EPA) stated that the endocrine disturbing component is an inner cell mediator that hinders the secretion or release of hormones present in the human body. These hormones are dependable to maintain body fluid balance, development process, and reproduction. Endocrine disruptors modulate the function of hormones by causing structural changes in cell surface/nuclear receptors viz. estrogen, androgen, progesterone, thyroid, and retinoids. Due to this alteration in down regulatory pathways may causes organal dysfunctioning, including dysfunction in the reproductive system. Therefore, an endocrine toxicant is a chemical compound. It can be natural or synthetic, which via environmental or unsuitable developmental contact, changes the hormonal or homeostatic structure. The group of molecules recognized as ETs is extremely varied and comprises artificial chemicals utilized as manufacturing solvents and their residues, such as polychlorinated biphenyls (PCBs), polybrominated biphenyls (PBBs), persistent organic pollutants (POPs), plastics, plasticizers like phthalate esters, pesticides like DDT, chlorpyrifos, methoxychlor, fungicides, and pharmaceutical products, e.g., diethylstilbestrol (DES). Naturally available chemicals such as phytoestrogens that are found in the food chain may also responsible for endocrine toxicity (Kuiper et al., 1998; Dickerson and Gore, 2007). The concentrations of urine were about 500 folds in genistein and daidzein higher in comparison to fed cow's milk (Cao et al., 2009). Thus, there is a higher need to study the potential of genistein and daidzein for endocrine interruption. The presence of these ETs or exerting any endocrine disturbing activities is very hard to predict due to their varied diversity, and these do not share any structural resemblance except generally having small molecular mass (less than 1,000 Daltons) components. However, in extremely broad conditions, ETs like POPs, PCBs, PBBs, and pesticides frequently having non-metallic compounds known as halogen group, such as Cl^- or Br. These compounds are frequently containing a phenolic compound in the form of a phenolic ring, i.e., considered to mimic natural steroids and allow ETs to act together with the receptors of chemical compounds as agonists or antagonists intended for similar responses. Toxic metals and semimetals might also possess DES properties, signifying that these components are ETs and more universal toxicants as well. Numerous classes of ETs work as anti-androgens and as agonists or antagonists' receptors of thyroid hormone, and androgenic ETs have been recognized. There are numerous sources of exposure to ETs and these can vary across the globe. The circumstance is continuously developing because the use of some ETs has been stopped many years ago and others in upcoming times, with major dissimilarities among countries. Various chronological instances of contaminant spills or toxicity from PCBs and POPs that explain a straight fundamental association between a toxicant and the demonstration of a reproductive as well as endocrine complications in humans or animals. Areas near industries are generally characterized by pollution from a broad variety of manufacturing chemicals that can percolate into land and surface water. Endocrine-disrupting chemicals (EDCs) enter in the food web and get accumulated at different tropic levels. However, organism are exposed to these toxicants through other routes like intake of contaminated water, intake of polluted air, foods in which EDCs gets accumulated, etc. due to this multiple exposures towards EDCs leads to the development of reproductive abnormalities. These compounds take much time to decompose. These cannot be metabolized or can be metabolized or these are able to worn-out into highly toxic molecules than the previous molecule. The compounds that were banned many years ago stay in higher amounts in the surroundings and feasible in every human or animal testing as a body part (Calafat and Needham, 2007; Porte et al., 2006). The generation of chemicals and chemical manufacturing industries has extremely increased with the increase in industrialization. A number of chemical compounds may be ETs because they disrupt the signaling pathways of endogenous hormones. There are various studies which showed the relation between the increase in contact with chemicals and the rise of a number of major diseases, for instance, conflicts in reproduction, cancers due to hormone problems, and several metabolic disorders, i.e., diabetes mellitus, weight gain, and cardiovascular diseases. Nonetheless, this relation and the mechanisms behind this relation is less understood till date. Up till now, various studies are there which show that ETs have various effects on the process of reproduction which are stimulated by hormones arbitrated mainly by estrogen receptors and the androgen receptors (ARs). Additionally, receptors of thyroid hormone and peroxisome proliferator-activated can target for the action ETs. There are various ways through which ETs affect these systems, for instance, by directly affecting receptor signaling or by the activation of other signaling pathways, particularly aryl hydrocarbon receptor (AhR) (Figure 6.1).

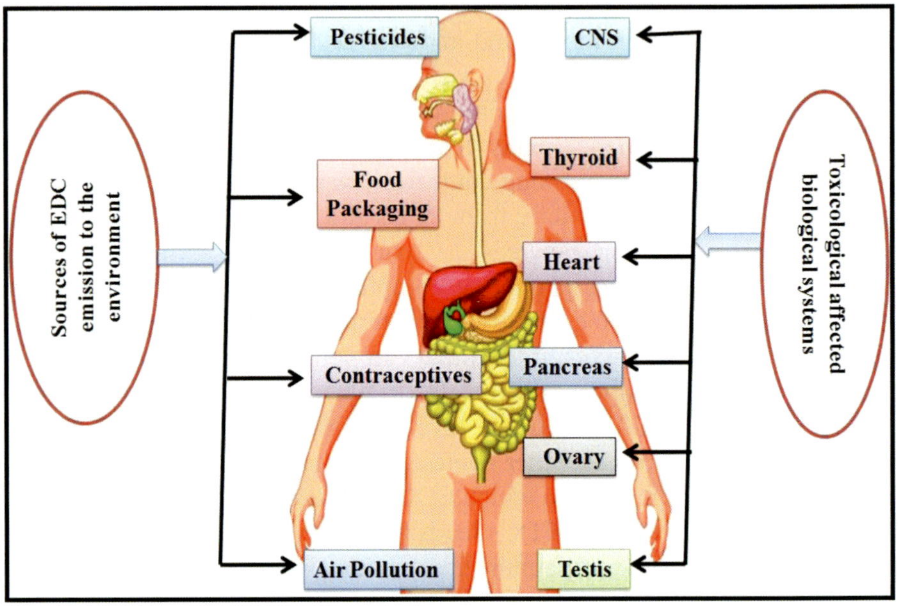

FIGURE 6.1 Effect of endocrine disrupting chemicals/toxicants on biological systems.

6.2 MECHANISTIC ASPECTS OF ENDOCRINE DISRUPTION

Endocrine disruption is referred to as any intervention exerted by an external substance with the endocrine system. Those mentioned external substances are actually known as ETs. Though ETs might affect every pathway of the endocrine system, mainly hormone receptor or nuclear receptor. The hormonal receptors signify relatives of transcription factors that are structurally associated (Gronemeyer et al., 2004). About 48 nuclear receptors have been found in mammals that are involved in all fundamental functions, for instance, development of fetal, body fluids balance, reproduction, and response to other xenobiotics components. The nuclear hormonal receptors belong to transcription factor are the receptors of steroid hormone such as receptors of estrogen, glucocorticoids, mineralocorticoids, and the receptors of thyroid hormone as well. A number of ETs may bind directly with these hormone receptors, therefore, increasing, or slow down the effect of a hormone. Nevertheless, other mechanisms are also described regarding the action of ETs particularly with the association of the receptors of hydrocarbon commonly aryl hydrocarbon. AhR stimulates all the cellular responses to xenobiotics contact. AhR comes under the category of "bHLH-PAS" proteins, and is universally articulated and well-preserved during evolution. Various kinds of organic substances like "polychlorinated dibenzodioxins, dibenzofurans, PCBs, PAHs such as 3-methylcholanthrene (3-MC), benzo[a] pyrene (BaP) and benzoflavone" trigger these AhR (aryl hydrocarbon) receptors (Poland and Knutson, 1982). Additionally, several other natural or endogenous substances have also been recognized as the activators of AhRs. The activation of AhRs or nuclear receptors is closely related at the transcription level (Ruegg et al., 2008). These receptors remain in the cytosol or nucleus of the cell as a composite with chaperones and co-repressors in the absence of ligand (Petrulis and Perdew, 2002; Picard, 2006). Due to the ligand binding, there is occurrence of various conformational changes and "dissociation of the repressive complex," "the recruitment of transcriptional co-activators," "receptor dimerization" (Lee et al., 2001; Hankinson, 2005). Nuclear receptors can homo-dimerize or heterodimerize with retinoid X receptors (RXRs), while aryl hydrocarbons dimerize with "AhR nuclear translocator (ARNT)" (Swanson, 2002; Claessens and Gewirth, 2004). On the other hand, several thyroid receptors (TR) are frequently be bounded by their recognition position lying on the deoxyribonucleic acid. The exchange of co-repressors with co-activators can be triggered due to the ligand binding which causes the activation of transcription (Wu and Koenig, 2000). The highly considered method of ETs /disruption is the direct binding to hormone receptors (Janosek et al., 2006; Ruegg et al., 2008). Ubiquitin-proteasome pathway is used for targeted degradation which control the level of nuclear hormones. In addition to this, the assembly of this complex is likely to be "AhR ligand-dependent" (Ohtake et al., 2007). It has been reported that AhR ligands have anti-estrogenic properties and these can reduce the level of estrogen receptors by the degradation of proteosome (Wang et al., 1993).

The activities of nuclear receptors and AhRs depend upon the co-activators of transcription. Various co-activators are shared by a lot of nuclear receptors and AhRs. Therefore, competition among the receptors of steroid hormone and AhRs for general type of transcriptional coregulators is a possible system in which AhRs (ions or neutral molecules) might disrupt the signaling of steroid hormones. Certainly, in mammalian cells, higher gene expression of the "co-activator-binding domain of AhR" and estrogen receptors cause impairment in the activities of receptors of estrogen and AhR receptors, respectively (Reen et al., 2002). Actually, competition between estrogen receptors and AhRs for ER and AhR for Ah receptor nuclear translocator protein showed to be partially liable for the anti-estrogenic characters of the dioxins (Brunnberg et al., 2003; Ruegg et al., 2007). Though, Ah receptor nuclear translocator protein is not measured a "classical transcriptional co-activator," it shares numerous characteristics of SRC type of transcriptional coregulators and has been shown to act as co-activator of the estrogens receptors (Brunnberg et al., 2003), particularly of estrogens receptors-β (Ruegg et al., 2007). Competition for general co-activators takes place also inside the family of nuclear receptor (Min et al., 2002).

6.3 DNA INTERFERENCE

A multifaceted relationship between AhRs, and estrogen receptors (ER) promotes ER target genes by reliable promoters. AhR gets activated and binds to the estrogen receptors which inhibit xenobiotic response elements (XREs). AhR cannot activate gene transcription, and slightly different composition than XREs (Safe and Wormke, 2003). Beside this, AhR can bind to XREs which closely binding to the certain promoter region of estrogen receptors. However, it can be necessary for activation of transcription of particular gene present on estrogen receptors (Hockings et al., 2006). Liganded AhR binding is similar to XREs which interferes negatively with the gene induction by estrogen receptors (Ohtake et al., 2003) and the ability of 3-methycholanthrene (3-MC) to activate estrogen receptors dependent transcription. It requires the biotransformation of metabolites among estrogenic activity. Therefore, 3-methycholanthrene (3-MC) was not metabolized and acted as antiestrogen related with dioxin (Swedenborg et al., 2008; Arcaro et al., 1999). Therefore, the effects of AhR and estrogen receptors having the capability of cells that converted into chemical substances like drugs, carcinogens, environmental pollutants, hydrocarbons, and pesticides into estrogenic chemical substances.

6.4 IMBALANCE OF METABOLIC HORMONES

Due to availability of hormones is helped by biosynthesis of metabolic hormones, transportation of hormones to the targeted cells, catabolism, and protein binding metabolic hormones. ETs interfere all the metabolic and catabolic processes (Baker et al., 1998; You et al., 2001 Boas et al., 2006). ETs mostly affected by catabolism of steroid hormone such as P450 enzymes

(CYP1A1, CYPIA2, CYP34A4, and CYP1A1). These are targeted genes of AhR and liable for 17β-estradiol hydroxylation (Tsuchiya et al., 2005). Due to disclosure of chemical substances, activation of enzyme happened which increased the catabolism of hormone. As a result, this xenobiotic affects/compromise hormone signaling. Thus, cytochrome 450 aromatase (CYP19) which converts testosterone to estradiol (by AhR directed gene). Initiation of activated AhR by ETs increased the degradation of steroid hormones and also increases the production of estradiol.

6.5 ARYL HYDROCARBON RECEPTOR (AHR)

AhR (Aryl hydrocarbon receptor) convoluted in the stimulation of the process of fatty acids, triglyceride synthesis from glucose or other substrates (Thackaberry et al., 2003; Sato et al., 2008). Besides, endocrine disruptors (EDs) (dioxin) have been increased the risk of diabetes mellitus (T2DM) (Fierens et al., 2003; Thackaberry et al., 2003). In type 2 diabetes, ARNT plays fundamental function and heterodimerization AhR (Gunton et al., 2005). Loss of ARNT manifestation causes impaired β-cells for insulin release. It is expected that initiation of AhR/ARNT by EDCs might be disturbs the stimulation of the lipolytic process and influences the intake of blood sugar in the liver tissues. In fact, group of AhR ligands, i.e., POPs cause or associated with metabolic syndrome (MetS) (Lee et al., 2007, 2006). Additionally, at a low level of dose revelation co-planar dioxin (PCB-77) and PCBs, enhances differentiation of adipocytes in AhR mediated and activates the discharge of inflammatory cytokines in adipose tissues. These types of dioxins accumulate in adipose tissue and developing obesity, atherosclerosis, and cardiovascular diseases (Arsenescu et al., 2008).

6.6 ESTROGEN RECEPTORS

Estrogen receptors (ER) take part in adipose metabolism and process of adipogenesis. The risk of obesity-related diseases is higher in postmenopausal women in comparison to men (Mattsson and Olsson, 2007). The lack of estrogen receptors increased body mass index (BMI) and insulin resistance (IR) (Heine et al., 2000; Takeda et al., 2003). Numerous environmental estrogens, such as octylphenol which are widely used as surfactant and highly impact on adipose tissue (Bonefeld-Jorgensen et al., 2007). Octylphenol mostly found in wastewater bodies. It is upregulated "resistin" and adipocytes specific hormone cause IR and decrease adipocytes differentiation (Lee et al., 2008; Legler and Brouwer, 2003; Meerts et al., 2001; Darnerud, 2003, 2008). These ETs or disrupting chemicals are mostly found in the environment. These compounds are basically lipophilic and deposited in the adipose tissue of the body (Darnerud, 2003; Lim et al., 2008; Hoppe and Carey, 2007; Howdeshell et al., 2003; Le et al., 2008; Vandenberg et al., 2007; Welshons et al., 2006).

There is scarcity of the scientific or systematic conformational evidences concerning effect of xenobiotics upon human health. On the other hand, Lang et al. (2008) published an epidemiology article on a US adult population which stated a connection amid high urinary levels of bisphenol-A (BPA) and cardiovascular disease, diabetes, and liver abnormalities. However, for the very first time, by using random sampling studies, we are able to correlate high BPA and various disorders in human beings. Furthermore, biologically significant dosages of BPA were latterly presented to pointedly constrain the adiponectin secretion from lipocytes and human explants (Hugo, 2008). Adiponectin is a significant adipocytokine used to protect human cells from various metabolic disorders, and this data derived from artificial insemination supported by the epidemiology report given by Lang et al. in 2008. In a different research, BPA managed at corporal appropriate amounts was showed to proliferate insulin amount in pancreatic b-cell. This appears to be estrogen receptor-reliant, and implicates indicating by a non-genomic or independent by gene transcription or protein synthesis passageway by extracellular regulated kinase 1/2 (Alonso-Magdalena et al., 2008). Even though the chief intermediaries of action of estrogen are the ERs (estrogen receptors), adjacently there are further ways by which estrogen signaling transmission takes place. Particularly, the non-conventional ER (estrogen receptor) GPR30 has in recent times been confirmed to be significantly convoluted in the impacts of estrogens on metabolic rate (Martensson et al., 2008). ER-GPR30 is membrane-enclosed and be a member of the class of combined G-protein receptors (Prossnitz et al., 2008). Mice with deficiency of ER-GPR30 agonize from hyperglycemia, decreased glucose forbearance, and higher blood pressure amid other warning sign (Martensson et al., 2008). Remarkably, previous studies have been revealed that GPR30 "G protein-coupled estrogen receptor 30" alike traditional receptors of estrogen, are enclosed and stimulated by numerous EDCs (endocrine disruptive chemicals), for example BPA (Thomas and Dong, 2006; Thomas et al., 2005). As per ER-GPR30 is identified in islets of the pancreas (Martensson et al., 2008), it possibly will moderately subsidize to the insulin raise after BPA disclosure.

6.7 RETINOID X RECEPTOR (RXRS)/PPARG

Peroxisome proliferator-activated receptors are lipo-regulatory and lipid-sensing receptors (Sugden and Holness, 2008), and PPARg (Peroxisome proliferator-activated receptor gamma) is a main controller in adipose tissue. A number of ligands, including fatty acids and eicosanoids, are responsible for the activation of this receptor, which formulates accommodating heterodimers along with RXR, and regulates involved genes in insulin motioning and glucose and fatty acid metabolic breakdown. Manmade PPARg agonists are in scientific usage to recover glucose forbearance in diabetes mellitus. Organotins like TBT (tributyltin chloride) were revealed to be capable of activating RXR by attaching to the ligand binding province of the receptors, in that way performing as agonists (Grun and Blumberg, 2006; Grun et al., 2006). Organotin complexes are chemical elements formed of tin, directly bound to numerous organic groups. Organotin were initially utilized in boat paint, however, their use is now banned. The toxicity of organotins in the aquatic environs consists of endocrine disorders, for example, "imposex" in gastropods. Organotins disclosure to human take place via consumption of fish and shellfish (Appel, 2004). Nowadays, organotins are highly in use to guard textiles and wood from the growth of fungus and as mediator in the manufacturing of plastic. Grun et al. (2006) stated that the impact of PPARg and RXR initiation through TBT which object the adipogenesis passageway. Specially, TBT was revealed to raise the quantity of distinguished adipo-cells inside the 3T3-L1 cell archetypal and to disturb regular adipose synthesis and accumulation in vertebrate progressive copies. Conflicting to the obesogenic assumption, it has been validated in monkeys and rodents that disclosure to PFCs (perfluorinated compounds), compounds that have been instituted to proliferation in serum testers in the former years, have strong an orexigenic impacts. PFCs, performing through PPARa, constrain weight gain and feeding behavior (Butenhoff et al., 2002; Asakawa et al., 2008). Remarkably, in recent times it has been revealed that ERb can perform as adverse controller of PPARg (Foryst-Ludwig et al., 2008). It declines transcriptional activity of ligand-induced PPARg and *in vitro* PPARg-induced adipogenesis. Moreover, ERb knock-out mice display upgraded insulin reaction when retained on a diet with high fat in contrast to wild type mice. These outcomes recommend that function of PPARg (Peroxisome proliferator-activated receptor gamma) cannot merely be influenced by endocrine disruptive chemicals sprightly relating with the receptor, but also by endocrine disruptive chemicals that regulate ERb activity.

6.8 ORGAN AND SYSTEMIC EFFECTS OF ENDOCRINE TOXICANTS (ETS)

Endocrine disruptive chemicals can impair ordinary physiological reactions correlated to the reproductive/multiplicative system. Furthermore, numerous *in vivo* and *in vitro* investigations are accessible about ovarian and testicular irregularities (Miller and Sharpe, 1998). Endocrine disrupting chemicals (EDCs) decrease amount along with eminence of sperms, as well as with an upturn in the incidence of prostate, testicular, and breast cancer (Toppari et al., 1996). In the wildlife species, these sexual faults have also been seen to a great extent, and environmental chemicals have been recounted to perform an important function in lieu of all these (Maffin et al., 2006). TDS (Testicular dysgenesis syndrome), reduced semen quality, and testicular carcinoma are the key complaints identified to be linked with a few environmental contaminants distressing the endocrine system of the body (Skakkebaek et al., 2001). DES brings some morphological and functional behavior alterations in the female reproductive system (Maffin et al., 2006). According to an *in-vitro* study conducted by Gore (2002) on two organochlorine pesticides (OCPs), i.e., chlorpyrifos and methoxychlor confirmed their impact on variation in the gonadotrophin hormone bio-synthesis in hypothalamic GT1-7 cells. Hence, contaminants object the entire structure of hormonal secretion in human beings. A study accompanied by 120 girl population in Belgium stated that PCBs have the destructive potential against various developmental stages of the reproductive system (Staessen et al., 2001). It was observed that children of pesticides exposed workers are also suffered from reproductive malfunctioning (Weidner et al., 1998). BPA, a well-known ED, percolates out as of plastic packing things resembling water carboys, reusable baby bottles and other domestic things. Therefore, BPA also places children at great risk of endocrine disruptive harms (Hunt et al., 2003). Correspondingly miscarriages observed in women are commonly due to aneuploidy. According to Sugiura-Ogasawara et al. (2005) the threat of miscarriage was elevated in women uncovered to BPA.

6.9 CARCINOGENICITY

A large number of chemicals that were used during the Second World War as weapons have been considered for their endocrine system disrupting properties inside the human body. An investigation on animals has been accompanied by around 48 endocrine disruptive chemicals registered in CDC (Center for Disease Control and prevention in the US) list, and the effects have revealed adjacent connection to developmental effects, mutagenicity, and carcinogenicity (Choi et al., 2004). A scientific electronic-based imitation analysis on animals exhibited that several EDCs have distressing impacts, including thyroid gland (TG) cancer, prostate cancer, and mutagenicity considerably more interrelated to cardio-vascular difficulties and MetS (Sot and Sonnenschein, 2010). Natural or synthetic estrogens can also act as carcinogenic and in further most instances of analogous tumors and poisonousness proliferate in the time of development of various different organs. Motherly source demonstrated to be chief way of cognate poisonousness (Birnbaum and Fenton, 2003). According to a study undertaken on newborns and youngsters residing in agronomic regions consuming domestic pesticides has demonstrated an augmented level of lymphoma and leukemia (Alexander et al., 2001). Different components like nitrosamines, alkenes, heterocyclic, and aromatic compounds exist in the tobacco smoke, which is the key reason behind the cancer of lung and alveoli. Moreover, carbamates are responsible to induce leukemia in newborns (Das et al., 2001). Correspondingly, atrazine is one of the broadly utilized insecticides used for regulating the weeds growth in the crops. The exposure of atrazine is linked with the mammary glands' cancer (Cooper et al., 2000). According to a report of the American Cancer Society, prostate cancer cases are growing quickly (Jemal et al., 2006). Benign prostate hyperplasia is the main difficulty of prostate cancer detected in 50% of men at the age of 60 years (Prins and Korach, 2008). All through the last 50 years, xenoestrogens (xenohormone that imitates estrogen) have been anticipated as the supreme possible source for increasing problem of breast cancer. It has been investigated that augmented levels of estrogen all through pregnancy, give rise to breast cancer and other endocrine interferences. DDT (dichlorodiphenyltrichloroethane), toxaphene, and further pesticides are having endocrine disturbing activities have been evidenced to prompt breast tumor.

6.10 EFFECT OF ENDOCRINE TOXICANTS (ETS) ON THE THYROID SYSTEM

Regular physiological processes, brain developmental stages, and metabolic control in the human body are regulated by thyroid hormones (THs). Disturbance or delayed release in THs can give rise to mental, developmental, and metabolic interruptions which ultimately caused damage of brain (WHO,

2013). Pharmacokinetics of thyroid system have been considered as one of the objectives for endocrine disruptive chemicals, including perchlorates, dioxins, BPA, flame retardants, pesticides, phytoestrogens, PCB and PAHs (Moriyama et al., 2002). There are many ways through which EDCs can affect the thyroid system. EDCs can perform as an agonist or antagonist as well modifying the THs passage or paths (Cheek et al., 1999). Organochlorines, a well-known active endocrine disrupting chemical have been testified to decrease the THs level, predominantly "T4" in several species. Further, organochlorines including alachlor and acetochlor also be a source for decreased levels of THs (T4) and proliferated UDPGT (uridine diphosphate glucuronosyl transferase) levels (Darnerud et al., 1996). It has been proved in various animal studies that PCBs have great harmony to thyroxin definite binding proteins, therefore greatly destruct customary THs physiology. Furthermore, this can change the growth and adrenal roles inside the human body (Green et al., 2005). Phthalate exposure results in impaired metabolic breakdown of THs in infants. After disclosure to altered forms of phthalates, histopathological harms have been detected in rats. Amid further chemicals, parabens are the usually utilized as additives in the food. Methyl parabens has been reported for declining activity of thyroid and a dosage reliant inhibition of iodide activities was detected throughout *in vitro* investigations (Scollon et al., 2004). Hence, there is a few lacunae that must be investigated to find out the link between parabens and problems of the thyroid system. In addition to other endocrine disruptive chemicals pesticides including DDT (dichlorodiphenyltrichloroethane) and HCB (hexachlorobenzene) have also revealed developing faults in the production of thyroid hormone. Dichlorodiphenyltrichloroethane declines the activity of THs while HCB through metabolic breakdown gives greatly noxious complexes, pointing THs. Numerous animal-based studies are accessible that upkeep toxicity of HCB on THs such as "T3" and "T4" (Rozman et al., 1986).

6.11 OBESITY AND DIABETES

In present, obesity mainly caused due to endocrine disruption that also leads to CVD (cardiovascular disorders) and diabetes (Newbold et al., 2008). Baillie-Hamilton (2002) reported that World Health Organization (WHO) find the way to explain obesity epidemics (in hypothesis) by announcing the world data on total obese patients and undernourished patients. However, some metabolic disorders such as hypertension, hyperinsulinemia, and dyslipidemia, etc., mainly occur due to obesity along with diabetes mellitus type-2 and CVDs that needs treatment to not increase the death rates (Maqbool et al., 2015). Various studies were reported (animal models) in which a significant increase was observed in mice weight after giving DES to "neonatal mice." Newbold et al. (2007) reported that the significant increase in abdominal body fat, also the significant increase of inflammatory biomarkers forms in the DES induced mice. In glucose metabolisms, "estrogen receptor-α" and "estrogen receptor-β" play an important role, in which estradiol, BPA, pesticides, and dioxins, etc., are similar in targeting ERα and ERβ (Le May et al., 2006). However, the mechanisms of homeostasis of glucose and insulin release got changed when EDCs attack all these receptors (Nadal et al., 2009). Although, various chemicals (natural and environmental) disturb these receptors that lead the changes or damage in the physiology of "glucose homeostasis" and pancreatic cells damage, that is the sign of diabetes (Ropero et al., 2008). A study of persistent organic carbon (OC) was demonstrated that shows the relation between obesity and diabetes. Moreover, the environment or surrounding around us can leads to diabetes in humans due to their high concentration (Lee et al., 2006; Remillard and Bunce, 2002). Alonso-Magdalena et al. (2010) reported that a small amount of dose, BPA leads to hyperinsulinemia that is interlinked with obesity and type-2 diabetes mellitus (T2DM). The proper regulation or control of glucose metabolism is regulated by a hormone (produced by α-cells), i.e., glucagon. α-cells and glucagon pathway are damaged also at low doses of BPA and DES (Alonso-Magdalena et al., 2005). Alonso-Magdalena et al. (2011) reported that few EDCs (BPA, dioxins, etc.), are mainly produced from adipose tissues when glucose homeostasis get disturbed due to higher deposition of it. As similar to the metabolic state, generally, they cause hyperinsulinemia during pregnancy and higher the concentration of estrogen concentration causes obesity as well as diabetes (Figure 6.2).

6.12 EFFECT ON THE NERVOUS SYSTEM

Various EDC having impact on the nervous system that further leads to behavioral and morphological changes with the help of various receptors. Frye et al. (2012) reported various censorious stages of development (perinatal, puberty periods, intrauterine, etc.), due to exposure towards EDCs. BPA shows their binding with "schizophrenia and neurotoxicological pathology" in different animal studies. Generally, all the neurotoxic problems are articulated by the sudden breakage in various pathways (physical development of organs, hormones, neurotransmitters, cellular anatomy, etc.) (Brown, 2009). Generally, estrogen shows neuroprotective effects. The sudden change or decrease in estrogen levels can cause the schizophrenia-like symptoms in males as well as in females (Lind et al., 2011), and it also may cause a mutated genetic makeup in the disturbing stage. Then it causes the Turner's and Klinefelter syndrome (Prior et al., 2000). De Lisi et al. (2005) were reported that the psychotic disorder occurs due to Klinefelter's syndrome demonstrated in genetic studies. It is also interlinked with the HPA axis (hypothalamic pituitary adrenal axis). Different animal studies show effect on hippocampal nerve due to higher concentration of glucocorticoid. Generally, during stress, the significant increase in glucocorticoid levels shows changes in the hypothalamic pituitary adrenal axis that leads to schizophrenia (Cotter and Pariante, 2002). However, EDCs affect different neurosteroids (such as allopregnanolone), which help in the regulation of "evoking psychiatric disorders" and also damage the brain that causes disorders (like schizophrenia and bipolar disorders). Generally, estrogen levels change when some variation forms in steroid levels in the body. In additionally, BPA also shows its presence that causes brain disorders (ADHD: attention deficit hyperactivity disorder) (Kimura-Kuroda et al., 2007; Masuo et al., 2004). All the disorders (neural, behavioral, and bipolar) are also found in children's, and infants because of BPA exposure by attacking on various hormonal pathways. Astrocytes also cause schizophrenia. LIF (leukemia inhibitory factor) demonstrated higher in various animal study models because of schizophrenia. Motor activity gets decreased by increase in LIF as similar as Alzheimer disease, autoimmune, and schizophrenia disorder. A metal mercury, mainly methyl mercury causes significant increase in oxidative stress of hippocampus (in brain) leads to neural problems in animal exposure. Few PCBs show negative impact at development stages, however, Aroclor 1221 shows significant positive effects on ER (estrogen receptor) at the time when demonstration of sexually dimorphic brain circuits occurs. In contrast to it, the evaluations of various endocrine disruptive chemicals with their toxic effects have been studied to decrease the effects on human beings. 2,3,8-tetrachlorodibenzo-p-dioxin is another endocrine disruptive chemical like other that shows negative effect on the brain (Fernandez et al., 2010; Head et al., 2008).

6.13 ACTION MECHANISM ON STEROID HORMONES

In steroids biosynthesis, various enzymes play their significant role and endocrine disruptive chemicals directly attack on them and physiology damage can lead to reproductive problems, cancers, nervous disorders, and growth abnormalities (Sanderson, 2006). Various organs (adrenal gland, brain, placenta, testis, etc.), help to regulate biosynthesis of steroid hormone in which CYP450 play an important role in it. In fish, the aromatase, CYP450 and CYP1A1 activity can be inhibited by Organotin that utilized in various industries, paints, and agriculture commonly. Altogether this gives rise to endocrine disrupting impact resembling "imposex" (development of penis in females) by suppressing aromatase in females (Fent, 2003). Certain by-products of DDT such as DDD [mitotane; 1,1-dichloro-2,2-bis(p-chlorophenyl) ethane], have been described to be adreno-toxic in dogs. Recently, different reporters have reported the involvement of benzene as convoluted in several non-cancerous condition effects containing problems of reproductive, cardiovascular system, nervous, immune, and endocrine (Bahadar et al., 2014). In additional research $MeSO_2$-DDE (Methylsulfonyl-DDE) has revealed adreno-toxic influence via control of P450c11-facilitated purposes in adreno-cortical cell lines Kin-8 and Y1. Subsequently harmfulness of everyday utilized elements cannot be eluded; we must to take severe quantities (Bahadar et al., 2014).

6.14 ENDOCRINE DISRUPTION (NUCLEAR RECEPTORS)

In EDCs, nuclear receptors are essentially ligand-inferable transcription aspects that control the appearance of particular genes convoluted in metabolic breakdown, differentiation/division, and sexual properties. Naturally occurring or artificial ligands are well-known to promote the activities of nuclear sensory receptors named co-activators (Rachez et al., 2000). Co-activators comprise p-160 family with integral histone acetyltransferase action, thyroid hormone receptor activator protein 220, vitamin D receptor-cooperating protein 205 (Lonard et al., 2004). Arise in nuclear receptors transcriptional activity has been detected in SERM preserved tissues. Different type of co-activators such as SRC (steroid receptor co-activator) blocks the other co-activators, and CBP (CREB binding protein) has been demonstrated in the degradation of proteins/amino acids. This occurs because of the modification of gene expressions and prevention of the action of co-activators. In SERM (selective estrogen-receptor modulator) preserved mouse cells, reaction to raloxifene and 4HT signs broad biological response of ligands, inducing nuclear receptors transcriptional activity. It has been studied that further down the stimulus of high calorie diet and high fat, these nuclear receptors become stimulated by organotin that is capable of promoting differentiation of adipocyte and susceptibility succeeding concerning obesity/heaviness (Grun and Blumberg, 2006). Disclosure of humans towards organotins is frequently detected from contaminated food bases, fungicides, and polluted water from textile businesses. Along with organotins, BPA and phthalates have also been demonstrated for exciting linked pathways of nuclear hormones and nuclear receptor's causing metabolic disorder (Grün and Blumberg, 2004; Golub and Doherty, 2004).

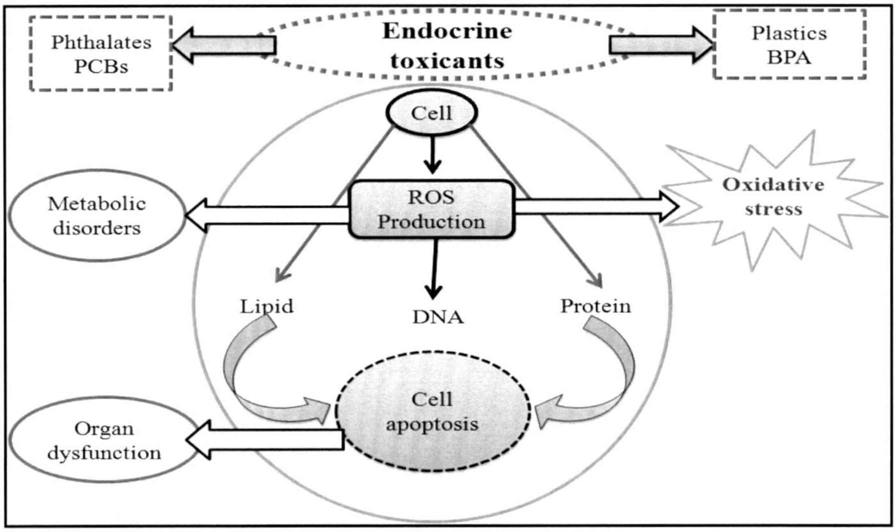

FIGURE 6.2 Effect of endocrine toxicants on cell causes metabolic disorders, organ dysfunction, and oxidative stress.

KEYWORDS

- **bisphenol-A**
- **endocrine disrupting chemicals**
- **estrogen receptors**
- **persistent organic pollutants**
- **polybrominated biphenyls**
- **polychlorinated biphenyls**

REFERENCES

Alexander, F. E., Patheal, S. L., Biondi, A., Brandalise, S., Cabrera, M. E., Chan, L. C., Chen, Z., et al., (2001). Transplacental chemical exposure and risk of infant leukemia with MLL gene fusion. *Cancer Res., 61*, 2542–2546.

Alonso-Magdalena, P., Laribi, O., Ropero, A. B., Fuentes, E., Ripoll, C., Soria, B., & Nadal, A., (2005). Low doses of bisphenol A and diethylstilbestrol impair Ca^{2+} signals in pancreatic α-cells through a nonclassical membrane estrogen receptor within intact islets of Langerhans. *Environ. Health Perspect., 113*, 969–977.

Alonso-Magdalena, P., Quesada, I., & Nadal, A., (2011). Endocrine disruptors in the etiology of type 2 diabetes mellitus. *Nat. Rev. Endocrinol., 7,* 346–353.

Alonso-Magdalena, P., Ropero, A. B., Carrera, M. P., Cederroth, C. R., Baquie, M., Gauthier, B. R., Nef, S., et al., (2008). Pancreatic insulin content regulation by the estrogen receptor ERα. *PloS One, 3*, 2069.

Alonso-Magdalena, P., Vieira, E., Soriano, S., Menes, L., Burks, D., Quesada, I., & Nadal, A., (2010). Bisphenol A exposure during pregnancy disrupts glucose homeostasis in mothers and adult male offspring. *Environ. Health Perspect., 118*, 1243–1250.

Appel, K. E., (2004). Organotin compounds: Toxicokinetic aspects. *Drug Metab. Rev., 36*, 763–786.

Arsenescu, V., Arsenescu, R. I., King, V., Swanson, H., & Cassis, L. A., (2008). Polychlorinated biphenyl-77 induces adipocyte differentiation and proinflammatory adipokines and promotes obesity and atherosclerosis. *Environ. Health Perspect., 116*, 761–768.

Asakawa, A., Toyoshima, M., Harada, K. H., Fujimiya, M., Inoue, K., & Koizumi, A., (2008). The ubiquitous environmental pollutant perfluorooctanoic acid inhibits feeding behavior via peroxisome proliferator-activated receptor-α. *Int. J. Mol. Med., 21*, 439–445.

Bahadar, H., Mostafalou, S., & Abdollahi, M., (2014). Current understandings and perspectives on non-cancer health effects of benzene: A global concern. *Toxicol. Appl. Pharmacol., 276*, 83–94.

Baillie-Hamilton, P. F., (2002). Chemical toxins: A hypothesis to explain the global obesity epidemic. *J. Altern. Complement. Med., 8,* 185–192.

Baker, M. E., Medlock, K. L., & Sheehan, D. M., (1998). Flavonoids inhibit estrogen binding to rat alpha-fetoprotein. *Proc. Soc. Exp. Biol. Med., 217*, 317–321.

Birnbaum, L. S., & Fenton, S. E., (2003). Cancer and developmental exposure to endocrine disruptors. *Environ. Health Perspect., 111*, 389–394.

Boas, M., Feldt-Rasmussen, U., Skakkebæk, N. E., & Main, K. M., (2006). Environmental chemicals and thyroid function. *Eur. J. Endocrinol., 154*, 599–611.

Bonefeld-Jørgensen, E. C., Long, M., Hofmeister, M. V., & Vinggaard, A. M., (2007). Endocrine-disrupting potential of bisphenol A, bisphenol A dimethacrylate, 4-n-nonylphenol, and 4-n-octylphenol *in vitro*: New data and a brief review. *Environ. Health Perspect., 115*, 69–76.

Brown, Jr. J. S., (2009). Effects of bisphenol-A and other endocrine disruptors compared with abnormalities of schizophrenia: An endocrine-disruption theory of schizophrenia. *Schizophr. Bull., 35,* 256–278.

Brunnberg, S., Pettersson, K., Rydin, E., Matthews, J., Hanberg, A., & Pongratz, I., (2003). The basic helix–loop–helix–PAS protein ARNT functions as a potent coactivator of estrogen receptor-dependent transcription. *Proc. Natl. Acad. Sci., 100*, 6517–6522.

Butenhoff, J., Costa, G., Elcombe, C., Farrar, D., Hansen, K., Iwai, H., Jung, R., et al., (2002). Toxicity of ammonium perfluorooctanoate in male cynomolgus monkeys after oral dosing for 6 months. *Toxicol. Sci., 69*, 244–257.

Cao, Y., Calafat, A. M., Doerge, D. R., Umbach, D. M., Bernbaum, J. C., Twaddle, N. C., Ye, X., & Rogan, W. J., (2009). Isoflavones in urine, saliva, and blood of infants: Data from a pilot study on the estrogenic activity of soy formula. *J. Expo. Sci. Environ. Epidemiol., 19*, 223–234.

Cheek, A. O., Kow, K., Chen, J., & McLachlan, J. A., (1999). Potential mechanisms of thyroid disruption in humans: Interaction of organochlorine compounds with thyroid receptor, transthyretin, and thyroid-binding globulin. *Environ. Health Perspect., 107*, 273–278.

Choi, S. M., Yoo, S. D., & Lee, B. M., (2004). Toxicological characteristics of endocrine-disrupting chemicals: Developmental toxicity, carcinogenicity, and mutagenicity. *J. Toxicol. Environ. Health. Part B, 7*, 1–23.

Claessens, F., & Gewirth, D. T., (2004). DNA recognition by nuclear receptors. *Essays Biochem., 40*, 59–72.

Cooper, R. L., Stoker, T. E., Tyrey, L., Goldman, J. M., & McElroy, W. K., (2000). Atrazine disrupts the hypothalamic control of pituitary-ovarian function. *Toxicol. Sci., 53*, 297–307.

Cotter, D., & Pariante, C. M., (2002). Stress and the progression of the developmental hypothesis of schizophrenia. *Br. J. Psychiatry., 181*, 363–365.

Darnerud, P. O., (2003). Toxic effects of brominated flame retardants in man and in wildlife. *Environ. Int., 29*, 841–853.

Darnerud, P. O., Morse, D., Klasson-Wehler, E., & Brouwer, A., (1996). Binding of a 3, 3′, 4, 4′-tetrachlorobiphenyl (CB-77) metabolite to fetal transthyretin and effects on fetal thyroid hormone levels in mice. *Toxicol., 106*, 105–114.

Das, P. C., McElroy, W. K., & Cooper, R. L., (2001). Alteration of catecholamines in pheochromocytoma (PC12) cells *in vitro* by the metabolites of chlorotriazine herbicide. *Toxicol. Sci., 59*, 127–137.

DeLisi, L. E., Maurizio, A. M., Svetina, C., Ardekani, B., Szulc, K., Nierenberg, J., Leonard, J., & Harvey, P. D., (2005). Klinefelter's syndrome (XXY) as a genetic model for psychotic disorders. *Am. J. Med. Genet. B Neuropsychiatr. Genet., 135*, 15–23.

Dickerson, S. M., & Gore, A. C., (2007). Estrogenic environmental endocrine-disrupting chemical effects on reproductive neuroendocrine function and dysfunction across the life cycle. *Rev. Endocr. Metab. Disord., 8*, 143–159.

Fernández, M., Paradisi, M., D'Intino, G., Del Vecchio, G., Sivilia, S., Giardino, L., & Calza, L., (2010). A single prenatal exposure to the endocrine disruptor 2, 3, 7, 8-tetrachlorodibenzo-p-dioxin alters developmental myelination and remyelination potential in the rat brain. *J. Neurochem., 115*, 897–909.

Fierens, S., Mairesse, H., Heilier, J. F. O., De Burbure, C., Focant, J. F. O., Eppe, G., De Pauw, E., & Bernard, A., (2003). Dioxin/polychlorinated biphenyl body burden, diabetes and endometriosis: Findings in a population-based study in Belgium. *Biomarkers, 8*, 529–534.

Foryst-Ludwig, A., Clemenz, M., Hohmann, S., Hartge, M., Sprang, C., Frost, N., Krikov, M., et al., (2008). Metabolic actions of estrogen receptor beta (ERβ) are mediated by a negative cross-talk with PPARγ. *PLoS Genet., 4*, 100–108.

Frye, C., Bo, E., Calamandrei, G., Calza, L., Dessì-Fulgheri, F., Fernández, M., Fusani, L., et al., (2012). Endocrine disruptors: A review of some sources, effects, and mechanisms of actions on behavior and neuroendocrine systems. *J. Neuroendocrinol., 24*, 144–159.

Golub, M. S., & Doherty, J. D., (2004). Triphenyltin as a potential human endocrine disruptor. *J. Toxicol. Environ. Health-B, 7*, 281–295.

Gore, A. C., (2002). Organochlorine pesticides directly regulate gonadotropin-releasing hormone gene expression and biosynthesis in the GT1-7 hypothalamic cell line. *Mol. Cell. Endocrinol., 192*, 157–170.

Green, R., Hauser, R., Calafat, A. M., Weuve, J., Schettler, T., Ringer, S., Huttner, K., & Hu, H., (2005). Use of di (2-Ethylhexyl) phthalate–containing medical products and urinary levels of mono (2-Ethylhexyl) phthalate in neonatal intensive care unit infants. *Environ. Health Perspect., 113*, 1222–1225.

Gronemeyer, H., Gustafsson, J. Å., & Laudet, V., (2004). Principles for modulation of the nuclear receptor superfamily. *Nat. Rev. Drug Discovery, 3,* 950–964.

Grün, F., & Blumberg, B., (2006). Environmental obesogens: Organotins and endocrine disruption via nuclear receptor signaling. *Endocrinol., 147*, 50–55.

Grun, F., Watanabe, H., Zamanian, Z., Maeda, L., Arima, K., Cubacha, R., Gardiner, D. M., et al., (2006). Endocrine-disrupting organotin compounds are potent inducers of adipogenesis in vertebrates. *Mol. Endocrinol., 20*, 2141–2155.

Gunton, J. E., Kulkarni, R. N., Yim, S., Okada, T., Hawthorne, W. J., Tseng, Y. H., Roberson, R. S., et al., (2005). Loss of ARNT/HIF1β mediates altered gene expression and pancreatic-islet dysfunction in human type 2 diabetes. *Cell, 122*, 337–349.

Hankinson, O., (2005). Role of coactivators in transcriptional activation by the aryl hydrocarbon receptor. *Arch. Biochem. Biophys., 433,* 379–386.

Head, J. A., Hahn, M. E., & Kennedy, S. W., (2008). Key amino acids in the aryl hydrocarbon receptor predict dioxin sensitivity in avian species. *Environ. Sci. Technol., 42*, 7535–7541.

Heine, P. A., Taylor, J. A., Iwamoto, G. A., Lubahn, D. B., & Cooke, P. S., (2000). Increased adipose tissue in male and female estrogen receptor-α knockout mice. *Proc. Natl. Acad. Sci., 97*, 12729–12734.

Hockings, J. K., Thorne, P. A., Kemp, M. Q., Morgan, S. S., Selmin, O., & Romagnolo, D. F., (2006). The ligand status of the aromatic hydrocarbon receptor modulates transcriptional activation of BRCA-1 promoter by estrogen. *Cancer Res., 66*, 2224–2232.

Hoppe, A. A., & Carey, G. B., (2007). Polybrominated diphenyl ethers as endocrine disruptors of adipocyte metabolism. *Obesity., 15*, 2942–2950.

Howdeshell, K. L., Peterman, P. H., Judy, B. M., Taylor, J. A., Orazio, C. E., Ruhlen, R. L., Vom, S. F. S., & Welshons, W. V., (2003). Bisphenol A is released from used polycarbonate animal cages into water at room temperature. *Environ. Health Perspect., 111*, 1180–1187.

Hugo, E. R., Brandebourg, T. D., Woo, J. G., Loftus, J., Alexander, J. W., & Ben-Jonathan, N., (2008). Bisphenol A at environmentally relevant doses inhibits adiponectin release from human adipose tissue explants and adipocytes. *Environ. Health Perspect., 116*, 1642–1647.

Hunt, P. A., Koehler, K. E., Susiarjo, M., Hodges, C. A., Ilagan, A., Voigt, R. C., Thomas, S., et al., (2003). Bisphenol A exposure causes meiotic aneuploidy in the female mouse. *Curr. Biol., 13*, 546–553.

Janošek, J., Hilscherová, K., Bláha, L., & Holoubek, I., (2006). Environmental xenobiotics and nuclear receptors—interactions, effects and *in vitro* assessment. *Toxicol. in Vitro., 20*, 18–37.

Jemal, A., Siegel, R., Ward, E., Murray, T., Xu, J., Smigal, C., & Thun, M. J., (2006). Cancer statistics, 2006. *Cancer J. Clin., 56,* 106–130.

Kimura-Kuroda, J., Nagata, I., & Kuroda, Y., (2007). Disrupting effects of hydroxy-polychlorinated biphenyl (PCB) congeners on neuronal development of cerebellar Purkinje cells: A possible causal factor for developmental brain disorders? *Chemosphere, 67,* 412–420.

Kuiper, G. G., Lemmen, J. G., Carlsson, B. O., Corton, J. C., Safe, S. H., Van, D. S. P. T., Van, D. B. B., & Gustafsson, J. A., (1998). Interaction of estrogenic chemicals and phytoestrogens with estrogen receptor β. *Endocrinol., 139*, 4252–4263.

Lang, I. A., Galloway, T. S., Scarlett, A., Henley, W. E., Depledge, M., Wallace, R. B., & Melzer, D., (2008). Association of urinary bisphenol A concentration with medical disorders and laboratory abnormalities in adults. *J. American Med. Assoc., 300*, 1303–1310.

Le May, C., Chu, K., Hu, M., Ortega, C. S., Simpson, E. R., Korach, K. S., Tsai, M. J., & Mauvais-Jarvis, F., (2006). Estrogens protect pancreatic β-cells from apoptosis and prevent insulin-deficient diabetes mellitus in mice. *Proc. Natl. Acad. Sci., 103*, 9232–9237.

Le, H. H., Carlson, E. M., Chua, J. P., & Belcher, S. M., (2008). Bisphenol A is released from polycarbonate drinking bottles and mimics the neurotoxic actions of estrogen in developing cerebellar neurons. *Toxicol. Lett., 176*, 149–156.

Lee, D. H., Lee, I. K., Porta, M., Steffes, M., & Jacobs, D. R., (2007). Relationship between serum concentrations of persistent organic pollutants and the prevalence of metabolic syndrome among non-diabetic adults: Results from the National Health and Nutrition Examination Survey 1999–2002. *Diabetol., 50,* 1841–1851.

Lee, D. H., Lee, I. K., Song, K., Steffes, M., Toscano, W., Baker, B. A., & Jacobs, D. R., (2006). A strong dose-response relation between serum concentrations of persistent organic pollutants and diabetes: Results from the National Health and Examination Survey 1999–2002. *Diab. Care, 29*, 1638–1644.

Lee, D. H., Lee, I. K., Song, K., Steffes, M., Toscano, W., Baker, B. A., & Jacobs, D. R., (2006). A strong dose-response relation between serum concentrations of persistent organic pollutants and diabetes: Results from the National Health and Examination Survey 1999–2002. *Diab. Care., 29,* 1638–1644.

Lee, J. W., Lee, Y. C., Na, S. Y., Jung, D. J., & Lee, S. K., (2001). Transcriptional coregulators of the nuclear receptor superfamily: Coactivators and corepressors. *Cell. Mol. Life Sci., 58,* 289–297.

Lee, M. J., Lin, H., Liu, C. W., Ding, W. H., & Kao, Y. H., (2008). Octylphenol stimulates resistin gene expression in 3T3-L1 adipocytes via the estrogen receptor and extracellular signal-regulated kinase pathways. *Am. J. Physiol. Cell Physiol., 294*, 1542–1551.

Legler, J., & Brouwer, A., (2003). Are brominated flame retardants endocrine disruptors? *Environ. Int., 29*, 879–885.

Lim, J. S., Lee, D. H., & Jacobs, D. R., (2008). Association of brominated flame retardants with diabetes and metabolic syndrome in the US population, 2003–2004. *Diab. Care., 31*, 1802–1807.

Lonard, D. M., Tsai, S. Y., O'Malley, B. W., (2004). Selective estrogen receptor modulators 4-hydroxytamoxifen and raloxifene impact the stability and function of SRC-1 and SRC-3 coactivator proteins. *Mol. Cell. Biol., 24*, 14–24.

Maffini, M. V., Rubin, B. S., Sonnenschein, C., & Soto, A. M., (2006). Endocrine disruptors and reproductive health: The case of bisphenol-A. *Mol. Cell. Endocrinol., 254*, 179–186.

Maqbool, F., Safavi, M., Bahadar, H., Rahimifard, M., Niaz, K., & Abdollahi, M., (2015). Discovery approaches for novel dyslipidemia drugs. *Curr. Drug Discov. Technol., 12,* 90–116.

Martensson, U. E., Salehi, S. A., Windahl, S., Gomez, M. F., Swärd, K., Daszkiewicz-Nilsson, J., Wendt, A., & Andersson, N., (2009). Deletion of the G protein-coupled receptor 30 impairs glucose tolerance, reduces bone growth, increases blood pressure, and eliminates estradiol-stimulated insulin release in female mice. *Endocrinol., 150*, 687–698.

Masuo, Y., Morita, M., Oka, S., & Ishido, M., (2004). Motor hyperactivity caused by a deficit in dopaminergic neurons and the effects of endocrine disruptors: A study inspired by the physiological roles of PACAP in the brain. *Regul. Pept., 123*, 225–234.

Mattsson, C., & Olsson, T., (2007). Estrogens and glucocorticoid hormones in adipose tissue metabolism. *Curr. Med. Chem., 14*, 2918–2924.

Meerts, I. A., Letcher, R. J., Hoving, S., Marsh, G., Bergman, A., Lemmen, J. G., Van, D. B. B., & Brouwer, A., (2001). *In vitro* estrogenicity of polybrominated diphenyl ethers, hydroxylated PDBEs, and polybrominated bisphenol A compounds. *Environ. Health Perspect., 109*, 399–407.

Miller, W. R., & Sharpe, R. M., (1998). Environmental oestrogens and human reproductive cancers. *Endocr.-Relat. Cancer., 5*, 69–96.

Min, G., Kim, H., Bae, Y., Petz, L., & Kemper, J. K., (2002). Inhibitory cross-talk between estrogen receptor (ER) and constitutively activated androstane receptor (CAR): CAR inhibits ER-mediated signaling pathway by squelching p160 coactivators. *J. Biol. Chem., 277*, 34626–34633.

Moriyama, K., Tagami, T., Akamizu, T., Usui, T., Saijo, M., Kanamoto, N., Hataya, Y., et al., (2002). Thyroid hormone action is disrupted by bisphenol A as an antagonist. *J. Clin. Endocrinol. Metab., 87,* 5185–5190.

Nadal, A., Alonso-Magdalena, P., Soriano, S., Quesada, I., & Ropero, A. B., (2009). The pancreatic β-cell as a target of estrogens and xenoestrogens: Implications for blood glucose homeostasis and diabetes. *Mol. Cell. Endocrinol., 304,* 63–68.

Needham, L. L., Calafat, A. M., & Barr, D. B., (2007). Uses and issues of biomonitoring. *Int. J. Hyg. Environ. Health., 210*, 229–238.

Newbold, R. R., Padilla-Banks, E., Jefferson, W. N., & Heindel, J. J., (2008). Effects of endocrine disruptors on obesity. *Int. J. Androl., 31,* 201–208.

Newbold, R. R., Padilla-Banks, E., Snyder, R. J., & Jefferson, W. N., (2007). Perinatal exposure to environmental estrogens and the development of obesity. *Mol. Nutr. Food Res., 51,* 912–917.

Ohtake, F., Baba, A., Takada, I., Okada, M., Iwasaki, K., Miki, H., Takahashi, S., et al., (2007). Dioxin receptor is a ligand-dependent E3 ubiquitin ligase. *Nature, 446*, 562–566.

Petrulis, J. R., & Perdew, G. H., (2002). The role of chaperone proteins in the aryl hydrocarbon receptor core complex. *Chem. Biol. Interact., 141*, 25–40.

Porte, C., Janer, G., Lorusso, L. C., Ortiz-Zarragoitia, M., Cajaraville, M. P., Fossi, M. C., & Canesi, L., (2006). Endocrine disruptors in marine organisms: Approaches and perspectives. *Comp. Biochem. Physiol. Part - C: Toxicol. Pharmacol., 143*, 303–315.

Prins, G. S., & Korach, K. S., (2008). The role of estrogens and estrogen receptors in normal prostate growth and disease. *Steroids, 73,* 233–244.

Prior, T. I., Chue, P. S., & Tibbo, P., (2000). Investigation of turner syndrome in schizophrenia. *Am. J. Med. Genet., 96,* 373–378.

Prossnitz, E. R., Arterburn, J. B., Smith, H. O., Oprea, T. I., Sklar, L. A., & Hathaway, H. J., (2008). Estrogen signaling through the transmembrane G protein-coupled receptor GPR30. *Annu. Rev. Physiol., 70*, 165–190.

Rachez, C., Gamble, M., Chang, C. P. B., Atkins, G. B., Lazar, M. A., & Freedman, L. P., (2000). The DRIP complex and SRC-1/p160 coactivators share similar nuclear receptor binding determinants but constitute functionally distinct complexes. *Mol. Cell. Biol., 20*, 2718–2726.

Reen, R. K., Cadwallader, A., & Perdew, G. H., (2002). The subdomains of the transactivation domain of the aryl hydrocarbon receptor (AhR) inhibit AhR and estrogen receptor transcriptional activity. *Arch. Biochem. Biophys., 408*, 93–102.

Remillard, R. B., & Bunce, N. J., (2002). Linking dioxins to diabetes: Epidemiology and biologic plausibility. *Environ. Health Perspect., 110*, 853–858.

Ropero, A. B., Alonso-Magdalena, P., Quesada, I., & Nadal, A., (2008). The role of estrogen receptors in the control of energy and glucose homeostasis. *Steroids., 73,* 874–879.
Rozman, K., Gorski, J. R., Rozman, P., & Parkinson, A., (1986). Reduced serum thyroid hormone levels in hexachlorobenzene-induced porphyria. *Toxicol. Letters., 30,* 71–78.
Rüegg, J., Hedengran-Faulds, M., Hase, M., Pongratz, I., & Gustafsson, J. Å., (2008). Coregulators in toxicology. In: *Nuclear Receptor Coregulators and Human Diseases* (pp. 517–537).
Safe, S., & Wormke, M., (2003). Inhibitory aryl hydrocarbon receptor– estrogen receptor α cross-talk and mechanisms of action. *Chem. Res. Toxicol., 16*, 807–816.
Sanderson, J. T., (2006). The steroid hormone biosynthesis pathway as a target for endocrine-disrupting chemicals. *Toxicol. Scis., 94*, 3–21.
Sato, S., Shirakawa, H., Tomita, S., Ohsaki, Y., Haketa, K., Tooi, O., Santo, N., et al., (2008). Low-dose dioxins alter gene expression related to cholesterol biosynthesis, lipogenesis, and glucose metabolism through the aryl hydrocarbon receptor-mediated pathway in mouse liver. *Toxicol. Appl. Pharmacol., 229*, 10–19.
Scollon, E. J., Carr, J. A., & Cobb, G. P., (2004). The effect of flight, fasting and p, p′-DDT on thyroid hormones and corticosterone in Gambel's white-crowned sparrow, Zonotrichia leucophrys gambelli. *Comp. Biochem. Physiol. Part-C: Toxicol. Pharmacol., 137*, 179–189.
Skakkebæk, N. E., Meyts, R. D., & Main, K. M., (2001). Testicular dysgenesis syndrome: An increasingly common developmental disorder with environmental aspects: Opinion. *Hum. Reprod., 16*, 972–978.
Soto, A. M., & Sonnenschein, C., (2010). Environmental causes of cancer: Endocrine disruptors as carcinogens. *Nat. Rev. Endocrinol., 6*, 363–370.
Staessen, J. A., Nawrot, T., Den, H. E., Thijs, L., Fagard, R., Hoppenbrouwers, K., Koppen, G., et al., (2001). Renal function, cytogenetic measurements, and sexual development in adolescents in relation to environmental pollutants: A feasibility study of biomarkers. *The Lancet, 357*, 1660–1669.
Sugden, M. C., & Holness, M. J., (2008). Role of nuclear receptors in the modulation of insulin secretion in lipid-induced insulin resistance. *Biochem. Soc. Trans., 36*, 891–900.
Sugiura-Ogasawara, M., Ozaki, Y., Sonta, S. I., Makino, T., & Suzumori, K., (2005). Exposure to bisphenol A is associated with recurrent miscarriage. *Hum. Reprod., 20*, 2325–2329.
Swanson, H. I., (2002). DNA binding and protein interactions of the AHR/ARNT heterodimer that facilitate gene activation. *Chem. Biol. Interact., 141,* 63–76.
Swedenborg, E., Rüegg, J., Hillenweck, A., Rehnmark, S., Faulds, M. H., Zalko, D., Pongratz, I., & Pettersson, K., (2008). 3-Methylcholanthrene displays dual effects on estrogen receptor (ER) α and ERβ signaling in a cell-type specific fashion. *Mol. Pharmacol., 73*, 575–586.
Takeda, K., Toda, K., Saibara, T., Nakagawa, M., Saika, K., Onishi, T., Sugiura, T., & Shizuta, Y., (2003). Progressive development of insulin resistance phenotype in male mice with complete aromatase (CYP19) deficiency. *J. Endocrinol., 176*, 237–246.
Thackaberry, E. A., Bedrick, E. J., Goens, M. B., Danielson, L., Lund, A. K., Gabaldon, D., Smith, S. M., & Walker, M. K., (2003). Insulin regulation in AhR-null mice: Embryonic cardiac enlargement, neonatal macrosomia, and altered insulin regulation and response in pregnant and aging AhR-null females. *Toxicol. Sci., 76*, 407–417.
Thomas, P., & Dong, J., (2006). Binding and activation of the seven-transmembrane estrogen receptor GPR30 by environmental estrogens: A potential novel mechanism of endocrine disruption. *J. Steroid Biochem. Mol. Biol., 102*, 175–179.

Thomas, P., Pang, Y., Filardo, E. J., & Dong, J., (2005). Identity of an estrogen membrane receptor coupled to a G protein in human breast cancer cells. *Endocrinol., 146*, 624–632.

Toppari, J., Larsen, J. C., Christiansen, P., Giwercman, A., Grandjean, P., Guillette, Jr. L. J., Jégou, B., et al., (1996). Male reproductive health and environmental xenoestrogens. *Environ. Health Perspect., 104*, 741–803.

Tsuchiya, Y., Nakajima, M., & Yokoi, T., (2005). Cytochrome P450-mediated metabolism of estrogens and its regulation in human. *Cancer Lett., 227*, 115–124.

Vandenberg, L. N., Hauser, R., Marcus, M., Olea, N., & Welshons, W. V., (2007). Human exposure to bisphenol A (BPA). *Reprod. Toxicol., 24*, 139–177.

Wang, X., Porter, W., Krishnan, V., Narasimhan, T. R., & Safe, S., (1993). Mechanism of 2, 3, 7, 8-tetrachlorodibenzo-p-dioxin (TCDD)-mediated decrease of the nuclear estrogen receptor in MCF-7 human breast cancer cells. *Mol. Cell. Endocrinol., 96*, 159–166.

Weidner, I. S., Møller, H., Jensen, T. K., & Skakkebæk, N. E., (1998). Cryptorchidism and hypospadias in sons of gardeners and farmers. *Environ. Health Perspect., 106*, 793–796.

Welshons, W. V., Nagel, S. C., & Vom, S. F. S., (2006). Large effects from small exposures. III. Endocrine mechanisms mediating effects of bisphenol A at levels of human exposure. *Endocrinol., 147*, 56–69.

World Health Organization, (2013). *State of the Science of Endocrine Disrupting Chemicals: An Assessment of the State of the Science of Endocrine Disruptors Prepared by a Group of Experts for the United Nations Environment Program (UNEP) and WHO* (p. 296).

Wu, Y., & Koenig, R. J., (2000). Gene regulation by thyroid hormone. *Trends Endocrinol. Metab., 11*, 207–211.

You, L., Sar, M., Bartolucci, E., Ploch, S., & Whitt, M., (2001). Induction of hepatic aromatase by p, p′-DDE in adult male rats. *Mol. Cell. Endocrinol., 178*, 207–214.

CHAPTER 7

Biochemical and Histopathological Alterations Caused by Endocrine Toxicants

NEHA JAIN

Department of Zoology, Guru Ghasidas Central University, Bilaspur, Chhattisgarh–495009, India, E-mail: drnehajain2016@gmail.com

ABSTRACT

In the present context, a wide array of health threats is posed by endocrine toxicants (ETs) or endocrine-disrupting chemicals (EDCs). These ETs are basically exogenous chemicals or a mixture of different chemicals which has the ability to interfere with any aspect of the endocrine hormone. EDCs contaminate the environment from a number of sources. The EDCs represent a wide range of compounds such as industrial chemicals, plasticizers, plastics, paints, filling agents, fuels, varnishes, and many other chemicals which are commonly found in our environment, food, and consumer products that alter the endocrine hormone secretion, synthesis, transport, and metabolism. Clinical evidence, epidemiological studies, and experimental models revealed that ETs, directly and indirectly, target the structure and functioning of different organs/tissues or cells of the animal/human body. The different mechanisms which target the normal system of an organism by different EDCs involve divergent pathways, i.e., through nuclear receptors, estrogen receptors, neurotransmitter receptors, and steroidal receptor activation. Since it has been well documented that the endocrine disturbances of different animals and the human body cause breast cancer, ovarian problem, eruption of thyroid, testicular carcinoma, Alzheimer's disease, schizophrenia, nerve

Environmental Endocrine Toxicants: Biology, Effects, and Management. Younis Ahmad Hajam, Rouf Ahmad Bhat, Khalid Rehman Hakeem, & Rajesh Kumar (Eds.)
© 2023 Apple Academic Press, Inc. Co-published with CRC Press (Taylor & Francis)

damage, and obesity. The EDC reported an elevation in oxidative stress, activity of enzymes, decreased the steroidogenesis and also altered the estrous cycle. Similarly, biochemical and histopathological alterations were reported after acute or chronic exposure of EDCs. Necessarily public must be made aware of the risks and use of EDCs in daily life. Thus, the major aim of this review chapter is to summarize the evidence regarding biochemical and histopathological alterations in the body and possible mechanisms.

7.1 INTRODUCTION

The Endocrine System of animals and human beings is a highly integrated physiological system that regulates the structure and functioning of cells/ tissue or organs of the body. The endocrine system controls over the other systems of a body through different glands those secrets hormones. These hormones act on target tissue or cells through various transduction pathways like receptors. The different glands of the endocrine system by which hormone secretion occurs include the pituitary, thyroid, hypothalamus, ovaries, adrenal, and testes. Internally hormone system changes constantly with reference to the environmental cues like photoperiod, temperature, food, and water availability, energy demands, and season or status of reproduction. In addition, aging and longevity are also influenced by the action of the endocrine hormone system. The signaling of endocrine system takes place in a feedback mechanism in which signal is initiated by the central nervous system (CNS) through the hypothalamus. The signaling of endocrine hormone acts by binding a hormone to a specific nuclear receptor transcription factors like estrogen receptor, androgen receptor (AR), thyroid receptor, glucocorticoid receptor, and the other receptors which are found in target tissues. In human, 48 receptors have been identified, and once a receptor is activated by a hormone, it alters the expression of specific genes, which in turn result in alteration of protein levels, thus changing the hormonal change in cellular function. Under normal conditions, the endocrine system acts normally, but if the endocrine hormone synthesis, secretion, transport, metabolism, binding action, or mechanism is altered by any chemical, it will result in the loss of normal structure and functioning of a cell/tissue or organ of the body. Simply such a deviation from normal endocrine functioning to abnormal through the interference of exogenous chemicals is called endocrine toxicity.

From the past few decades, it has been reported that many exogenous substances have the ability to interfere with the hormone synthesis, transport or hormone action mechanism thus could induce toxic effects. This gave rise to the identification and documentation of endocrine toxicants (ETs) as a new class of toxic agent's (Zoeller et al., 2014). These substances which alter the normal functioning of the endocrine system normal functioning of endocrine system are known as endocrine disrupting chemicals (EDCs). Mostly, EDCs include exogenous chemicals which have the ability to interfere with any aspect of hormonal action. EDCs produce adverse effects on growth, development, neurological disorder, genetic alteration, immune stress, and abnormal growth pattern. It has been reported that some EDCs cause nonmonotonic dose responses, suggesting that low doses of EDCs may have significant unanticipated impacts on some endpoints. This alteration of normal endocrine functioning has brought the tissues or cells under physiological stress, which in turn alter the tissues or cells biochemically and histopathological. There is a need to address the vary minute influences of the endocrine disruptors (EDs) on the biochemical, histological, physiological, and genetic parameters of an organism. With this, there is also a need to advance the knowledge and information on endocrine disruption mechanisms. Therefore, the major aim of this review chapter is to summarize the evidence regarding biochemical and histopathological alterations in the body and its possible mechanism. More importantly, this chapter will encompass the research challenges and gaps in the existing knowledge and future research perspectives on endocrine toxicity.

7.2 ENDOCRINE DISRUPTING CHEMICALS (EDCS)

It has been reported from last few decades that some exogenous substances or chemicals have the ability to interfere the hormone levels or activity and could cause toxic effects. Research on these substances/chemicals identified them as "endocrine disruptors" or "endocrine disrupting chemicals (EDCs)." In scientific literature, this terminology was first used in 1993 (Zoeller et al., 2014). Endocrine society has defined EDCs as: an exogenous (non-natural) chemical, or mixture of chemicals, that interfere with any respect of hormone action. Recently WHO has defined EDCs as; These are the substances or a mixture (natural or synthetic) that alter the function(s) of the endocrine system and consequently cause adverse health effects in an intact organism, progeny, or (sub)population. ETs or EDCs show similarity to a number of chemical classes, which aren't known by their chemical structure or by their utilization, but they are identified by their mode of action mechanism. Further, the EDCs have been classified by their origin (Herman et al., 2020) as:

- Synthetic endocrine disrupting chemicals (S-EDCs); and
- Natural endocrine disrupting chemicals (N-EDCs).

Both of these interact with endocrine receptors and disturb hormonal balance. There are lakhs of natural and synthetic chemicals of which thousands may be EDCs. A short list of the representative EDCs and their application is provided in Table 7.1. Human and other animals come in contact to endocrine disrupting chemicals (EDCs) by different routes, i.e.:

- Consumption through the skin of organism;
- By inhalation;
- By transfer from one generation to another through reproduction if women have EDCs in their body.

TABLE 7.1 Some Common Category of Endocrine Disrupting Chemicals (EDCs) with Their Use

Sl. No.	Category/Use	Examples of EDCs
1.	Pesticides	Majority of fungicides, insecticides, herbicides, rodenticides, plant growth hormones and then bio-pesticides
2.	Children products	Cadmium, zinc, lead, phthalates
3.	Food contact materials	Bisphenol, phthalates, phenol
4.	Electronics and building materials	Brominated flame retardants, polychlorinated biphenyls
5.	Personal care products, medical tubing	Phthalates
6.	Antibacterial	Triclosan
7.	Textiles, clothing	Perfluorochemicals
8.	Cosmetics and spray	Cream, deodorants, fresheners, etc.

7.3 ACTION MECHANISM OF ENDOCRINE DISRUPTING CHEMICALS (EDCS)

The endocrine system of any organism constitutes a network of hormones, which coordinates various functions of tissues and cells. This hormonal network possesses a number of loops of stimulation and retroaction in a cascade in order to set different physiological parameters and functions for normal health of an organism. Under normal conditions fundamentally the hormone signaling remains normal. EDCs can interfere the hormone signaling and alter the physiological, biochemical, genetic, and histopathological parameters of tissues or cells of an organism. This interference of EDCs with endocrine system occurs through various mechanisms, either binding directly to the receptor of the hormone, or indirectly elevate/inhibit the concentration of endogenous hormones. The basic mechanisms behind the alteration of endocrine hormones by the exposure of EDCs are given below:

i. The most accepted mechanism universally is the direct one, in which EDCs interacts with the hormone receptor, which in turn increase or decrease (Falco et al., 2014) the synthesis of hormone thus alters the pathway of target cells (Yves et al., 2017).
ii. The second one is an indirect mechanism in which the concentration of endogenous active hormones is stimulated or inhibited of either their synthesis or degradation (Golub and Doherty, 2004; Mazur et al., 2015) or availability.

7.4 BIOCHEMICAL ALTERATION BY EDC

Many technological approaches like transcriptomics, proteomics, mass spectroscopy and metabolomics have been used by the toxicologists to reveal the influence of EDCs on the biochemical parameters of tissue or cells and its associated mechanism (Barakat and Beltagy, 2015; Collette et al., 2010; Hamid et al., 2020a, b; Mathieu-Denoncourt et al., 2015). Growing evidence suggests that EDC exposure disturb hormonal activity, which in result alter the physiological parameters of cells or tissues. However, the effect and magnitude of EDCs fluctuate with the species, age, sex, environment, and trophic level. Recently it has been well reported that the EDCs significantly alters the biochemical properties like glycogen, protein, enzymes, carbohydrates, secondary metabolites, etc., of tissues or cells (Bhoi and Patole, 2019; Pallavi et al., 2016; Abhijith et al., 2016; Asgar et al., 2017; Hussein et al., 2019; Rohankar and Dar, 2019; Al-Ghanim et al., 2020; Woo and Chung, 2020; Ghayyur et al., 2021; Dawood et al., 2020a,b; Ghelichpour et al., 2020). The alterations in the biochemistry of tissues or cells are guided by the hormonal functions. Thus, in the current text, we will discuss the influence of the EDCs on biochemical properties regulated by hormonal activities. EDCs have the ability to interfere with the biochemical or physiological parameters of the hypothalamus-pituitary-gonadal axis, but it's reported that every axis of endocrine system might be the target for every EDCs and their action isn't limited to a single axis

(Lauretta et al., 2019). Over the past few years effect of EDCs in alteration of physiological parameters and functioning has been a very challenging task for the scientific community (Backhaus et al., 2000; Brosche and Backhaus, 2010; Cedergreen, 2014; Ding et al., 2017). However, researchers across the globe successfully reported the influence of EDCs on the biochemical parameters, which are regulated by the endocrine system.

Several EDCs act on the endocrine (hormonal gland) systems, where they influence the endocrine axes; as a result, it alters the secretion of hormones from its anterior and posterior lobe (Fujimoto, 2001; Soriano-Guillen and Argente, 2019; Sen and Sellix, 2016). These hormones are responsible for the growth, development, and reproduction. During the exposure of EDCs, the inhibition or stimulation of these hormones occur, which ultimately changes the functioning of tissue or cell. For example, growth hormones (GHs) in addition to the growth-promoting role, it plays an essential role in protein, carbohydrate, and fat metabolism. Furthermore, GH modulates the activities of different target organs like liver, kidney, bone, cartilage, skeletal muscle, and adipose cells. Any change in the secretion of GH hormone under exposure of the EDCs can cause the biochemical changes in the cells or tissues of different organs. These statements were proved by the reports published (Maradonna and Carnevali, 2018; Shabir et al., 2019), who found that the EDCs alters the carbohydrate and lipid concentrations in target organs. Similarly, the other hormones alter the biochemical activity of the target organs under the influence of EDCs. Once EDCs enter into the body of a human or any other animal, these EDCs are accumulated in the kidney, liver, and pancreas, where they cause deleterious effects on protein, lipid, and carbohydrate metabolism through alteration in the relevant enzyme activity. Phthalates inhibit the enzymatic activity of 11 β-hydroxysteroid dehydrogenase, which is responsible for inactivation of endogenous glucocorticoids (Chapman et al., 2013). Different EDCs increase the enzymatic activity which is responsible for the catabolism of different amino acids (Moyer et al., 2017). EDCs also influence the endocrine organs and cause alteration in the level of hormones that are involved in glucose regulation (Dogan and Can, 2011; Heindel et al., 2015; Sabir et al., 2019).

The exposure of EDCs in human and other animals promote the accumulation of lipids in cells and alter the rate of reactive oxygen species (ROS) like malondialdehyde (MDA) and Superoxide dismutase (SOD), thus disturbing the balance of oxidative stress (Maradonna and Carnevali, 2018). It is reported that lipid accumulation was enhanced by the activation of SREBP-1c and PPARa signaling pathway (Zhang et al., 2017). Similarly in zebrafish, the exposure of EDCs disturbed the metabolic processes in the liver by bringing alterations in the hormonal pathway of the endocrine system, which resulted in alteration of five major biological pathways, i.e.:

- Transcription factor network of FOXA2 and FOXA3;
- Pathway of different metabolic processes;
- Alteration in the metabolism of amino acids and derivatives;
- Alteration in the lipid and lipoprotein metabolism; and
- Alteration in triacylglycerol, fatty acid, and ketone body metabolism (Huff et al., 2018).

Kopp et al. (2017) revealed that the disturbance in the circadian rhythms and hormonal pathway was regarded as one of the potential mechanisms by which EDCs influence lipid metabolism. It was observed that the protein content was decreased on the exposure of different EDCs (Choudhary, 2000; Khare and Singh, 2002; Tripathi and Singh, 2003; Kumar et al., 2004; Mishra et al., 2005; Muley et al., 2007; Tiwari and Singh, 2009; Ghelichpour, 2017; Bhoi and Patole, 2019). The reduction in the level of protein and stimulation in the level of amino acids through EDCs exposure directs a high protein hydrolytic activity due to elevation of enzyme protease activity (Muley et al., 2007), which might be a result of disturbance in the hormonal activity. Further, reduction in the total protein content under EDCs exposure may also be due to the destruction of tissues, imbalance of cellular fraction, and a consequent impairment in the synthetic machinery of proteins.

Various diseases in human by EDC-induced biochemical and physiological alteration have been reported by many authors (Shimano and Sato, 2017; Maradonna and Carnevali, 2018; Ayisi et al., 2018; Cheng and Guo, 2018; Walker et al., 2021). As the endocrine system of human and other animals regulate a number of physiological and biochemical mechanisms. The identification of the EDCs capable of perturbing these systems may provide very essential insights into the development of EDC induced metabolic diseases.

7.5 HISTOPATHOLOGICAL ALTERATION AFTER EXPOSURE TO EDCS

Endocrine signaling via the hypothalamus results in the production of hormones which result in maintaining the structure and functioning of different organs of humans and other animals. These hormones produced by the hypothalamus act on the pituitary gland, thyroid gland (TG), pineal gland, parathyroid gland, ovaries, pancreas, testes, adrenal gland, gastrointestinal gland, etc., in the target organs. During the production of a specific hormone, it activates a specific receptor, which alters the expression of specific genes, that gene is responsible for the production of a specific protein for a specific function. So, a specific hormonal pathway is followed for the structure and function of a specific organ. Like the cell/tissue metabolism, physiology, growth, reproduction, etc., are governed by a group of endocrine or exocrine hormones, similarly the anatomy/histology of an organ are structured by the functioning of specific hormones. The different endocrine hormones which are responsible for the anatomical/histological arrangements of different organs of humans and other animals is presented in Table 7.2.

Any disturbance in the hormone signaling pathway results in the alteration of expression of specific genes, resulting in the altered protein levels, which changes the cellular function. The EDCs are responsible for causing endocrine toxicity including alteration in receptor activity pathway. These chemicals alter the ability of the hormone to bind and activate their specific receptors leading to the disruption of the normal signaling cascade. Since the structure and functioning of cells, tissues or organs depend upon the specific hormone production. Thus, change in the anatomy of cells, tissues, or organs due to EDCs are a subject of concern in the present context.

Histology or histopathology is regarded as a promising field of research in toxicology is a promising field of research in toxicology, which provides a clear picture of toxic effects of EDCs and other xenobiotics in vital structure and function of living organisms (Wester and Canton, 1991; Schwaiger et al., 1992; Dutta, 1996; Munoz et al., 2015). It shows the clear picture of histo-architecture of different organs of human and other animals like liver, kidney, heart, reproductive organs, breasts, ovary, testis, or muscles under EDCs exposure, and this helps us to identify the deviation of histo-architecture of specific organ from its normal structure. Therefore histo-pathological alterations can be used as best indicators to monitor the effects of EDCs and other toxicants in human and other animals, and are reported to be the reflection of the overall health status of an organism (Meyers and Hendricks, 1985; Schwaiger et al., 1997; Thophon et al., 2003; Munoz et al., 2015; Drishya et al., 2016; Olaniya, 2020). Further histological alteration of different organs under exposure of EDCs are now widely used as biomarkers to evaluate the health of an organism (Wester and Canton, 1991; Hinton et al., 1993; The et al., 1997; Maurya and Malik, 2016a, b, c; Pereira et al., 2017).

TABLE 7.2 Some Endocrine Hormones that Regulate the Anatomy of Different Organs of Human and Other Animals

SL. No.	Endocrine Hormone	Regulating Anatomy/Histology
1.	Anti-diuretic hormone (ADH), urodilatin, endothelin, natriuretic, parathyroid hormone (PTH)	Kidney
2.	Oxytocin, prolactin-releasing hormone (PRH), prolactin (PRL), estradiol, estrogen, progesterone	Breasts
3.	Thyrotropin-releasing hormone (TRH)	Every organ
4.	Human growth hormone (hGH)	Growth and repair
5.	Thyroid-stimulating hormone (TSH).	Hormone production
6.	Luteinizing hormone (LH), estrogen, and progesterone, gonadotropin-releasing hormone (GnRH), follicle-stimulating hormone (FSH)	Male and female reproductive parts
7.	Melanocyte-stimulating hormone (MSH)	Skin
8.	Thyroxine (T4) and triiodothyronine (T3)	Thyroid gland
9.	Parathyroid	Bones, blood, and muscles
10.	Secretin	Pancreas
11.	Motilin	Gastrointestinal tract
12.	Cholecystokinin, gastric inhibitory polypeptide	Liver, gall bladder
13.	Hydrocortisone, aldosterone, corticosterone, epinephrine, adrenaline, norepinephrine	Heart and blood vessel development

Liver is also a prime target organ for various types of EDC stress and other contaminants (Salamat and Zarie, 2012). Lipid peroxidation is the most important mechanism involved in EDCs induced toxicity (Maradonna and Carnevali, 2018). Though any damage to the liver of fish eventually causes severe physiological and histological disruptions (Pereira et al., 2017). Sharma and Jindal (2020) reported that EDCs exposure bring different histological alterations in liver, i.e., foamy cells, congestion of blood, lymphocytes infiltration, necrosis, melanomacrophage aggregation, pyknotic nuclei, blood sinusoid dilation, hypertrophy, and vacuolation. As this histo-architecture of the liver is regulated by a specific set of hormones through a specific receptor pathway, the EDCs might have disturbed this pathway which resulted in the alteration in the histology of the liver. Similarly, in kidney under exposure of EDCs, Moniruzzaman et al. (2017) reported several histological alterations such as hypertrophy, renal corpuscles, blood vessels, kidney tubules, ruptured collecting duct, ruptured renal corpuscle, hepatopoietic tissue, vacuolation, Bowman's space, degenerated hepatopoietic tissue, intra-cytoplasmic vacuoles, degenerated kidney tissue, narrowing of tubular lumen and degeneration of glomerular cells under exposure of profenofos pesticides. The tissue damage in kidney, like vacuolar degeneration, necrosis, and atrophy were reported under the exposure of EDCs (Tilak and Yacobu, 2002). Similar findings were reported in by many authors (Rahman et al., 2002; Banerjee and Bhattacharya, 1994; Das and Mukherjee, 2000; Dawood et al., 2020a, b; Faris et al., 2020). Sodium fluoride has significantly altered the gonadosomatic index in fish (Agniwanshi et al., 2014). It might be due to the histological alteration in the reproductive organs. Similarly, the historical alteration could take place by the regulation of hormonal activity under ECDs exposure.

7.6 CONCLUSION

The endocrine system is a highly integrated physiological system that regulates metabolism, growth, development, reproduction, and response to stress, among other functions. Hormones produced by endocrine glands regulates functions of body. From the last few decades, many studies have reported that some exogenous chemicals have the ability to interfere with the level and action of hormones; thus, they could induce toxic effects. These influences of exogenous have led to the identification of EDCs. EDCs target humans and other animals in routine life-used products, environment, and occupational exposure. The physiological, biochemical, anatomical, and structural alterations in cells, tissues, or organs by EDCs exposure in humans and other animals coincide with the following characteristics: (i) EDCs alter cell proliferation, cell death, or nutrient supply, (ii) EDCs are genotoxic, (iii) EDCs are immunosuppressive agents, (iv) EDCs induce epigenetic alterations, and (v) EDCs induce oxidative stress. In addition, EDCs are able to modulate endocrine hormone activity through receptor-mediated effects. The major targets of EDCs are nuclear receptors, estrogen receptors, neurotransmitter receptors, steroidal receptor activation, and oxidative stress. These actions on the endocrine system bring biochemical and histological changes in different tissues or organs of humans and other animals. It is further recommended that there should be an increase in understanding of the mechanism and effects of EDCs, which may include updating basic and clinical research in this area, developing the precautionary principles, and increase the involvement of individual researchers and scientific community in communicating or implementing changes in public policy and awareness. EDC targets humans and other animals in routine life-used products, environment, and occupational exposure. The major targets of EDCs are nuclear receptors, estrogen receptors, neurotransmitter receptors, steroidal receptor activation, and oxidative stress. The acute and chronic EDCs exposure to animals and human beings causes biochemical and histopathological alterations. This biochemical and histopathological alteration of EDCs disturbs the normal functioning of the animals and human beings. The disturbance of the endocrine system in the human body causes breast cancer, ovarian problem, and eruption of thyroid, testicular carcinoma, Alzheimer's disease, schizophrenia, nerve damage, and obesity.

KEYWORDS

- **biochemical changes**
- **endocrine disrupting chemicals**
- **growth hormones**
- **histopathological alterations**
- **malondialdehyde**
- **natural endocrine-disrupting chemicals**

REFERENCES

Abhijit, B. D., Ramesh, M., & Poopal, R. K., (2016). Responses of metabolic and antioxidant enzymatic activities in gill, liver and plasma of *Catla catla* during methyl parathion exposure. *J. Basic Appl. Zool., 77,* 31–40.

Agniwanshi, S., Shedpure, M., & Jain, N., (2014). Effect of sodium fluoride on body weight gain and gonadosomatic index in freshwater Catfishes. *J. Ind. Pollut. Control., 30*(2), 339–343.

Al-Ghanim, K. A., Shahid, M., Vijayaraghavan, P., Al-Misned, F. A., Young, O. K., & Hak-Jae, K., (2020). Sub-lethal effect of synthetic pyrethroid pesticide on metabolic enzymes and protein profile of non-target Zebrafish, Danio rerio. *Saudi J. Biol. Sci., 27*(1), 441–447.

Ayisi, C. L., Yamei, C., & Zhao, J. L., (2018). Genes, transcription factors and enzymes involved in lipid metabolism in fin fish. *Agri. Gene., 7*, 7–14.

Backhaus, T., Altenburger, R., Arrhenius, Å., Blanck, H., Faust, M., Finizio, A., Gramatica, P., et al., (2003). The BEAM-project: Prediction and assessment of mixture toxicities in the aquatic environment. *Cont. Shelf Res., 23,* 1757–1769.

Barakat, K. M., & Beltagy, E. A., (2015). Bioactive phthalate from marine *Streptomyces ruber* EKH2 against virulent fish pathogens. Egypt. *J. Aquat. Res., 41*, 49–56.

Bhattacharjee, D., & Das, S., (2013). Toxicity of organochlorine pesticide,- Lindane, to fish: A review. *J. Chem. Pharm. Res., 5,* 90–96.

Bhoi, S. S., & Patole, S. S., (2019). Biochemical changes of freshwater fish, *Channa marulius* (Ham Buch) exposed to 3/4th Sub lethal concentration of cypermethrin and fenvalerate. *Int. J. Life Sci., 13,* 14–18.

Bishop, P. L., & Willett, C. E., (2014). The use and acceptance of other scientifically relevant information (OSRI) in the U.S. environmental protection agency (EPA) endocrine disruptor screening program, birth defects. *Res. B Dev. Reprod. Toxicol., 101*, 3–22.

Borgert, C. J., Mihaich, E. M., Quill, T. F., Marty, M. S., Levine, S. L., & Becker, R. A., (2011). Evaluation of EPA's tier 1 endocrine screening battery and recommendations for improving the interpretation of screening results. *Regul. Toxicol. Pharmacol., 59*, 97–411.

Brosche, S., & Backhaus, T., (2010). Toxicity of five protein synthesis inhibiting antibiotics and their mixture to limnic bacterial communities. *Aquat. Toxicol., 99*, 457–465.

Cedergreen, N., (2014). Quantifying synergy: A systematic review of mixture toxicity studies within environmental toxicology. *PLoS One, 9*, 96580.

Chapman, K., Holmes, M., & Seckl, J., (2013). 11β-hydroxysteroid dehydrogenases: Intracellular gate-keepers of tissue glucocorticoid action. *Physiol Rev.*, *93*, 1139–1206.

Chaudhari, R. T., (2000). Impact of heavy metals on some physiological aspects of the freshwater bivalve, Parresia cylindrical. *J. Indian Poll. Con., 18*, 93–97.

Cheng, X., Li, J., & Guo, D., (2018). SCAP/SREBPs are central players in lipid metabolism and novel metabolic targets in cancer therapy. *Curr. Top Med. Chem., 18*, 484–493.

Collette, T. W., Teng, Q., Jensen, K. M., Kahl, M. D., Makynen, E. A., Durhan, E. J., Villeneuve, D. L., et al., (2010). Impacts of an anti-androgen and an androgen/anti-androgen mixture on the metabolite profile of male fathead minnow urine. *Environ. Sci. Technol., 44*, 6881–6886.

Das, B. K., & Mukherjee, S. C., (2003). Toxicity of cypermethrin in *Labeo rohita* fingerlings: Biochemical, enzymatical and haematological consequences. *Comp. Biochem. Physiol., 134*, 109–121.

Dawood, M. A., AbdEl-Kader, M. F., Moustafa, E. M., Gewaily, M. S., & Abdo, S. E., (2020a). Growth performance and hemato-immunological responses of *Nile tilapia* (*Oreochromis niloticus*) exposed to deltamethrin and fed immunobiotics. *Environl. Sci. Pollu. Res.*, 1–10.

Dawood, M. A., Abdo, S. E., Gewaily, M. S., Moustafa, E. M., SadAllah, M. S., AbdEl-Kader, M. F., Hamouda, A. H., et al., (2020b). The influence of dietary β- glucan on immune, transcriptomic, inflammatory and histopathology disorders caused by deltamethrin toxicity in Nile tilapia (Oreochromis niloticus). *Fish Shellfish Immunol., 98*, 301–311.

Ding, K., Lu, L., Wang, J., Wang, J., Zhou, M., Zheng, C., Liu, J., et al., (2017). *In vitro* and in silico investigations of the binary-mixture toxicity of phthalate esters and cadmium (II) to *Vibrio qinghaiensis* sp.-Q67. *Sci. Total Environ., 580*, 1078–1084.

Dogan, D., & Can, C., (2011). Endocrine disruption and altered biochemical indices in male *Oncorhynchus* mykiss in response to dimethoate. *Pestic. Biochem. Phys., 99*, 157–161.

Drishya, M., Ambikadevi, A., & Aswin, B., (2016). Histopathological changes in the gills of fresh water fish, *Catla catla* exposed to electroplating effluent. *Int. J. Fish. Aquat., 4,* 13–16.

Dutta, H. M., (1996). A composite approach for evaluation of the effects of pesticides on fish. In: Munshi, J. S. D., & Dutta H. M., (eds.), *Fish Morphology* (p. 334). New York, Science Publishers Inc.

Falco, M. D., Sellitti, A., Sciarrillo, R., Capaldo, A., Valiante, A., Lachetta, G., Forte, M., & Laforgia, V., (2014). Nonylphenol effects on the HPA axis of the bioindicator vertebrate, Podarcis sicula lizard, *Chemosphere., 104,* 190–196.

Faris, S., K., (2020). Short-time effects of malathion pesticide on functional and histological changes of liver and kidney in female mice. *Pal J. Biol. Science., 23*, 1103–1112.

Fujimoto, N., (2001). Effect of endocrine disruptors on the pituitary glands. *J. Toxicol. Pathol., 14*, 65–69.

Ghayyur, S., Fiaz, K. M., Tabassum, S., Saleem, A. M., Sajid, M., Badshah, B., Azhar, K. M., et al., (2021). A comparative study on the effects of selected pesticides on hemato-biochemistry and tissue histology of freshwater fish *Cirrhinus mrigala* (Hamilton, 1822). *Saudi J. Biol. Sci., 28,* 603–611.doi.org/10.1016/j.sjbs.2020.10.049.

Ghelichpour, M., Mirghaed, A. T., Hoseini, S. M., & Jimenez, A. P., (2020). Plasma antioxidant and hepatic enzymes activity, thyroid hormones alterations and health status of liver tissue in common carp (*Cyprinus carpio*) exposed to lufenuron. *Aquaculture, 516*, 734634.

Ghelichpour, M., Taheri, M. A., Mirzargar, S. S., Joshaghani, H., & Ebrahimzadeh, M. H., (2017). Plasma proteins, hepatic enzymes, thyroid hormones and liver histopathology of *Cyprinus carpio* (Linnaeus, 1758) exposed to an oxidizing pesticide, indoxacarb. *Aquac. Res., 48*, 5666–5676.

Golub, M., & Doherty, J., (2004). Triphenyltin as a potential human endocrine disruptor. *J. Toxicol. Environ. Health B Crit. Rev., 7*, 281–295.

Hamid, N., Junaid, M., & Pei, D. S., (2020b). Individual and combined mechanistic toxicity of sulfonamides and their implications for ecological risk assessment in the three gorges reservoir area (TGRA). *China. J. Hazard. Mater., 382,* 121106.

Hamid, N., Junaid, M., Manzoor, R., Jia, P. P., & Pei, D. S., (2020a). Prioritizing phthalate esters (PAEs) using experimental *in vitro/vivo* toxicity assays and computational in silico approaches. *J. Hazard. Mater., 398*, 122851.

Hamid, N., Muhammad, J., & De-Sheng, P., (2021). Combined toxicity of endocrine-disrupting chemicals: A review. *Ecotoxicol. Environ. Saf., 215*, 112136. doi.org/10.1016/j.ecoenv.2021.112136.

Heindel, J. J., Newbold, R., & Schug, T. T., (2015). Endocrine disruptors and obesity. *Nat. Rev. Endocrinol., 11*, 653.

Herman, A., et al., (2020). Human exposure to synthetic endocrine disrupting chemicals (S-EDCs) is generally negligible as compared to natural compounds with higher or comparable endocrine activity: How to evaluate the risk of the S-EDCs? *Arch. Toxicol.*, *94,* 2549–2557. doi: 10.1007/s00204-020-02800-8.

Hinton, D. E., & Lauren, D. J., (1990). Liver structural alterations accompanying chronic toxicity: Potential biomarkers of exposure. In: McCartney, J. F., & Shugart, L. R., (eds.), *Biomarkers of Environmental Contamination* (pp. 15–57). Boca Raton, FL: Lewis Publ.

Huff, M., Da Silveira, W. A., Carnevali, O., Renaud, L., & Hardiman, G., (2018). Systems analysis of the liver transcriptome in adult male zebrafish exposed to the plasticizer (2-Ethylhexyl) Phthalate (DEHP). *Sci Rep.*, *8*, 2118.

Hussein, M. M., Elsadaawy, H. A., El-Murr, A., Ahmed, M. M., Bedawy, A. M., Tukur, H. A., Swelum, A. A., & Saadeldin, I. M., (2019). Endosulfan toxicity in Nile tilapia (*Oreochromis niloticus*) and the use of lycopene as an ameliorative agent. Comparative biochemistry and physiology part C: *Toxicol. Pharmacol.*, *224*, 108573.

Khare, A., & Kingh, S., (2002). Impact of malathion on protein content in the freshwater fish, *Clarias batrachus*. *J. Ecotoxicol. Environ. Monit.*, *12*, 129–132.

Kopp, R., Martinez, I. R., Legradi, J., & Legler, J., (2017). Exposure to endocrine disrupting chemicals perturbs lipid metabolism and circadian rhythms. *J. Environ. Sci.*, *62,* 133–137.

Kumar, K., Patri, P., & Pandey, A. K., (2004). Haematological and biochemical responses in the freshwater air-breathing teleost, *Anabas testudineus* (Bloch) exposed to mercury. *J. Ecophysiol. Occup. Hlth.*, *4*, 97–108.

Maradonna, F., & Carnevali, O., (2018). Lipid metabolism alteration by endocrine disruptors in animal models: An overview. *Front. Endocrinol. (Lausanne).*, *9,* 654. doi: 10.3389/fendo. 2018.00654.

Maurya, P. K., & Malik, D. S., (2016a). Distribution of heavy metals in water, sediments and fish tissue (*Heteropneustis fossilis*) in Kali River of western U.P. India. *Int. J. Fish. Aquat. Stud.*, *4*, 208–15.

Maurya, P. K., & Malik, D. S., (2016b). Accumulation and distribution of organochlorine and organophosphorus pesticide residues in water, sediments and fishes, *Heteropneustis fossilis* and *Puntius ticto* from Kali River, India. *J. Toxicol. Environ. Health Sci.*, *8,* 30–40.

Maurya, P. K., & Malik, D. S., (2016c). Bioaccumulation of xenobiotics compound of pesticides in riverine system and its control technique. A critical review. *J. Ind. Pollut. Control.*, *32*, 580–594.

Mazur, C. S., Marchitti, S. A., & Zastre, J., (2015). P-glycoprotein inhibition by the agricultural pesticide propiconazole and its hydroxylated metabolites: Implications for pesticide-drug interactions. *Toxicol. Lett.*, *232*, 37–45.

Meyers, T. R., & Hendricks, J. D., (1985). *Histopathology* (pp. 283–331). Hemisphere Publishing Corporation, Washington, USA.

Mishra, D. K., Bohidar, K., & Pandey, A. K., (2005). Effect of sublethal Carbary and cartap exposure on protein metabolism of the teleost, *Channa punctatus* (Bloch). *J. Inland Fish. Soc.*, *37*, 30–34.

Moniruzzaman, M., Khan, M. M., & Islam, M. S., (2017). Effects of profenofos induced histopathology and recovery patterns in silver barb (*Barbonymus gonionotus*). *Progress. Agric.*, *28,* 240–248.

Moyer, B. J., Rojas, I. Y., Kerley-Hamilton, J. S., Nemani, K. V., Trask, H. W., Ringelberg, C. S., et al., (2017). Obesity and fatty liver are prevented by inhibition of the aryl hydrocarbon receptor in both female and male mice. *Nutr Res.*, *44*, 38–50.

Muley, D. V., Karanjkar, D. M., & Maske, S. V., (2007). Impact of industrial effluents on the biochemical composition of freshwater fish, *Labeo rohita*. *J. Environ. Biol.*, *28,* 245–249.

Muñoz, L., Weber, P., Dressler, V., Baldisserotto, B., & Vigliano, F. A., (2015). Histopathological biomarkers in juvenile silver catfish (*Rhamdia quelen*) exposed to a sublethal lead concentration. *Ecotoxicol. Environ. Saf., 113*, 241–247.

Olaniyi, K., (2020). Histopathology studies of selected organs of *Hemichromis* fasciatus [1] inhabiting igun gold mining and opa reservoirs, South Western Nigeria: A comparative study. *Int. J. Adv. Life Sci. Technol., 4,* 1–10.

Pallavi, S., Ajay, S., & Pandey, A. K., (2016). Pesticides toxicity in fishes: Biochemical, physiological and genotoxic aspects. *Biochem. Cell. Arch., 16*, 199–218.

Rahman, M. Z., Hossain, Z., Mollah, M. F. A., & Ahmed, G. U., (2002). Effect of dazinon 60 EC on *Anbas testudineus*, Channa *punctatus* and *Barbodes gonionotus*. *Naga., 25*, 8–12.

Reif, D. M., Martin, M. T., Tan, S. W., Houck, K. A., Judson, R. S., Richard, A. M., Knudsen, T. B., et al., (2010). Endocrine profiling and prioritization of environmental chemicals using tox cast data. *Environ. Health Perspect., 118*, 1714–1720.

Rohankar, P. H., & Dar, F. A., (2019). Biochemical change in protein of freshwater fish Channa punctatus, in response to treatment with glyphosate. *Int. J. Life Sci.*, 110–112.

Sabir, S., Muhammad, S. H. A., Fareeha, F., Uzma, S., Malik, H. M., & Kanwal, R., (2019). Role of cadmium and arsenic as endocrine disruptors in the metabolism of carbohydrates: Inserting the association into perspectives. *Biomed. Pharmacother., 114*, 108802. doi.org/10.1016/j.biopha.2019.108802.

Salamat, N., & Zarie, M., (2012). Using of fish pathological alterations to assess aquatic pollution: A review. *WJFMS, 4,* 223–231.

Schwaiger, J., Bucher, F., Ferling, H., et.al., (1992). A prolonged toxicity study on the effect of sublethal concentrations of bis(tri-n-butylin)oxide (TBTO): Histopathological and histochemical findings in rainbow trout (*Oncorhynchus mykiss*). *Aquat. Toxicol., 23*, 31–48.

Schwaiger, J., Wanke, R., Adam, S., Pawert, M., Honnen, W., & Triebskorn, R., (1997). The use of histopathological indicators to evaluate contaminant-related stress in fish. *J. Aquat. Ecosyst. Stress Recovery, 6*, 75–86. doi: 10.1023/ A:1008212000208.

Sen, A., & Sellix, M. T., (2016). The circadian timing system and environmental circadian disruption: From follicles to fertility. *Endocrinology., 157*, 3366–3373. doi: 10.1210/en.2016-1450.

Sharma, R., & Jindal, R., (2020). Assessment of cypermethrin induced hepatic toxicity in *Catla catla*: A multiple biomarker approach. *Environ. Res., 184*, 109359.

Shimano, H., & Sato, R., (2017). SREBP-regulated lipid metabolism: Convergent physiology—Divergent pathophysiology. *Nat. Rev. Endocrinol., 13*, 710–730.

Soriano-Guillen, L., & Argente, J., (2019). Central precocious puberty, functional and tumor-related. *Best Pract. Res. Clin. Endocrinol. Metab*. doi: 10.1016/j.beem.2019.01.003.

The, S. J., Adams, S. M., Hinton, D. E., et al., (1997). Histopathological biomarkers in feral fresh water fish populations exposed to different types of contaminant stress. *Aquat. Toxicol., 37*, 51–70.

Thophon, S., Kruatrachue, M., Upathan, E. S., Pokethitiyook, P., Sahaphong, S., & Jarikhuan, S., (2003). Histopathological alterations of white seabass, Lates calcarifer in acute and subchronic cadmium exposure. *Environ. Pollut., 121*, 307–320.

Tilak, & Yacobu., (2003). Toxicity and effect of fenvalerate on fish *Ctenophargodon idella. J. Ecotoxicol. Environ. Monitoring., 12*, 09–15.

Tiwari, S., & Singh, A., (2009). Changes in some biochemical parameters in the liver and muscle of *Colisa fasciatus* due to toxicity of ethanolic extract of *Nerium indicum* (Lal Kaner) latex. *Nat. Prod. Radi., 8,* 48–54.

Tripathi, P. K., & Singh, A., (2003). Toxic effects of dimethoate and carbaryl pesticides on protein metabolism of the freshwater snail *Lymnaea acuminata*. *Bull. Environ. Contam. Toxicol., 70,* 146–152.

Walker, C., Garza, S., Paadopoulos, V., & Culty, M., (2021). Impact of endocrine-disrupting chemicals on steroidogenesis and consequences on testicular function. *Mol. Cell. Endocrinol., 527,* 111215. doi.org/10.1016/j.mce.2021.111215.

Wester, P. W., Canton, J. H., et. al., (1991). The usefulness of histopathology in aquatic toxicity studies. *Comp. Biochem. Physiol. C. Comp. Pharmacol. Toxicol., 100*, 115–117.

Willett, C. E., Bishop, P. L., & Sullivan, K. M., (2011). Application of an integrated testing strategy to the U.S. EPA endocrine disruptor screening program, *Toxicol. Sci., 123*, 15–25.

Woo, S. J., & Chung, J. K., (2020). Effects of trichlorfon on oxidative stress, neurotoxicity, and cortisol levels in common carp, *Cyprinus carpio* L., at different temperatures. *Comp. Biochem. Physiol. C. Comp. Pharmacol. Toxicol., 229*, 1–9.

Yves, C., (2017). Endocrine disruptor compounds (EDCs) and agriculture: The case of pesticides. *C. R. Biol., 340,* 406–409.

Zhang, W., Shen, X. Y., Zhang, W. W., Chen, H., Xu, W. P., & Wei, W., (2017). The effects of di 2- ethyl hexyl phthalate (DEHP) on cellular lipid accumulation in HepG2 cells and its potential mechanisms at the molecular level. *Toxicol. Mech. Methods., 27*, 245–252.

Zoeller, R. T., Bergman, A., Becher, G., et al., (2014). A path forward in the debate over health impacts of endocrine disrupting chemicals. *Environ. Health, 13*, 118.

CHAPTER 8

Polychlorinated Biphenols and Their Impact on the Endocrine System

DIKSHA[1], SEEMA RAI[2], YOUNIS AHMAD HAJAM[3], and RAJESH KUMAR[4]

[1]*Division Zoology, Department of Biosciences, Career Point University, Hamirpur, Himachal Pradesh–176041, India*

[2]*Department of Zoology, Guru Ghasidas Vishwavidayalaya, Bilaspur, Chhattisgarh–495001, India*

[3]*Department of Life Sciences and Allied Health Sciences, Sant Baba Bhag Singh University, Padhiana, Jalandhar, Punjab, India*

[4]*Department of Biosciences, Himachal Pradesh University, Shimla, Himachal Pradesh–171005, India, E-mail: drkumar83@rediffmail.com*

ABSTRACT

Polychlorinated biphenyls (PCBs) are organic carbon (OC) as similar as dichloro diphenyl trichloroethane (DDT). It is produced by replacing hydrogen (H) with chlorine (Cl), i.e., tetrachloride, pentachloro, hexachloro, heptachloro, octachloro, nonachloro, and decachloro biphenyl. PCBs are lipophilic in nature due to which they accumulate in the soil as well as in water bodies for a long time period and further in the food chain. The other sources of PCBs are evaporation of paints, downright leakages into streams and ocean dumping. PCBs have high boiling point, non-flammability, and having chemical stability so due to these they used in electrical work commercially and in various industries (like as plasticizers in paints, plastics, various rubber products, dyes, carbon less copy paper, etc.). PCBs have been linked to chronic effects (immune system damage, decrease in pulmonary

Environmental Endocrine Toxicants: Biology, Effects, and Management. Younis Ahmad Hajam, Rouf Ahmad Bhat, Khalid Rehman Hakeem, & Rajesh Kumar (Eds.)
© 2023 Apple Academic Press, Inc. Co-published with CRC Press (Taylor & Francis)

function, bronchitis, and hormones interference that leads to cancer. There are different models to study endocrine effects on a few endocrine system aspects. Various disorders recognized that are linked with their interaction with ES (endocrine system), such as predominantly developmental toxicity, immunotoxicity, metabolic disorders, thyroid disorders, disorder in the reproductive system (both male and female). Therefore, this chapter summarizes endocrine system disrupting ability, sources, and their effects and mechanism of action.

8.1 INTRODUCTION

Polychlorinated biphenyls (PCBs) are organic carbon (OC) as similar as Di-chloro diphenyl trichloroethane (DDT) (Nemerow and Agardy, 1998). It produces by replacing hydrogen (H) with chlorine (Cl), i.e., "tetrachloro, pentachloro, hexachloro, heptachloro, octachloro, nonachloro, and decachlorobiphenyl" (Wu et al., 2018; Zhang et al., 2015; Zhu et al., 2011; Javorska et al., 2011; Windham et al., 2010; Niu et al., 2006). PCBs have empirical formula C12H10-xClx (x = 1–10). However, there are about 209 PCB congeners, in which chlorine atom position and varies in each congener. They are colorless and odorless crystal. Polychlorinated bisphenols have fire retardant properties when mixed with some materials. In the 20th century, these polychlorinated bisphenols proven as valuable products or chemicals in industries. There are so many products produced from polychlorinated bisphenol's such as capacitors, microwaves, AC, fluid-cooled motors, switches, voltage regulators, vacuum pumps, waxes, inks, electrical cables, adhesives, dyes, paints, pesticides, electrical insulating materials, sealants, etc., in Figure 8.1.

Since the 1970s, polychlorinated bisphenols have been banned in different countries. Now a days maximum materials contain polychlorinated bisphenols from the starting those releases into the surroundings (Hu and Hornbuckle, 2010; Hu et al., 2010; Shanahan et al., 2015). PBCs belong to persistent organic pollutants (POPs) family that were utilized in whole world and commercially produced by USA in industries (Erickson and Kaley, 2011). PCBs are not reacted easily with other materials and their thermal stability; due to these properties, they are utilized as coolants and lubricants (as transformers and capacitors). PCBs are having various specific properties as mentioned in Table 8.1 Also, PCBs have dielectric property.

PCBs are the mixture of various PCBs congeners and in which Aroclor is one of them. The varieties of Aroclor are mainly it depends upon the degree of chlorination. Aroclor 1254 is one of the examples that is the mixture of mono-heptachlorinated biphenyl congeners having approx. 54% chlorine content. Homolog congeners are that PCBs congeners having the same number of Chlorine (Cl) atoms and their (chlorine) different positions are known as isomers. The physicochemical properties vary broadly and these PCBs have low vapor pressure and water solubility. PCBs have the ability to dissolve in some organic solvents, oils, and fats. They cannot degrade easily, but they can destroy through some chemicals and some biochemical processes and thermal processes that occur metabolically or intentionally. Due to their high thermal stability property, these mechanisms are very difficult to proceed. These processes (environmental and metabolic) are very slow as compared to other substances.

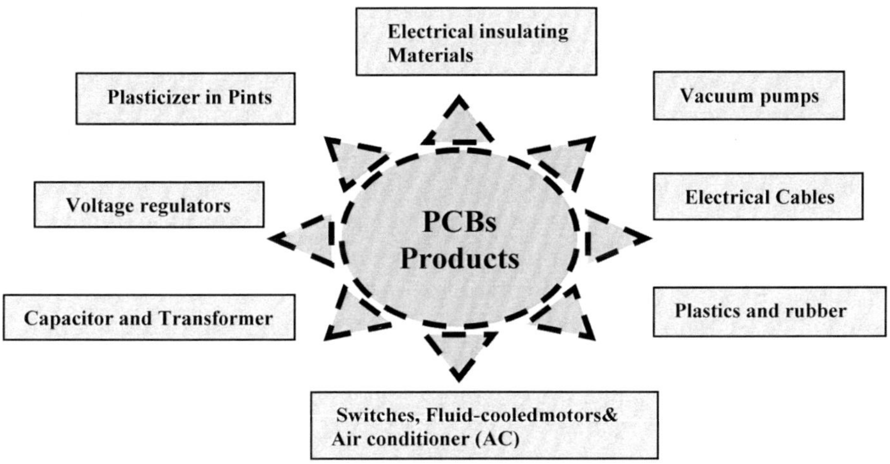

FIGURE 8.1 Different products produce from PCBs.

TABLE 8.1 Properties of Polychlorinated Biphenyls (PCBs)

SL. No.	Properties of Polychlorinated Biphenyls (PCBs)
1.	PCBs have high boiling point and non-flammable
2.	Having chemical stability and high thermal stability
3.	Low vapor pressure and water stability
4.	Having ability to dissolve in some organic solvents (oils and fats)
5.	Having ability to survive in biotic and abiotic environments
6.	It has dielectric properties
7.	It cannot degrade easily

Till dichlorodiphenyltrichloroethane is used to prevent malaria by applying it on mosquitoes, it was utilized as pesticides. On the other hand, brominated derivatives are used as "flame retardants" (furniture, electronic devices). Fluorinated compounds have property to repel water and oil. For manufacturing the food containers, textile coatings, leather, cement, cosmetics, etc., latter substances are utilized (Kissa, 2001). However, DDT and polychlorinated biphenols were banned for 50 years ago, but there is the presence of them in the environment still by accumulating with the food chains. Generally, humans are exposed to POPs through diet, breathing, and through dermal absorption (Trudelet al., 2011; Harrad et al., 2010). Various authors reported the POPs in peoples (fetal blood) that is the major concern of human health at present (Muller, 2015, 2017). Some evidence was reported for two decades that express the mutual relationship between polychlorinated biphenols and in chronic disease (endocrine disorders, metabolic disorders, and CVD) (Ahmed, 2013; Lee et al., 2011; Perkins et al., 2016). Various lab experiments showed the inflammatory effect on endothelial cells (endothelial damage and atherosclerosis development) (Hennig et al., 2002; Yilmaz et al., 2012). Meanwhile, there are no studies declared from epidemiological studies. There are two studies on the "Anniston population" in which high PCB serum levels were detected with high Bp and hypertension (Goncharov et al., 2010, 2011). In US, the high level of hypertension was detected having high PCB levels in general population (Everett et al., 2008; Ha et al., 2009). In the case of contaminated fish consumption no significant combination of interaction was recognized in the other studies (Raymond et al., 2016; Valera et al., 2013). In the general population, hypertension problem was detected by the consumption of PCBs contaminated rice (Akagi and Okumura, 1985). When more than one OH groups are added in PCBs then it is characterized as OH-PCBs. However, about 837 mono-hydroxyl polychlorinated biphenols are determined in the surroundings. One of the best methods to calculate the acid-base dissociation constant of mono-OH-PCBs was silico methods and the range was about 4 to 10 (Grimm et al., 2015). There are about 19 polychlorinated biphenols having 3 or 4 ortho position of Cl that able to stabilize atropisomers to "racemization" under various physiological conditions and it was determined by PCBs atropisomers isolation through "liquid chromatography" and diastereomeric intermediation (Haglund, 1996; Püttmann et al., 1989). Kania-Korwel and Lehmler reported that in about 19 "chiral PCBs" found in the mixture of PCBs and that are produced worldwide. These Chiral PCBs were demonstrated in human as well as in animals to their release in the surroundings (Kania-Korwel and Lehmler, 2016; Lehmler and Robertson, 2001; Lehmler et al., 2010). These chiral PCBs are recognized in excess about in humans and in animals (wildlife) because of preferential interaction of them or some biological processes PCBs transportation (Lehmler et al., 2010). Lauby-Secetan et al. (2016) reported that PCBs were classified by IARC (as class 1) human carcinogens. OH-PCBs play an important role in it as initiators, promoters, and as co-carcinogens. Initiation is the first step in it. There are some studies in which OH-PCBs initiating activity evaluated. Espandiari and co-workers made an experiment and observed that PCB3, PCB15, PCB52, and PCB 77 shows positive results and PCB-12 and PCB 38 was negative in modified rat liver foci assay helps to evaluate initiation activity of compounds (Espandiari et al., 2003). Now the second step is promotion, by inhibiting GJIC (gap junction intercellular communication), it leads to a promotion mechanism in which all nondioxin (PCBs and OH-PCBs) show positive effect in *in-vitro* gap junction intercellular communication assay (Hamers et al., 2011; Machala et al., 2004) The PCB congeners having 4 or more than 4 chlorine atoms with no substitute at ortho position that helps to exhibits dioxin-like (DL) toxicity is known as DL PCBs (Zhang et al., 2015; Seegal et al., 2011). Due to their valuable properties, they are used in different chemical industries broadly to produce products (heat transfer, rubber, and plastic products, hydraulic machines, dyes, and pigments (Zhu et al., 2011, 2012). About 1.5 million metric tons of PCBs were evaluated broadly, out of which about 10% remains in the environment (Wolska et al., 2012). Various administrative authorities raised PCB's effects on the environment (including toxic effect, persistency, and its bioaccumulation). In history, PCBs products were produced having commercial names like in USA 'Aroclor,' in Germany 'Chlophen,' in Japan 'Kanechlor' and in Italy 'Fenclor' (Hu et al., 2011).

In the last few decades, the major attention showed towers the toxicity of PCBs that is exactly to the PCDFs (polychlorinated dibenzofurans) and PCDDs (polychlorinated dibenzo-p-dioxins). PCB-77, PCB-126 and PCB-169 exhibit DL toxicity because absence of chlorine (as substitute) and ortho position (Carro et al., 2018). DL PCB toxicity is as nearly similar to TCDD (5,2,3,7,8-tetrachlorodibenzo-p-dioxin). The PCBs have ability to survive in both environments, i.e., biotic and abiotic due their high stability (Zeng et al., 2013). Due to their various physical and chemical properties they validate as essential organic compounds utilized broadly. But also, they are dangerous for the environment due to some specific properties of them. In which dielectric property is one of the significant properties of polychlorinated biphenols. With scientific and public concern research indicates that

PCBs were undesirable (Randoll et al., 2014). Batang et al. (2016) reported that PCBs accumulation into food chain can affects "ecological integrity" and hum health. However, these PCBs were banned for many years but still used, mostly in developing countries (Mahmood et al., 2014; Men et al., 2014). Moreover, PCBs utilization and its occurrence in whole world is reported but in Bangladesh (a state of India) it is not recognized yet because these have been not produced their ever. But some tools and equipment (capacitors, lubricating oils, etc.), imported to Bangladesh. In the 20th century, PCBs were commercially synthesized for different industries also. The provision of PCB pollution shows multiple health effects that raise many questions and health concern. PCBs totally transformed into phenols due to "enzymatic biotransformation" in flora and fauna. Also, by the hydrolysis of sulfated and glucuronidase metabolites into "OH-PCBs" that shows toxic effects through different mechanisms.

8.2 SOURCE OF PCBS

The major cause of the occurrence of polychlorinated biphenols (PCBs) are the leakage of PCBs in the environment and waste incineration. Also, the utilization of PCBs in various industries through worldwide, having no proper disposal arrangements creates environmental issues (contamination). There is no any specific designed landfill area that can use for the management of PCB wastes. Mostly the PCB waste was dumped into those landfills having no proper management. They were utilized broadly due to their physicochemical properties. In USA, PCBs are marketed as Aroclor trademark that identified as 4-digits number. First two digit shows carbon number and other two-digits shows the chlorine percentage in that mixture. Like USA, different trademark was named for the PCBs (Kaneclor in Japan, Fenclor in Italy, Pyralene in France, etc.) (Watts et al., 1998) (Table 8.2).

TABLE 8.2 Different Trademarks Used in Different Countries

SL. No.	Trademark of PCBs	Name of Country
1.	Kaneclor	Japan
2.	Fenclor	Italy
3.	Pyralene	France
4.	Aroclor	USA
5.	Chlophen	Germany

In US, since 1976, its manufacturing, processing was banned under "Toxic Substances Control Act." In between 1930 and 1979, about 600 million kg PCBs were utilized by North America only (Laukers, 1986). About 15% of that release into the environment (either legally or illegally). PCBS are soluble in lipids and becomes accumulated in soil and water for prolonged time. Also, the PCBs disposal negligence and not considered is another source by which PCBs are recognized in the environment. In 1966, PCBs were recognized firstly, when DDT (Dichloro diphenyl trichloro-ethane) quantification was considered in the samples (Huang et al., 2011). It considered as common pollutant globally that identified in "air, soil, sediments, water, milk, fish, wildlife, plants, human adipose tissues and in blood samples" and that depends upon the degree of chlorination that helps to influence the PCBs adsorption (Figure 8.2).

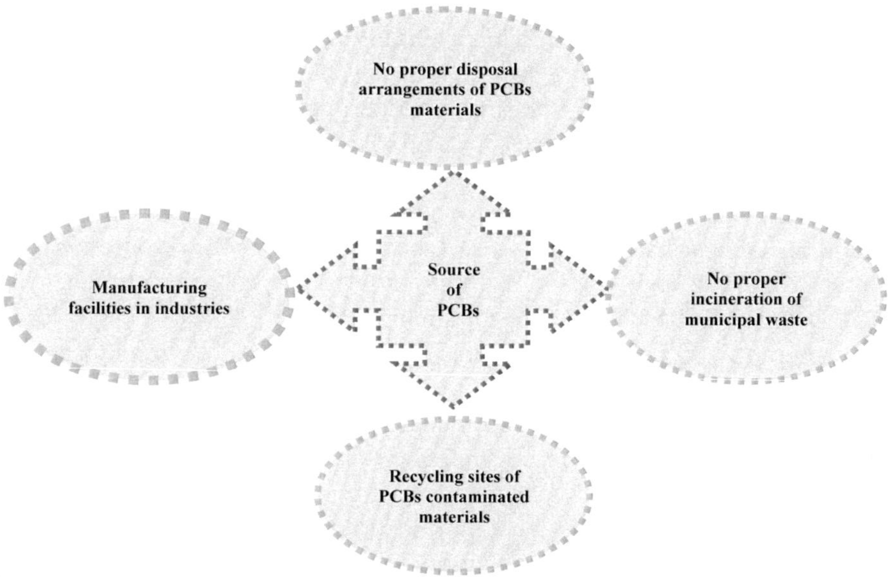

FIGURE 8.2 Various sources of PCBs.

In soil, it is the major source of PCBs in the "terrestrial ecosystem" because of the hydrophobic chemicals (Cabrerizo et al., 2011; Armitage et al., 2006). Their potency and solubility remain limited because they are hydrophobic in nature and have slow desorption kinetics (Luthy et al., 1997; Xing and Pignatello, 1997). The slow increase in their potency is because of DOC (dissolved organic carbon) and OC (mobile organic carbon) (Moeckel et al., 2008), which play an important role in driving the infiltration fluxes of "hydrophobic chemicals" (Enell et al., 2016; Persson et al., 2008). A study was considered by Rylander et al. (2005) in which they evaluated that in type-2 diabetes patients have more effected by PCBs. Also, different studies determined the relation in between type-2 diabetes and PCB exposure (Codru et al., 2007; Vasiliu et al., 2006; Wang et al., 2005). In an experimental study, the association of diabetes mellitus and polychlorinated biphenols-153 concentrations and p, p00DDT (dichloro diphenyl trichloroethylene) in 544 serum sample (among Swedish fishermen's Wives) (Rignell-Hydborn et al., 2007). Similar studies were reported by Lee et al. (2006, 2007) (the dose-response striking forms in between serum conc. In different POPs and in the prevalence of diabetes. A study was also conducted by Turyk et al. (2009) on a cohort of Great Lakes sport fish consumers that was established in 1990 and followed through 2005. This study cohort the relationship in between POPs (PCBs, body burdens, health effects, Sprague Dawley, Pregnant Sprague Dawley rats) by giving them PCB 153 orally from GD10 through GD16. In Male and female rats of 1–3 weeks, a dose was given to them that showed a decrease in the T3 and T4 plasma concentrations. There was no change in thyroid hormone levels reported by Kobayashi et al. (2008). Induction of pro-inflammatory mediators in liver and brains, within 24 hr., the strongest expression was recorded in the following PCB class: II-P (intra and infragy-ramidal), GD (gestation day), PND (postnatal day), VP (vasopressin), LH (luteinizing hormone), OH-PCBs (hydroxylated polychlorinated biphenyls), TCDD (2,3,7,8-tetrachlorodibenzopdioxin), PCDF (2,3,4,7,8-pentachlorod-ibenzofuran), ROS, A1254 (Aroclor 1254), T3 (tri-iodothyronine) and T4 (thyroxine).

Phytoremediation in Polychlorinated biphenols: phytoremediation is a cheap process because there has no energy consuming tool. It is a process that has been significantly considered an alternative source to prevent or to remove organic pollutants present in soils, water, and in sediments (Meagher, 2000). "Plant-mediated bioremediation" is the most significant technique of phytoremediation (Macek et al., 2000). This process is summarized in two different steps. One is the biodegradation of pollutants in soil and uptake into plant tissue (through roots) due to plant enzyme transformation either volatilization into the atmosphere (Harvey et al., 2002). Bio-stimulation and bio-augmentation are the two important types of bioremediation techniques. There are some verities of plant that helps in utilization of PCBs, i.e., *Medicago sativa* (alfalfa), *Lespadeza cuneate* (Chinese bush clover), *Phalaris arundinacea* (reed canary grass), *Salix alaxensis* (Alaska willow), *Lathyrus sylvestris* (everlasting pea), *Cucurbitaceae* (cucurbits), *Picea glauca* (white spruce), *Panicum virgatum* (switchgrass) and *Sparganium* (bur-reed) and about 27 various weeds (Table 8.3) (Slater et al., 2011; Ficko et al., 2010; Chekol et al., 2004).

TABLE 8.3 Some Varieties of Plant That Helps in Utilization of PCBs

SL. No.	Scientific Name of Plants	Common Name of Plants
1.	*Medicago sativa*	Alfalfa
2.	*Lespadeza cuneate*	Chinese bush clover
3.	*Phalaris arundinacea*	Reed canary grass
4.	*Salix alaxensis*	Alaska willow
5.	*Lathyrus sylvestris*	Everlasting pea
6.	*Cucurbitaceae*	Cucurbits
7.	*Picea glauca*	White spruce
8.	*Panicum virgatum*	Switchgrass
9.	*Sparganium*	Bur-reed

M. sativa shows positive results in the first two years in the field scale phytoremediation (Tu et al., 2011). *Sparganium* helps to promote oxidation (low-chlorinated PCBs) through "rhizodegradation." *Panicum virgatum* helps in degradation of high as well as low congener mixture (Meggo et al., 2013). Cucurbitaceae helps to decrease PCBs concentration in sludge (38.63%) and in sediment (27.38%) (Wyrwicka et al., 2014). For the treatment of PCBs, various *in-situ* technologies impact soils as well as sediments was conducted since a few decades (Tyagi et al., 2011; Sowers and May, 2013). In water these PCBs are found on the surface of sediments. Disposal sites and atmospheric transportation are the primary way that confirms the PCBs entry into the soil. They have long-term effects and life. In surface water sediments, there is a significant accumulation of PCBs. Adsorption of PCBs is promoted by organic water, clay surface and macroparticles dispersion. Mostly in dry season, PCBs discharged into air having high PCBs concentration. However, the deposition of PCBs occurs in the rainy season and during rainfall. If it is not then PCBs settle down on the surface of soil. PCB concentration is high where human activity performed or near the shorelines (in water). However, PCBs are absorbed by human as well as in animals through the skin, lungs, GI (gastrointestinal tract). In the human body, they accumulate into the different muscles and adipose tissues transported through the bloodstream. PCBs cause different health effects which depend upon the age, sex, and body area. PCBs are carcinogenic reported through various animal studies. In animals,

PCBs cause mild liver damage if PCBs eaten by them for a short time period even some of them will die. Also in seabirds, they affect them and increases mass mortality rate. Cui et al. (2015) reported that UP-PBCs (Unintentionally produced PCBs are shows significant concern as environmental pollution than that of IP-PBCs (intentionally produced PCBs). Mostly, the cause of the increase of PCBs concentration is due to increase in the unintentionally produced PCBs in the soil (Cui et al., 2017). PCBs are the major concern due to their toxic nature that affects the environment and further entered into our food chains and food webs (Alegria et al., 2016). After banning, it still presents in the environment and leads to as a threat for humans as well as wildlife. Some studies show that these compound exposure causes "neurotoxicity, genotoxicity, carcinogenicity, and reproductive toxicity" (Lusnic et al., 2016; Elnar et al., 2015). They entered into the aquatic environment through different ways such as wastewater from industries, MSW, deposition, and atmospheric transport, etc. (Yang et al., 2012; Totten et al., 2006; Zhou et al., 2001). EDCs (endocrine-disruption chemicals) defined as an exogenous substance that affects function of endocrine system and also affect in intact organisms, its progeny (United N.E.P., 2012). World Health Organization (2002) defines endocrine-disruption chemicals that identify the relationship between endocrine activity (EA) and the adverse effects shown in intact organisms that help to identify EDC. It is also recommended by the Scientific Committee of the European Food Safety Authority. There are three criteria concluded such as: (a) adverse effect in an intact organism; (b) EA; (c) plausible causal relationship between the two. EDCs is continuously increased and causes as a concern publicly having two main reasons behind this concern (a) consumer exposure to EDCs; (b) linkage in between EDC to different disease since decades (Katsikantami et al., 2013; Buha et al., 2018; Petrakis et al., 2017; Sifakis et al., 2017; Bourguignon et al., 2016; Heindel, 2015; Trasande et al., 2015; Vandenberg et al., 2012). EDC toxicity is recognized that have "safe threshold" concept utilized in assessing the safety (Beausoleil et al., 2017). EDCs are act as NMDR (nonmonotonic dose-response), low-dose effects and developmental vulnerability is essential for translating findings towards EDCs to human health (Gore et al., 2015).

8.3 PCBS EFFECT ON ENDOCRINE SYSTEM

Since 1929, PCBs were utilized in industries due to their physicochemical properties (Alcocket al., 1994). In which its electrical insulating properties, they used to manufacture electrical equipment (Merkel et al., 1999).

A NIEHS (National Institute of Environmental Health Science) workshop on "Characterizing the Effects of Endocrine Disruptors on Human Health at Environmental Exposure Levels" held in Raleigh (North Carolina) in 1998 to discuss the methods and data needed to prevent EDs. PCBs cause adverse effect on human as well as on wildlife. PCBs have been linked to chronic effects (immune system damage, decrease in pulmonary function, bronchitis, and hormones interference that leads to cancer (Schecter et al., 2006). There are different models to study endocrine effects on a few endocrine system aspects. *In-vitro* models are suitable to study aspects of the endocrine system in isolation. They are also useful as "prescreening tools" for the identification of putative endocrine-disruptive chemicals and also allow to analyze different number of compounds in less time. There is a large-scale data on related to the endocrine effects of polychlorinated biphenols. PCBs affects verities of endocrine system such as estrogen and androgen system (Lee Kuhre, 1995), thyroid hormone system (Watts, 1998), retinoid system, corticosteroid system and different endocrine pathways. AhR (arylthdrocarbon) receptor is responsible that shows dioxin effects (non-ortho and mono-ortho) many other pathways by cross-talk of arylthdrocarbon (AhR) and ARNT (arylthdrocarbon nuclear transport protein) transcription factor with "Per-Arnt-Sim" family of helix-loop-helix proteins (Laukers, 1986). However, PCBs interaction along with thyroid and estrogen hormonal systems also reported in previous studies. Various effects observed, such as endocrine effects on the endocrine components (hormones, enzymes, carrier proteins, receptors, endocrine glands) and some effects associated with the endocrine system (neurodevelopment, reproduction, immune effects, and induction of ES Tumors). Various disorders recognized that are linked with their interaction with ES (endocrine system), such as predominantly developmental toxicity (Ruel et al., 2019), immunotoxicity (Kuwatsuka et al., 2014), metabolic disorders (Hofe et al., 2014), thyroid disorders, disorder in the reproductive system (both male and female) (Eskenazi et al., 2017; Jirsová et al., 2010; Carmichael et al., 2003). Ah-receptor dependent and independent mechanisms (liver hypertrophy) caused by PCB-induced effect. Also, "neuro-developmental effects" by which in steroid hormones some changes occur and lead to thyroid hormone disruption. Immunological effects and cancer are also caused by PCB-induced effects. In both animal studies and in *in-vitro* studies showed endocrine-disrupting effects of polychlorinated biphenols (PCBs). Mostly it effects on thyroid and reproductive functions. However, all other disorder also a wide concern for humans as well as in wildlife. In this, we discussed about all the endocrine disorders caused by PCB with their mode of action and mechanisms.

8.3.1 THYROID DISRUPTION

PCBs are broadly spread pollutant which is man-made that further leads to toxic effects. These PCBs are correlated with the POCs (persistent organochlorine compounds), i.e., "dibenzo-p-dioxins, dibenzo-p-furans, and organochlorine pesticides (OCPs)." Due to these epidemiological studies are very difficult that specify the compounds suitable for the biological effects. They are lipid soluble in nature. The major dietary source for humans is animal fats. Brouwer et al. (1998) reviewed a study in which they show that in animal studies, PCBs exposure causes thyroid hormone disturbances. PCBs exposure shows a decrease in T4 levels. Three mechanisms, i.e., direct effect on thyroid gland (TG) (cause decreases thyroid hormone level), reduction in thyroid hormone levels (causes induction in UDP-glucuronosyltransferases) and binding of TTR (transthyretin) (causes displacement of the natural ligands). Mostly in adults, T4 is transported through "thyroglobulin" but TTR mediates T4 transportation across the blood-brain barrier and fetal transported through the placenta (Southwell et al., 1993; Schreiber et al., 1995). Some PCBs metabolites (dibenzo-p-dioxins, and dibenzo-p-furans) shows a comparative TTR binding. Transplacental exposure towards polychlorinated biphenols and towards dioxins causes "neuro-developmental disruption of offspring" in different animal studies (Brouwer et al., 1995). In few epidemiological studies, the PCBs maternal diet and dioxin exposure shows negative effects on "infant psychomotor development" (Gladen and Rogan, 1991; Jacobson and Jacobson, 1996; Patandin et al., 1996; Walkowiak et al., 2001; Grandjean et al., 2001). However, in Dutch epidemiologic studies, polychlorinated biphenols and dioxin exposure shows deficiencies (infant-psychomotor a development) and circulations changes in TSH and in thyroid hormones (THs) (Koopman-Esseboom et al., 1994). It was discussed that PCBs and dioxins exposure and neurodevelopment disruption can be mediated by disrupted thyroid hormone homeostasis. Zoeller (2010) reviewed that the polychlorinated levels showed connected by elevated decreases the level of THs. They do not affect THS (thyroid stimulating hormone) levels and TG. There are various biological sources of variations that affect the relation of polychlorinated biphenols and THs levels. Polychlorinated biphenols (PCBs) and TH (thyroid hormone) level relation shows negative effects by birth delivery mechanisms (Herbstman et al., 2008).

However, various animal studies show adverse relation in polychlorinated biphenols and thyroid hormone level of circulation. Thyroid function regulated at hypothalamic and pituitary levels (Boas et al., 2012). A good response of feedback loop shows TSH levels elevated. It results in the compensatory changes in hypoplasia of thyroid tissue and histopathological in animal studies (Martin and Klaassen, 2010). Several evidences in analytical methods were evident for the chemical studies in biological tissues. It was very difficult to identify relevant marker effects of polychlorinated biphenols on the thyroid functions and action mechanisms. All type of studies (animal study, human, *in-vitro*, and computational) were conducted for this. Polychlorinated congeners are metabolized into OH-PCBs (hydroxylated metabolites). The structure of thyroid hormone and hydroxylated metabolites are same. It results into disruption of thyroid homeostasis (by interaction of TH receptors) (Kitamura et al., 2005). Polychlorinated biphenols also disturb thyroid receptors (THs) action. It shows that higher-chlorinated polychlorinated biphenols have thyroid receptor-agonistic action and the polychlorinated biphenols having low chlorine atoms show TR-agonist (Bai et al., 2018). *In-vitro* and *in-vivo* studies show positive results towards binding of polychlorinated biphenols and hydroxylated metabolites towards T4 and TTR. In human T4 is transported through TBG (thyroxin-binding globulin). T4 and TTR binding play an important role in blood-brain barrier transport and altering brain levels of thyroxin hormone in humans. Thyroid hormone-regulated gene expression that associated along with polychlorinated biphenols (PCBs) and hydroxylated metabolites (OH-PCBs). Induction of hepatic UDPGTs (uridine diphosphate glucuronyle transferases) is another mechanism through AhR activation by coplanar polychlorinated biphenols 126 congeners (Fisher et al., 2006). However, Ortho-substituted polychlorinated biphenols congeners produce AhR-dependent effects through CYP enzyme induction. Acute endocrine disorder effects of two different polychlorinated congeners (in 5 weanling female Sprague-Dawley rates) towards ortho-polychlorinated biphenols congeners (caused in T4 level without T3) changes their effects of ortho-congeners on "feedback method of HPT axis (Khan et al., 2002). This mechanism includes the interaction to THs in with TR (causes thyromimetic effects) and helps to prevent thyroxinemia. Ca^{2+} helps to regulate hormone secretion, its interference is important in it. Ortho-substituted polychlorinated polyphenols have direct action towards the "ryanodine-sensitive Ca^{2+}" (RyRs) in the skeleton and muscles (Wong et al., 1997). It suggested its potential role in polychlorinated biphenols and RyRs interaction in distributed TSH (release from the pituitary gland). PCBs and EDCs mixture are one of the major concerns of "thyroid disruption" of polychlorinated biphenols that becomes synergism in b/w chemicals. The thyroid function causes a significant health effect (low dose of EDCs) (Buha et al., 2013; Chen et al., 2016). Low dose exposure of EDCs mixture becomes real-life human scenarios (Bopp et al., 2018). There are many studies links of animal as well as human studies towards PCBs effects on male and female reproductive system. Polychlorinated biphenyls (PCBs) show estrogenic and anti-estrogenic activities that impacts upon different receptor levels and affect steroidogenesis modulation. Recently, in *in-vitro* studies, most of the studies show PCBs effects on Leydig cell homeostasis. All these study work helps to improve the molecular targets shows dose-dependent effects and some molecular mechanisms that were already identified. Some *in-vitro* studies show Aroclor 1254 exposures. In Leydig cells of mouse shows lipid peroxidation with ROS and antioxidant enzyme system increases and decrease, respectively and also decreases the cell viability. It occurs due to decreases in activities of 3β and 17β-hydroxysteroid dehydrogenase (shows irregular synthesis of testosterone) (Aydin and Erkan, 2017). In male rats, shows Aroclor 1254 exposure at the time of altered testosterone, estradiol lactation revealed altered testicular architecture (TA) and ABP levels in TIF (testicular intestinal fluids) and in serum also. However, different authors recognized a decrease in FSH levels, ABP (Androgen-binding proteins), ERβ (estrogen receptor β) and AR (androgen receptor) (Sugantha Priya et al., 2017). All finding shows "mechanistic supports of Sertoli cell dysfunctions (SCD) (as consequences of polychlorinated biphenols exposures." Recently *in-vitro* studies recognized the significant role for ER (endocrine regulation) on "enantioselective activity of mono-congener" towards nuclear receptors. It shows some evidences of the complexity of PCBs reproductive toxicity (Pěnčíková et al., 2018). Specifically, +ve and –ve PCB 136 were showed contrasting effects on AR activation, also various vigor that considered inducing pregnant X receptor-mediated gene expression. Furthermore, they showed close similarity of constitutive AR-dependent gene expression induction and also "suppression of AR-mediated gene expression." However, irregular physiology of reproductive system was shown in rodents, i.e., inter-generational and trans-generational effects of polychlorinated biphenols (Pocar et al., 2012).

Recently, in *in vitro*, in fetal sheep cells, DNA methylation and genomic integrity is reported towards Aroclor 1254 exposure (Table 8.4) (Anzalone et al., 2018).

8.3.2 PCBS EFFECT ON PANCREAS

POPs (persistent organic pollutants) are different chemicals which show lipophilic nature. They have the ability to accumulate in fat and further able to degradation (Abelsohn et al., 2002). POPs have the ability to accumulate in

TABLE 8.4 Endocrine Disruption Such as Thyroid Disruption, Effect on Adrenal Gland, and Effect on Pancreas

SL. No.	Source of PCBs	Effect on Endocrine System	Endocrine Disruption	Disorders/Diseases
1.	Plastics, contaminated water, contaminated soil, industries, pesticides, fungicides, insecticides, recycling sites of PCBs contaminated materials, canned food	Thyroid disruption	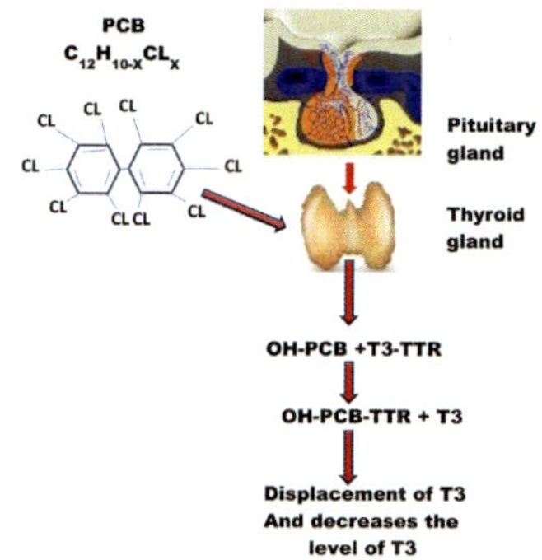	Disrupt thyroid hormone homeostasis, effect on level of thyroid hormones and thyrotropin (Kyle et al., 2019)
2.		Effect on adrenal gland	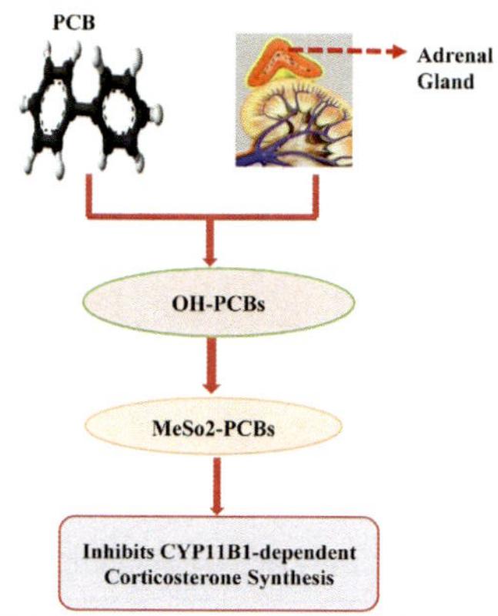	Addison's diseases, adrenal gland tumors, congenital adrenal hyperplasia (CAH), Cushing's syndrome (Lauretta, 2019)

TABLE 8.4 *(Continued)*

SL. No.	Source of PCBs	Effect on Endocrine System	Endocrine Disruption	Disorders/Diseases
3.		Effect on Pancreas	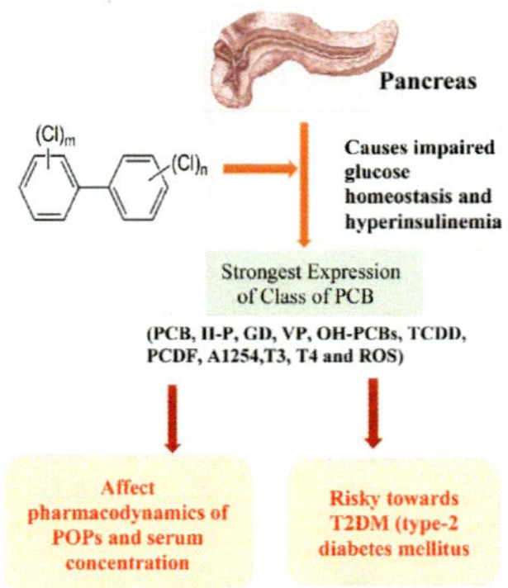	Acute pancreatitis, loss of functioning, diabetes, digestion problems, etc.

human adipose tissues. It is a slow process but causes long-term effect. Low dose exposure leads to type-2 diabetes. Different epidemiological studies report the relation between type-2 diabetes and in POPs (PCBs, dioxins, and OCPs including DDT). Obesity is not linked with type-2 diabetes in people having low POP levels (Lee, 2006) studied in a cross-sectional study in US general population. It also plays a significant role in pathogenesis of type-2 diabetes (Lee et al., 2006). An experiment determined the exposure of rats towards mixed POPs, hepatosteatosis, and insulin resistance (IR) (Ruzzin et al., 2010). There are many reasons that help to explore the interest in between POPs and type-2 diabetes. The first one is aging that is associate with insulin secretory defects that cause a high risk of type two diabetes (Steenland et al., 1999). And secondly, body composition change that occurs during aging (increase in fat, central redistribution of adipose tissues) (Harris, 2002) and these changes also affects pharmacodynamics of POPs and serum concentration.

It shows that POPs are risky to the T2DM (type-2 diabetes mellitus) (Song et al., 2016; Ngwa et al., 2015; Lasram et al., 2014). In epidemiological studies, type of POPs, PCBs were associated with diabetes (Singh and Chan, 2017; Lee et al., 2014; Cave et al., 2010; Vasiliu et al., 2006). Some pathological effects and underlining mechanisms lead to exposure to polychlorinated biphenols analyzed in animal models. PCB-77 and PCB-126 with 50 mg/kg and 1.6 mg/kg, respectively treated on mice and it resulted in impaired glucose homeostasis (Baker et al., 2013). Female C57B/6 mice exposed to 36 mg/kg/wk Aroclor 1254 for 20 weeks showed hyperinsulinemia and IR (Gray et al., 2013). Our previous work exhibited that sub-chronic exposure to Aroclor 1254 (0.5, 5, 50, and 500 μg/kg) for 60 days caused impaired glucose homeostasis and hyperinsulinemia in male mice, the pathological mechanisms included the down-regulation of the insulin receptor signaling pathway in both liver and skeletal muscle (Zhang et al., 2015) and disturbance in the function of β-cells (Xi et al., 2019). Nevertheless, the alteration of α-cells and β-cells is still not well understood. In US (1977), they completely stop working on manufacturing of polychlorinated biphenols when they find out PCBs presence in the environment that causes various disease in both animal as well as in humans. In recent days there are no specific source of manufacturing of PCBs except some activities and processes performed by industrial levels. The PCBs release in the environment on the commercial levels some process leads to environmental fate such as volatilization, biochemical transformation, and preferential bioaccumulation and all these processes mainly due to degree of chlorination of PCBs molecules

(ATSDR, 2000). Mainly in soil and in sediments, mostly absorption of the higher chlorinated polychlorinated biphenols congeners occurs. Ivanciuc et al. (2006) reported that these polychlorinated biphenols also found in the food chain due to their lipophilicity and stability. However, they specially embedded in fatty tissues. This bioaccumulation of polychlorinated biphenols found in humans that further leads to relevance towards human health. Many researchers determine the concentration of serum polychlorinated biphenols-153 in Swedish fishermen and their wives, in which about 196 are men and about 184 were women's (Rylander et al., 2005). All the participants take a part gave answered the question on different topics also (diabetes, medication, diseases, etc.), and the the results of the concentration (PCB-153) serum determines the relation with the diabetes type-2 in different variables. If there was an increase of about 100 ng polychlorinated biphenols 53/g lipids then it relates or of 1.16 (95%) chlorine 1.03–1.32 (p ¼ 0.03). It shows the relationship between diabetes mellitus and polychlorinated biphenols. Many other workers were determining the relation between type-2 diabetes and polychlorinated biphenols (Chen et al., 2008; Codru et al., 2007). Also, Rignell-Hydborn et al. (2007) demonstrated the relation between diabetes and PCB-153 concentration and p, p0-DDT in the sample of 544 serum samples in Swedish fishermen's wives in cross-sectional studies. Lee et al. (2006, 2007a, b) also studied this work. In this the striking dose-response formed in between serum concentration in about six POPs and diabetes. Turyk et al. (2009) were also study on Cohort of Great Lakes Sport Fish consumer. In the studied cohort, the relation in between polychlorinated biphenols (different PCBs, body burden species forms health issues) that established in 1990 till 2005. In female rats (Sprague Dawley) which was pregnant, PCB-153 was given orally from GD10 by GD16 (0, 16, or 64 mg/kg/day). In 1–3 weeks, offspring (male and female) the dose was given, and it results that it decreases the T3 and T4 plasma concentration and there was no change was demonstrated in TSH level. Different researchers reported the pro-inflammatory expression, PCB-104 and PCB-153 were injected in mice (150 mmol/kg b.w.) induction in different organs (liver, lungs, and brain) (Kobayashi et al., 2008). Sipka et al. (2008) studied the strongest expression of proinflammatory within 24 hr. independent of the class of polychlorinated biphenols (PCBs), intra, and infragyramidal (II-P), gestation day (GD), postnatal day (PND), vasopressin (VP), per os (p.o.), LH, body weight (B.W.), OH-PCBs, 2,3,7,8-tetrachlorodibenzo-dioxin (TCDD), 2,3,7,8-pentachlorodibemzofuran (PCDF), Aroclor 1254 (A1254), T3, ROS and T4). Type-2 diabetes was not accumulated with monoPCB-118 and ortho-PCB-118. Langerhans islets are composed of four types of functional cells (including β cell and α-cell) (Bonner-Weir et al., 2015). Insulin is secreted by β cell and glucagon is from α-cell that plays a significant role in glycemia regulation. In these β cell forms as plastic during T2DM development (to adapt insulin requirements) (Cnop et al., 2011). In few cases β cells do not able to compensate to fulfill the metabolic demand. In that case hyperglycemia and T2DM forms that also accompanied through β cell and functions (Rhodes, 2005). There are different studies that reported that there was an alteration in pancreas α-cell functioning and its morphology during IR stage (in animal and human). Sudden increase in α-cell mass and in hypoglucagonemia observed in those people having T2DM (Ellenbroek et al., 2017). Obesity and IR were interacting (associated) when glucagon level is increased (Færch et al., 2016). A-cell mass was low in people with T1DM (Bonnet-Serrano et al., 2018). In mice induced with a "high fat diet, Pancreatic α-cell mass and area" reduced along the time "islet compensatory stage and hypoglucagonemia" observed in fed mice (Merino et al., 2015). Alteration of both these cells (β cell and α-cell) is associated with apoptosis, proliferation, and size of cell, neogenesis from ductal progenitor cells (Merino et al., 2015; Chesnokova et al., 2009). Aroclor 1254 exposure results in increased and decreased in β cell and α-cell, respectively (Zhang et al., 2015). Also, it further investigated for its heterogeneous adaptation of both cells. In regulation of plasma glucose homeostasis and islet cell function Estrogen and androgen play a significant role. Mainly estrogen is helpful to maintain the insulin sensitivity under normal condition that further useful in β cell function. The estrogen level or concentration (low and high) is also helpful to promote IR and T2DM level (Nadal et al., 2009). ERαand ERβis the estrogen receptor leading as a key molecule that involving in "modulating pancreatic β-cell function" (Alonso-Magdalena et al., 2013; Liu et al., 2009). E2 (17β-estradiol) increases the survival of β cell and control of glucose homeostasis in different animal models by ERα and ERβ signaling (Le May et al., 2006; Alonso-Magdalena et al., 2008, 2013). In men, if testosterone levels become decreased, then it causes serious risk for the T2DM development (Ding et al., 2006; Fukui et al., 2007). Li et al. (2017) reported that the decreased level of testosterone can increases the risk factor for IR and T2DM in adults' men. In few clinical studies have determined that low serum testosterone reduces the sensitivity of insulin and that helps to promote IR and T2DM level (independent of age and body fat mass) (Xia et al., 2015; Yeap et al., 2009). β-cell receptor androgen receptor (AR) increases the β-cells functioning by Camp dependent pathway (Navarro et al., 2016). Although, polychlorinated biphenols are the EDs that can also disturb the sex hormone level and their receptors.

8.3.3 PCBS EFFECT ON ADRENAL

It is a gland belongs to the "neuroendocrine hypothalamus-pituitary-adrenal axis" (regulated by crosstalk mechanisms between its components). It represents the primary hormonal response towards homeostatic challenge by steroidogenic tissue and chromaffin tissue of different "embryological derivations" (Cavaliere et al., 2018; Herman et al., 2016). Steroidogenic tissue and chromaffin tissue derived from mesoderm mesenchymal and embryonic neural crest cells, respectively (De Falco et al., 2002). Adrenal gland is mainly composed of "peripheral cortex and inner medulla in mammals." In this, the peripheral cortex is made up of "steroidogenic tissue" that further forms mineral corticoid, glucocorticoids, and androgen hormones. They are controlled by HPA axis and kidney. CRH and ACTH help to control the glucocorticoids secretions specifically. For the inhibition of adrenocorticotropic hormones (ACTH) and corticotropic-releasing-hormones (CRH) these glucocorticoids show negative feedback loop from hypothalamus and pituitary (Manojlović-Stojanoski et al., 2006; Bornstein et al., 1997; Nussdorfer, 1996). Adrenal gland plays an important role in basal physiological demands (balance salt and water) to predict reproduction, development, and metabolic regulation in animals (Esposito and Varriale, 2015). It is not easy to find out the physical changes in "unpredictable stressors (as predation, illness, etc.)" (De Falco et al., 2004, 2002). Since 1937, PBCs were associated with liver injury when some fatal jaundice cases were demonstrated to PCB and "chlorinated naphthalene" (Drinker et al., 1937). Maroni et al. (1981) reported dose-dependent liver enzyme elevation and hepatomegaly in PCB exposed electrical worker. In US, PCB was produced by Monsanto (as trade name Aroclor) including dielectric fluids and hydraulic fluids. MSDS M00018515) a Monsanto material safety data sheet) reported that in animal study PCBs causes liver injury prolonged and repeatedly if its exposure is of sufficient degree and time. PCBs are chlorine substituted biphenyl ring. They are thermodynamically stable in nature. Before banning of PCB about 130 to 209 PCB congeners were produced commercially. PCBs are the major source of the environmental pollution (including food supply, breast milk in the air). In 2017, PCBs were ranked five on "the Agency for Toxic Substance and Disease Registry's Substance Priority List." In 2003–2004, a survey was conducted by NHANES

(National Health and Nutrition Examination Survey) in which PCB content was detected in the blood of the participants. PCB congeners were classified as doxin like and non-doxin like. They have the ability to regulate AhR (aryl hydrocarbon receptor) (Safe et al., 1985). DL and NDL types of PCBs were along with NAFLD (in cohort study) (Cave et al., 2010). They are very concentrated in adipose tissues and in liver. The congener with high chlorine is mostly resistant towards environment and the hepatic metabolism leads their bioaccumulation. They also affect bile acids, microbiome, and fecal metabolism because they undergo entero-hepatic circulation (Hoffman et al., 2019; Petriello et al., 2018). The NAFLD exceeds 25% at global level. It may cause liver damage and liver-related transplantation (Younossi et al., 2016). Nonalcoholic fatty liver disease (NAFLD) forms "pathologic spectrum" that form steatosis towards steatohepatitis either having fibrosis or not and hepatocellular carcinoma (Wahlang et al., 2013). There are two hit hypotheses to explain the effect of NAFLD with other substance that further leads to progressive liver disease. NAFLD may result in the hepatic manifestation of metabolic syndrome (MetS). Therefore, the association of PCB (having endocrine, metabolic, and signaling disrupting chemicals) and NAFLD is very surprising (Hardesty et al., 2018, 2019; Heindel et al., 2017). In the recent, PCB resulted towards and alter organokines implicated in NAFLD pathogenesis (including FGF-21: fibroblast growth factor-21; FGF-15: fibroblast growth factor-15; glucagon peptide-1 and adiponectin (Hardesty et al., 2017, 2019; Shi et al., 2019; Gadupudi et al., 2019; Clair et al., 2018; Petriello et al., 2018; Cheng et al., 2018; Wahlang et al., 2017, 2019). MDC (metabolism disrupting chemicals) are the chemicals that are utilized to promote obesity, diabetes, NAFLD and lipid and glucose ($C_6H_{12}O_6$) metabolism that may require second "hit" (Heindel et al., 2017). Non-doxin-like PCBs (PCB 153 and Aroclor 1260 PCB) is not good diet-induced NAFLD. They did not affect in mice (having normal diet) (Hardesty et al., 2019; Hardesty et al., 2019; Wahlang et al., 2013, 2014, 2017; Wu H et al., 2017; Al-Eryani L et al., 2015; Shi et al., 2012). However, doxin-like PCBs affects both NAFLD in rodents "fed with normal diet" and in abnormal diet that induces NAFLD (Chi et al., 2019; Petriello et al., 2018; Kania-Korwel et al., 2017; Gadupudi et al., 2016; Al-Eryani et al., 2015). Recently in the last few decades, hormonal, and health effects are resulting in concerns of contamination in food, food products and in the environment. All the substances refer towards ED that defined as "an exogenous substance that affects functioning of the endocrine system may cause adverse health effects in intent organisms, its population and progent." POPs, i.e., PCBs, PBDE (polybrominated diphenyl ether) and PFAs (polyfluoroalkylated) substances are highly stable chemically, having slow metabolism and excretion. POPs are lipophilic in nature that do not able to accumulate in adipose tissues and also enable to pas lipid bilayer of CM (cell membrane). Therefore, POPs acts as "hormonally active agents" and also considered as EDC (endocrine disrupting chemicals) (Gore et al., 2015).

The EDC exposure also causes hormonal impact HPT (hypothalamic pituitary-thyroid) axis and HPG (hypothalamic-pituitary-gonadal axis) (Bloom et al., 2014; Dallaire et al., 2009; Schell et al., 2009; Pelletier et al., 2002; Vested et al., 2014; Joensen et al., 2013; Grandjean et al., 2012; Haugen et al., 2011; Giwercman et al., 2006). There is a slight investigation formed that considered the effect of exposure towards persisting organic pollutants (POPs) on HPA (hypothalamic-pituitary-adrenal) axis. However, a negative association shown in PFOS (prenatal perfluorooctane sulfonate) and cord blood of cortisol level (Goudarzi et al., 2017). In animal and in-Virto studies suggested DE (disruptive effects) in the HPA axis (Wilson et al., 2016; Verboven et al., 2010; Zimmer et al., 2009; Li and Wang, 2005; Johansson et al., 1998).

8.3.4 REPRODUCTIVE ENDOCRINE (RD) DISRUPTION

Various studies were reported (animal as well as human studies) exposure towards PCBs in both male as well as in females were introduced. Although a mechanical data still needed to support it. It shows both activities (estrogenic and anti-estrogenic) that affects different receptor levels and steroidogenesis. Hypothalamus and pituitary gland are responsible for the function regulation in female reproductive system (through LH and FSH secretion). These hormones are helpful for the ovarian and uterine function regulation (including GH (Gonadal hormones synthesis and secretions) (Yilmaz et al., 1996, 1998). EDCs have the ability to accumulate with reproductive system that affects female reproductive organ functioning and structure (Gore et al., 2015). EDCs causes various disorders such as hormonal imbalance, fertility decreased or infertility causes PCOS, cancer, puberty dilation, adverse birth outcomes (Guven et al., 1999; Safe, 2013; Yesildaglar et al., 2010). It exhibits the toxic effects, due to which it helps in the development and growth of reproductive tissues when interfere along the hormone physiology. They interfere with hormones binding with ER and androgen receptor that further shows agonistic and antagonistic effect (Sifakis et al., 2017). Bisphenol-A (BPA) has bind with ERα and ERβ that shows estrogenic activity (Bloom et al., 2016; Le Fol et al., 2017). EDCs affect some hormone sensitive diseases in females such as breast cancer, endometrial cancer, and also an ovarian cancer. Additionally, EDCs also act on enzymes found in "steroidogenesis and metabolisms." Phthalates inhibits the synthesis of testosterone in Leydig cells and further responsible for anti-androgenic activities (Foster, 2005). Organochlorinated pesticides and dioxin helps in induction of CYP1 gene expression when bind with AhR. Tetrachloride dibenzo-p-Dioxin shows interfere along steroidogenesis in the ovary that leads to low steroid level (Karman et al., 2012). In methoxychlor format albolite same effects were observed by different workers (Harvey, 2015). Phthalates have ability to inhibit CYP1B1 expression. It also stimulates AhR (Zachow and Uzumcu, 2006). The endocrine disruption chemicals show adverse effects on female reproductive system that leads to folliculogenesis (Sifakis et al., 2017). BPA decreases the fertility rate and its level accumulated along with antral follicles and oocytes receiving fertility treatment decreases in women (Souter et al., 2013). Attachment of this plasticizer has shown with PCOS (polycystic ovarian syndrome), decreases implantation failure risk, decreases fertilization and endometriosis (Kandaraki et al., 2011; Ehrlich et al., 2012). In humans, due to BPA, phenols, and Phthalate's miscarriages, premature birth and behavioral deficits have been linked with them (Sugiura-Ogasawara et al., 2005; Philippat et al., 2012, 2017). During the pre-natal and peri-natal period, the EDCs exposure greatly affect the reproductive system than that of adult life. The endocrine disruptive chemicals exposure shows importance (as type and adverse effect on reproductive function).

As female reproductive system, in male reproductive system, hypothalamus, anterior pituitary (AP) hormones and androgen play an important role in regulation of male reproductive system functioning, in which androgen provides "driving force development of the genitalia" (Gore et al., 2015; Yilmaz et al., 2001, 2006). There are different studies that showed the EDCs effects on the male reproductive system. It shows sperm count reduction, mortality rate and abnormal sperm morphology (Safe, 2013; Bonde et al., 2016). In North Europe, especially in Danish men, showed decreased sperm count and the semen quality (Jensen et al., 2002; Jorgensen et al., 2012). The testosterone disruption through "Leydig cells in young Faroese men" accumulated along serum PCBs levels (Petersen et al., 2018). Also, BPA and phthalates shows effect on motility, sperm count, spermatozoa DNA damage and on testosterone levels (Hauser et al., 2008; Mendiola et al., 2010). EDCs exposure also showed a decrease in semen quality, testicular cancer, etc. (Nordkap et al., 2012). Androgenic and anti-androgenic chemicals show significant role in maintained of health of male reproductive system that reported in different studies (Gore et al., 2015). There are the mechanisms that adversely affect endocrine disruptive chemicals on reproductive functions that inhibit "5α-reductase and aromatase" (Sifakis et al., 2017).

Recently, the detrimental effect of PCBs towards Leydig cell homeostasis is focused in *in vitro* studies. It was mainly focused on increasing the molecular targets (considering dose depended) of the effects and a well proved method against other known mechanisms. In *in vitro* studies, lipid peroxidation was demonstrated with an increase in ROS and decrease antioxidant enzyme system in Leydig cells of mouse towards Aroclor 1242. Escobar-Morreale et al. (2017) reported decreased 3β and 17 β-hydroxysteroid dehydrogenase activity that indicates testosterone synthesis impaired. However, another study demonstrates the effects of PCB 103 and 106 on Leydig cells of mouse, that revels the dose specific and type-specific increase in "estrogen-related receptor type α expression" that helps to enhance Ca (calcium) level and sex steroid secretion and also altered Mitochondria ultra-structure (without change in its membrane potential) (Kjaer et al., 2017). The male rat was exposed towards the A1254 (Aroclor 1254) in the time of lactation demonstrated in a study altered TA, T (testosterone), E (estradiol) and ABP (androgen binding proteins) levels (in both serum and testicular interstitial fluid). Saboor Aftab et al. (2013) was also demonstrated a dose-dependent decrease in gene expression levels of FSH (follicle stimulating hormones), ABP, ERβ (estrogen receptor β) and AR (androgen receptor). All the studies and findings provided mechanical support of SCD. Taylor et al. (2015) also demonstrates the complexity of polychlorinated biphenols reproductive toxicity by studying on "enantioselective action of single congener along with nuclear receptors (having importance for ER). Although increase and decrease of PCB 136 showed +ve effects on ER activation and numerous "potency to induced pregnane X receptor-mediated gene expression." The induction of constitutive receptor-dependent gene expression and suppression of AR-mediated gene expression are the same. Various researchers reported the identification of impaired reproductive physiology and transgenerational effects of PCB in rodents (Dhindsa et al., 2018; Brambilla et al., 2009). However, some epigenetic methods and mechanisms were demonstrated to mediate transmission effects in different generations in relation to PBCs (Crawford et al., 2007). In the *in-vitro* model of pregnancy exposure towards A1254 (Aroclor 1254), the DNA altered methylation and genomic integrity of fetal cells was demonstrated by Glass et al. (1977). Molecular mechanisms and MOAs of PCBs induced RD (reproductive endocrine) (Table 8.5).

TABLE 8.5 Endocrine Disruption in Reproductive System

	Reproductive Disruption		
SL. No.	**PCBs Effects on Endocrine System**	**Endocrine Disruption**	**Disorders/Disease**
1.	Male reproductive system	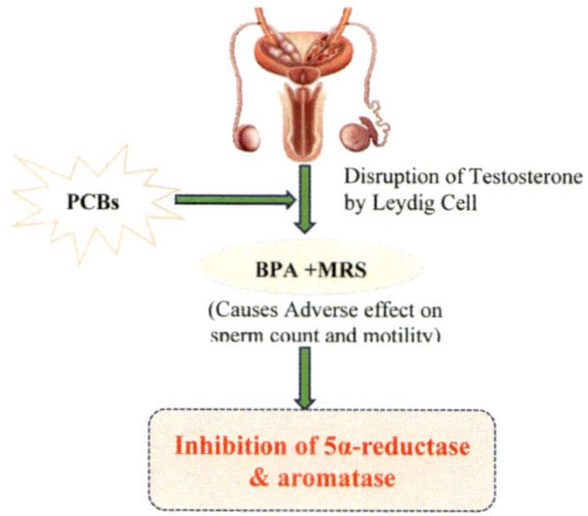	Decrease sperm count, delayed puberty, prostate disease, testicular cancer, etc. (Bayram et al., 2019)
2.	Female reproductive system	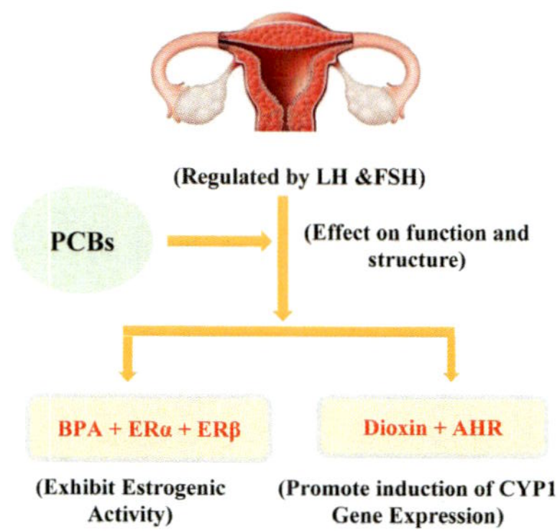	Breast cancer, prostate cancer, malformation in fetus, disrupted hormone action, irregular cyclicity, decreased fertility, infertility, polycystic ovarian syndrome, endometriosis, precocious or delayed puberty and adverse birth outcomes, etc. (Bayram et al., 2019)

KEYWORDS

- **dioxin-like**
- **endocrine system**
- **persistent organic pollutants**
- **polychlorinated biphenyls**
- **reproductive system**
- **trichloroethane**

REFERENCES

Abelsohn, A., Gibson, B. L., Sanborn, M. D., & Weir, E., (2002). Identifying and managing adverse environmental health effects: 5. persistent organic pollutants. *cmaj., 166*, 1549–1554.

Ahmed, R. G., (2016). Early weaning PCB 95 exposure alters the neonatal endocrine system: Thyroid adipokine dysfunction. *J. Endocrinol., 219*, 205–215.

Akagi, K., & Okumura, M., (1985). Association of blood pressure and PCB level in yusho patients. *Environ. Health Perspect., 59*, 37–39.

Alcock, R. E., Halsall, C. L., Harris, C. A., Johnston, A. E., Lead, W. A., Sanders, G., & Jones, K. C., (1994). Contamination of environmental samples prepared for PCB analysis. *Environ. Sci. Technol., 28*, 1838–1842.

Alegria, H., Martinez-Colon, M., Birgul, A., Brooks, G., Hanson, L., & Kurt-Karakus, P., (2016). Historical sediment record and levels of PCBs in sediments and mangroves of jobs bay, Puerto Rico. *Sci. Total Environ., 573*, 1003–1009.

Al-Eryani, L., et al., (2015). Identification of environmental chemicals associated with the development of toxicant-associated fatty liver disease in rodents. *Toxicol. Pathol., 43*, 482–497.

Alonso-Magdalena, P., Ropero, A. B., Carrera, M. P., Cederroth, C. R., Baquié, M., Gauthier, b. r., nef, s., et al., (2008). *Pancreatic Insulin Content Regulation by the Estrogen Receptor ER Alpha, 3*, 2069.

Alonso-Magdalena, P., Ropero, A. B., García, A. M., Sergi, S., Iván, Q., Muhammed, S. J., Albert, S., et al., (2013). Antidiabetic actions of an estrogen receptor β selective agonist. *Diabetes, 62*, 2015–2025.

Anzalone, D. A., Sampino, S., Czernik, M., Iuso, D., & Ptak, G. E., (2018). Polychlorinated biphenyls (PCBs) alter DNA methylation and genomic integrity of sheep fetal cells in a simplified in sitro model of pregnancy exposure. *Toxicol. Vitro,* 39–46.

Armitage, J. M., Hanson, M., Axelman, J., & Cousins, I. T., (2006). Levels and vertical distribution of PCBs in agricultural and natural soils from Sweden. *Sci. Total Environ., 371*, 344–352.

Aydin, Y., & Erkan, M., (2017). The toxic effects of polychlorinated biphenyl (Aroclor 1242) on tm3 Leydig cells. *Toxicol. Ind. Health*, 636–645.

Bai, X., Yan, L., Ji, C., Zhang, Q., Dong, X., Chen, A., & Zhao, M. A., (2018). Combination of ternary classification models and reporter gene assays for the comprehensive thyroid hormone disruption profiles of 209 polychlorinated biphenyls. *Chemosphere., 210,* 312–319.

Baker, N. A., Michael, K., Victoria, E., Fang, J., Wei, Y., Arnold, S., Manjula, S., Morris, A. J., et al., (2013). Coplanar polychlorinated biphenyls impair glucose homeostasis in lean c57bl/6 mice and mitigate beneficial effects of weight loss on glucose homeostasis in obese mice. *Environ. Health Perspect., 121*, 105.

Batang, Z. B., Alikunhi, N., Gochfeld, M., Burger, J., Al-Jahdali, R., Al-Jahdali, H., Aziz, M. A., et al., (2016). Congener-specific levels and patterns of polychlorinated biphenyls in edible fish tissue from the central red seacoast of Saudi Arabia. *Sci. Total Environ., 572*, 915–925.

Beausoleil, C., Beronius, A., Bodin, L., Bokkers, B. G. H., Boon, P. E., Burger, M., Cao, Y., et al., (2017). *Review of Non-Monotonic Dose-Responses of Substances for Human Risk Assessment, EFSA Support* (p. 13). Public.

Bloom, M. S., Jansing, R. L., Kannan, K., Rej, R., & Fitzgerald, E. F., (2014). Thyroid hormones are associated with exposure to persistent organic pollutants in aging residents of upper Hudson river communities. *Int. J.Hyg. Environ. Health., 217*, 473–482.

Bloom, M. S., Mok-Lin, E., & Fujimoto, V. Y., (2016). Bisphenol A and ovarian steroidogenesis. *Fertil. Steril., 106*, 857–863.

Boas, M., Feldt-Rasmussen, U., & Main, K. M., (2012). Thyroid effects of endocrine disrupting chemicals. *Mol. Cell. Endocrinol., 355,* 240–248.

Bonde, J., Flachs, E. M., Rimborg, S., Glazer, C. H., Giwercman, A., Ramlau-Hansen, C. H., Hougaard, K. S., et al., (2016). The epidemiologic evidence linking prenatal and postnatal exposure to endocrine disrupting chemicals with male reproductive disorders: A systematic review and meta-analysis. *Hum. Reprod. Update., 23,* 104–125.

Bonner-Weir, S., Sullivan, B. A., & Weir, G. C., (2015). Human islet morphology revisited: Human and rodent islets are not so different after all. *J. Histochem. Cytochem., 63*,604–612.

Bonnet-Serrano, F., Diedisheim, M., Mallone, R., & Larger, E., (2018). *Decreased A-Cell Mass and Early Structural Alterations of the Exocrine Pancreas in Patients with Type 1 Diabetes: An Analysis Based on the nPOD Repository, 13*, 0191528.

Bopp, S. K., Barouki, R., Brack, W., Dalla, C. S., Dorne, J. L. C. M., Drakvik, P. E., Faust, M., et al., (2018). Current EU research activities on combined exposure to multiple chemicals. *Environ. Int., 120,* 544–562.

Bornstein, S. R., Ehrart-Bornstein, M., & Scherbaum, W. A., (1997). Morphological and functional studies of the paracrine interaction between cortex and medulla in the adrenal gland. *Microsc. Res. Tech., 36*, 520–533.

Bourguignon, J. P., Slama, R., Bergman, Å., Demeneix, B., Ivell, R., Kortenkamp, A., Panzica, G. C., et al., (2016). Science-based regulation of endocrine disrupting chemicals in Europe: Which approach?, *Lancet Diabetes. Endocrinol., 16*, 30121–30128.

Boyle, A. W., Silvin, C. J., Hassett, J. P., Nakas, J. P., & Tanenbaum, S. W., (1992). Bacterial PCB biodegradation. *Biodegradable, 3*, 285–298.

Brambilla, D. J., Matsumoto, A. M., Araujo, A. B., & McKinlay, J. B., (2009). The effect of diurnal variation on clinical measurement of serum testosterone and other sex hormone levels in men. *J. Clin. Endocrinol. Metab., 94*, 907–913.

Brouwer, A., Ahlborg, U. G., Van, D. B. M., Birnbaum, L. S., Boersma, R. E., Bosveld, B., Denison, M. S., et al., (1995). Functional aspects of developmental toxicity of polyhalogenated aromatic hydrocarbons in experimental animals and human infants. *Eur. J. Pharmacol., 293*, 1–40.

Brouwer, A., Morse, D. C., Lans, M. C., Schuur, A. G., Murk, A. J., Klasson-Wehler, W., Bergman, Å., & Visser, T. J., (1998). Interactions of persistent environmental organohalogens with the thyroid hormone system: Mechanisms and possible consequences for animal and human health. *Toxicol. Ind. Health.*, *14*, 59–84.

Buha, A., Antonijević, B., Bulat, Z., Jaćević, V., Milovanović, V., & Matović, V., (2013). The impact of prolonged cadmium exposure and co-exposure with polychlorinated biphenyls on thyroid function in rats. *Toxicol. Lett.*, *221,* 83–90.

Buha, A., Matovic, V., Antonijevic, B., Bulat, Z., Curcic, M., Renieri, E. A., Tsatsakis, A. M., et al., (2018). Overview of cadmium thyroid disrupting effects and mechanisms. *Int. J. Mol. Sci.*, *19,* 1501.

Cabrerizo, A., Dachs, J., Moeckel, C., Ojeda, M. J., Caballero, G., Barceló, D., & Jones, K. C., (2011). Factors influencing the soil–Air partitioning and the strength of soils as a secondary source of polychlorinated biphenyls to the atmosphere. *Environ. Sci. Technol.*, *45*, 4785–4792.

Carmichael, S. L., Shaw, G. M., Nelson, V., Selvin, S., Torfs, C. P., & Curry, C. J., (2003). Hypospadias in California: Trends and descriptive epidemiology. *Epidemiol.*, 94, 499–510.

Carro, N., Garcia, I., Ignacio, M., & Mouteira, A., (2018). Polychlorinated dibenzo-p-dioxins and dibenzofurans (PCDD/Fs) and dioxin-like polychlorinated biphenyls (Dl-PCBS) in bivalve mollusk from Galician rias (N.W., SPAIN), *Chemosphere,* 782–792.

Cavaliere, G., Viggiano, E., Trinchese, G., De Filippo, C., Messina, A., Monda, V., Valenzano, A., et al., (2018). Feeding a high-fat diet induces hypothalamic oxidative stress and inflammation, and prolonged hypothalamic AMPK Activation in rat animal model. *Front. Physiol.*, *9,* 818.

Cave, M., Appana, S., Patel, M., Falkner, K. C., McClain, C. J., & Brock, G., (2010). Polychlorinated biphenyls, lead, and mercury are associated with liver disease in American adults: NHANES 2003–2004. *Environ. Health. Perspect.*, *118*, 1735–1742.

Chekol, T., Vough, L., & Chaney, R. L., (2004). Phytoremediation of polychlorinated biphenyl-contaminated soils: The rhizosphere effect. *Environ. Int.*, *30*, 799–804.

Chen, H. Y., Ko, Y. C., & Lee, C. C., (2008). Relationship between insulin sensitivity and exposure to dioxins and polychlorinated biphenyls in pregnant women. *Environ. Res.*, *107*, 245–253.

Chen, H., Liu, Z., Zhang, X., Jia, X., Li, Q., Su, Q., & Wang, W., (2016). Assessment of synergistic thyroid disrupting effects of a mixture of EDCs in ovariectomized rats using factorial analysis and dose addition. *Toxicol. Res.*, *5,* 1585–1593.

Cheng, S. L., Li, X., Lehmler, H. J., Phillips, B., Shen, D., & Cui, J. Y., (2018). Gut microbiota modulates interactions between polychlorinated biphenyls and bile acid homeostasis. *Toxicol. Sci.*, *166*, 269–287.

Chesnokova, V., Wong, C., Zonis, S., Gruszka, A., Wawrowsky, K., Ren, S. G., Benshloo, A., & Yu, R., (2009). Diminished pancreatic beta-cell mass in securin-null mice is caused by beta-cell apoptosis and senescence. *Endocrinology.*, *150*, 2603–2610.

Chi, Y., Lin, Y., Lu, Y., Huang, Q., Ye, G., & Dong, S., (2019). Gut microbiota dysbiosis correlates with a low dose PCB126-induced dyslipidemia and non-alcoholic fatty liver disease. *Sci. Total Environ.*, *653*, 274–282.

Chi, Y., Wang, H., Lin, Y., Lu, Y., Huang, Q., Ye, G., & Dong, S., (2019). Gut microbiota characterization and lipid metabolism disorder found in PCB77-treated female mice. *Toxicol.*, *420*, 11–20.

Clair, H. B., Pinkston, C. M., Rai, S. N., Pavuk, M., Dutton, N. D., Brock, G. N., Prough, R. A., et al., (2018). Liver disease in a residential cohort with elevated polychlorinated biphenyl exposures. *Toxicol. Sci., 164*, 39–49.

Cnop, M., Igoillo-Esteve, M., Hughes, S. J., Walker, J. N., Cnop, I., & Clark, A., (2011). Longevity of human islet α-and β-cells. *Diabetes Obes. Metab., 13*, 39–46.

Codru, N., Schymura, M. J., Negoita, S., Rej, R., & Carpenter, D. O., (2007). Akwesasne task force on the environment. Diabetes in relation to serum levels of polychlorinated biphenyls and chlorinated pesticides in adult Native Americans. *Environ. Health Perspect., 115*, 1442–1447.

Crawford, E. D., Barqawi, A. B., O'Donnell, C., & Morgentaler, A., (2007). The association of time of day and serum testosterone concentration in a large screening population. *BJU Int., 100*, 509–513.

Cui, S., Fu, Q., Ma, W. l., Song, W. W., Liu, L. Y., & Li, Y. F., (2015). A preliminary compilation and evaluation of a comprehensive emission inventory for polychlorinated bi-phenyls in China. *Sci. Total Environ., 533*, 247–255.

Dallaire, R., Dewailly, E., Pereg, D., Dewailly, É., Pereg, D., Dery, S., & Ayotte, P., (2009). Thyroid function and plasma concentrations of polyhalogenated compounds in Inuit adults. *Environ. Health Perspect., 117*, 1380–1386.

De Falco, M., Laforgia, V., Valiante, S., Virgilio, F., Varano, L., & De Luca, A., (2002). Different patterns of expression of five neuropeptides in the adrenal gland and kidney of two species of frog. *Histochem. J., 34*, 21–26.

De Falco, M., Sciarrillo, R., Virgilio, F., Fedele, V., Valiante, S., Laforgia, V., & Varano, L., (2004). Annual variations of adrenal gland hormones in the lizard *Podarcis sicula. J. Comp. Physiol. A. Neuroethol. Sens. Neural. Behav. Physiol., 190*, 675–681.

De Koning, E. J., (2017). Pancreatic α-cell mass in obesity. *Diabetes Obes. Metab., 19*, 1810–1813.

Dhindsa, S., Ghanim, H., Batra, M., & Dandona, P., (2018). Hypogonadotropic hypogonadism in men with diabesity. *Diabetic Care, 41*, 1516–1525.

Ding, E. L., Song, Y., Malik, V. S., & Liu, S., (2006). Sex differences of endogenous sex hormones and risk of type 2 diabetes: A systematic review and meta-analysis. *JAMA, 295*, 1288–1299.

Drinker, C. K., Warren, M. F., & Bennett, G. A., (1937). The problem of possible systemic effects from certain chlorinated hydrocarbons. *J. Ind. Hyg. Toxicol., 19*, 283–311.

Ehrlich, S., Williams, P. L., Missmer, S. A., Flaws, J. A., Berry, K. F., & Calafat, A. M., (2012). Urinary bisphenol A concentrations and implantation failure among women undergoing *In vitro* fertilization. *Environ. Health Perspect., 120*, 978–983.

Elnar, A. A., Desor, F., Marin, F., Soulimani, R., & Nemos, C., (2015). Lactational exposure to low levels of the six indicator non-dioxin-like polychlorinated biphenyls induces DNA damage and repression of neuronal activity, in juvenile male mice. *Toxicol., 328*, 57–65.

Enell, A., Lundstedt, S., Arp, H. P. H., Josefsson, S., Cornelissen, G., Wik, O., & Berggrenkleja, D., (2016). Combining leaching and passive sampling to measure the mobility and distribution between porewater, Doc, and colloids of native oxy-Pahs, N-Pacs, and Pahs in historically contaminated soil. *Environ. Sci. Technol., 50*, 11797–11805.

Erickson, M. D., & Kaley, 2nd R. G., (2011). Applications of polychlorinated biphenyls. *Environ. Sci. Pollut. Res. Int., 18*, 135–151.

Escobar-Morreale, H. F., Santacruz, E., & Luque-Ramírez, (2017). Prevalence of 'obesity-associated gonadal dysfunction' in severely obese men and women and its resolution after bariatric surgery: A systematic review and meta-analysis. *Hum. Reprod. Update., 23*, 390–408.

Eskenazi, B., Rauch, S. A., Tenerelli, R., Huen, K., Holland, N. T., Lustig, R. H., Kogut, K., et al., (2017). In utero and childhood Ddt, Dde, Pbde and Pcbs exposure and sex hormones in adolescent boys: The chamacos study. *Int. J. Hyg. Environ. Health., 220,* 364–372.

Espandiari, P., Glauert, H. P., Lehmler, H. J., Lee, E. Y., Srinivasan, C., & Robertson, L. W., (2003). Polychlorinated biphenyls as initiators in liver carcinogenesis: Resistant hepatocyte model. *Toxicol. Appl. Pharmacol., 186*, 55–62.

Esposito, T., Lucariello, A., Hay, E., Contieri, M., Tammaro, P., Varriale, B., Guerra, G., De Luca, A., & Perna, A., (2019). Effects of curcumin and its adjuvant on Tpc1 thyroid cell line. *Chem. Biol. Interact., 305*, 112–118.

Everett, C. J., Mainous, A. G., Frithsen, I. L., Player, M. S., & Matheson, E. M., (2008). association of polychlorinated biphenyls with hypertension in the 1999–2002 national health and nutrition examination survey. *Environ. Res., 108*, 94–97.

Færch, K., Vistisen, D., Pacini, G., Torekov, S. S., Johansen, N. B., Witte, D. R., Jonsson, A., et al., (2016). Insulin resistance is accompanied by increased fasting glucagon and delayed glucagon suppression in individuals with normal and impaired glucose regulation. *Diabetes., 65*, 3473–3481.

Ficko, S. A., Rutter, A., & Zeeb, B. A., (2010). Potential for phytoextraction of PCBs from contaminated soils using weeds. *Sci. Total Environ., 408*, 3469–3476.

Fisher, J. W., Campbell, J., Muralidhara, S., Bruckner, J. V., Ferguson, D., Mumtaz, M., Harmon, B., et al., (2006). Effect of PCB 126 on hepatic metabolism of thyroxine and perturbations in the hypothalamic-pituitary-thyroid axis in the rat. *Toxicol. Sci., 90,* 87–95.

Foster, P. M., (2005). Mode of action: Impaired fetal Leydig cell function effects on male reproductive development produced by certain phthalate esters. *Crit. Rev. Toxicol., 35*, 713–719.

Fukui, M., Soh, J., Tanaka, M., Kitagawa, Y., Hasegawa, G., Yoshikawa, T., Miki, T., & Nakamura, N., (**2007**). Low serum testosterone concentration in middle aged men with type 2 diabetes. *Endocr. J., 54*, 871–877.

Gadupudi, G. S., Elser, B. A., Sandgruber, F. A., Li, X., Gibson-Corley, K. N., & Robertson, L. W., (2018). Pcb126 inhibits the activation of AMPKCREB signal transduction required for energy sensing in the liver. *Toxicol. Sci., 163*, 440–453.

Gadupudi, G. S., Klaren, W. D., Olivier, A. K., Klingelhutz, A. J., & Robertson, L. W., (2016). PCB-126-induced disruption in gluconeogenesis and fatty acid oxidation precedes fatty liver in male rats. *Toxicol. Sci., 149*, 98–110.

Giwercman, A. H., Rignell-Hydbom, A., Toft, G., Rylander, L., Hagmar, L., Lindh, C., Pedersen, H. S., et al., (2006). Reproductive hormone levels in men exposed to persistent organohalogen pollutants: A study of Inuit and three European cohorts. *Environ. Health Perspect., 114*, 1348–1353.

Gladen, B. C., & Rogan, W. J., (1991). Effects of perinatal polychlorinated biphenyls and dichloro diphenyl dichloroethene on later development. *J. Pediatr., 119*, 58–63.

Glass, A. R., Swerdloff, R. S., Bray, G. A., Dahms, W. T., & Atkinson, R. L., (1977). Low serum testosterone and sex-hormone-binding-globulin in massively obese men. *J. Clin. Endocrinol. Metab., 45,* 1211–1219.

Goncharov, A., Bloom, M., Pavuk, M., Birman, I., & Carpenter, D. O., (2010). Blood pressure and hypertension in relation to levels of serum polychlorinated biphenyls in residents of Anniston, Alabama. *J. Hypertens., 28*, 2053–2060.

Goncharov, A., Pavuk, M., Foushee, H. R., & Carpenter, D. O., (2011). Blood pressure in relation to concentrations of PCB congeners and chlorinated pesticides. *Environ. Health Perspect., 119*, 319–325.

Gore, A. C., Chappell, V. A., Fenton, S. E., Flaws, J. A., Nadal, A., Prins, G. S., Toppari, J., & Zoeller, R. T., (2015). Executive summary to Edc-2: The endocrine society's second scientific statement on endocrine-disrupting chemicals. *Endocr. Rev., 36,* 1–150.

Goudarzi, H., Araki, A., Itoh, S., Sasaki, S., Miyashita, C., Mitsui, T., Nakazawa, H., Nonomura, K., & Kishi, R., (2017). The association of prenatal exposure to perfluorinated chemicals with glucocorticoid and androgenic hormones in cord blood samples: The Hokkaido study. *Environ. Health Perspect., 125*, 111–118.

Grandjean, P., Grønlund, C., & Kjær, I. M., (2012). Reproductive hormone profile and pubertal development in 14-year-old boys prenatally exposed to polychlorinated biphenyls. *Reprod. Toxicol., 34*, 498–503.

Grandjean, P., Weihe, P., Burse, V. W., Needham, L. L., Storrhansen, E., Heinzow, B., Debes, F., et al., (2001). Neurobehavioral deficits associated with PCB in 7-year-old children prenatally exposed to seafood neurotoxicants. *Neurotoxicol. Teratol., 23*, 305–317.

Gray, S. L., Shaw, A. C., Gagne, A. X., & Chan, H. M., (2013). Chronic exposure to PCBs (Aroclor 1254) exacerbates obesity-induced insulin resistance and hyperinsulinemia in mice. *J. Toxicol. Environ. Health, Part A, 76*, 701–715.

Grimm, F. A., He, X., Teesch, L. M., Lehmler, H. J., Robertson, L. W., & Duffel, M. W., (2015). Tissue distribution, metabolism, and excretion of 3,3′dichloro-4′-sulfooxy-biphenyl in the rat. *Environ. Sci. Technol., 49*, 8087–8095.

Ha, M. H., Lee, D. H., Son, H. K., Park, S. K., & Jacobs, D. R., (2009). "Association between serum concentrations of persistent organic pollutants and prevalence of newly diagnosed hypertension: Results from the national health and nutrition examination survey 1999–2002." *J. Human Hypertens., 23*, 274–286.

Haglund, P., (1996a). Enantioselective separation of polychlorinated biphenyl atropisomers using chiral high-performance liquid chromatography. *J. Chromatogr., 724*, 219–228.

Hamers, T., Kamstra, J. H., Cenijn, P., Pencikova, K., Palkova, L., Simeckova, P., Vondracek, J., et al., (2011). *In vitro* toxicity profiling of ultrapure non-dioxin-like polychlorinated biphenyl congeners and their relative toxic contribution to PCB mixtures in humans. *Toxicol. Sci., 121*, 88–100.

Hardesty, J. E., Al-Eryani, L., Wahlang, B., Falkner, K. C., Shi, H., Jin, J., Vivace, B. J., et al., (2018). Epidermal growth factor receptor signaling disruption by endocrine and metabolic disrupting chemicals. *Toxicol Sci., 162*, 622–634.

Hardesty, J. E., Wahlang, B., Falkner, K. C., Clair, H. B., Clark, B. J., Ceresa, B. P., Prough, R. A., & Cave, M. C., (2017). Polychlorinated biphenyls disrupt hepatic epidermal growth factor receptor signaling. *Xenobiotica., 47*, 807–820.

Hardesty, J. E., Wahlang, B., Falkner, K. C., Shi, H., Jin, J., Wilkey, D., Merchant, M., et al., (2019). Hepatic signaling disruption by pollutant polychlorinated biphenyls in steatohepatitis. *Cell Sign., 53*, 132–139.

Hardesty, J. E., Wahlang, B., Falkner, K. C., Shi, H., Jin, J., Zhou, Y., Wilkey, D. W., et al., (2019). Proteomic analysis reveals novel mechanisms by which polychlorinated biphenyls compromise the liver promoting diet-induced steatohepatitis. *J. Proteome. Res., 18*, 1582–1594.

Harrad, S., Goosey, E., Desborough, J., Abdallah, M. A. E., Roosens, L., & Covaci, A., (2010). Dust from U.K. primary school classrooms and daycare centers: The significance of dust as a pathway of exposure of young U.K. children to brominated flame retardants and polychlorinated biphenyls. *Environ. Sci. Technol., 44*, 4198–4202.

Harris, T. B., (2002). Invited commentary: Body composition in studies of aging: New opportunities to better understand health risks associated with weight. *Am. J. Epidemiol., 156*, 125, 126.

Harvey, C. N., Chen, J. C., Bagnell, C. A., & Uzumcu, M., (2015). Methoxychlor and its metabolite HPTE inhibit camp production and expression of estrogen receptors alpha and beta in the rat granulosa cell *In vitro*. *Reprod. Toxicol., 51*, 72–78.

Harvey, P. J., Campanella, B. F., Castro, P. M., Harms, H., Lichtfouse, E., Schäffner, A. R., Smrcek, S., & Werck-Reichhart, D., (2002). Phytoremediation of polyaromatic hydrocarbons, anilines and phenols. *Environ. Sci. Pollu. Resea., 9*, 29–47.

Haugen, T. B., Tefre, T., Malm, G., Jönsson, B. A., Rylander, L., Hagmar, L., Bjørsvik, C., et al., (2011). Differences in serum levels of Cb-153and P,P'-Dde, and reproductive parameters between men living south and north in Norway. *Reprod. Toxicol., 32*, 261–267.

Hauser, R., (2008). Urinary phthalate metabolites and semen quality: A review of a potential biomarker of susceptibility. *Int. J. Androl., 31*, 112–117.

Heindel, J. J., Balbus, J., Birnbaum, L., Brune-Drisse, M. N., Grandjean, P., Gray, K., Landrigan, P. J., et al., (2015). Developmental origins of health and disease: Integrating environmental influences. *Endocrinol., 156,* 3416–3421.

Heindel, J. J., Blumberg, B., Cave, M., Machtinger, R., Mantovani, A., Mendez, M. A., Nadal, A., et al., (2017). Metabolism disrupting chemicals and metabolic disorders. *Reprod. Toxicol., 68*, 3–33.

Hennig, B., Meerarani, P., Slim, R., Toborek, M., Daugherty, A., & Silverstone, A. E., (2002). Proinflammatory properties of coplanar PCBs: *In vitro* and *in vivo* evidence *Toxicol. Appl. Pharmacol., 181,* 174–183.

Herbstman, J. B., Sjödin, A., Apelberg, B. J., Witter, F. R., Haiden, R. U., Patterson, D. G., Panny, S. R., et al., (2008). Birth delivery mode modifies the associations between prenatal polychlorinated biphenyl (PCB) and polybrominated diphenyl ether (PBDE) and neonatal thyroid hormone levels. *Environ. Health Perspect., 116,* 1376–1382.

Hofe, C. R., Feng, L., Zephyr, D., Stromberg, A. J., Hennig, B., & Gaetke, L. M., (2014). Fruit and vegetable intake, as reflected by serum carotenoid concentrations, predicts reduced probability of polychlorinated biphenyl-associated risk for type 2 diabetes: National health and nutrition examination survey 2003–2004. *Nutr. Res., 34,* 285–293.

Hoffman, J. B., Flythe, M. D., & Hennig, B., (2019). Environmental pollutant mediated disruption of gut microbial metabolism of the prebiotic inulin. *Anaerobe., 55*, 96–102.

Hu, D., & Hornbuckle, K. C., (2010). Inadvertent polychlorinated biphenyls in commercial paint pigments. *Environ. Sci. Technol., 44*, 2822–2827.

Hu, D., Martinez, A., & Hornbuckle, K. C., (2008). Discovery of non-Aroclor PCB (3,dichlorobiphenyl) in Chicago air. *Environ. Sci. Technol., 42*, 7873–7877.

Hu, D., Martinez, A., & Hornbuckle, K. C., (2011). Sedimentary records of non-Aroclor and Aroclor PCB mixtures in the Great Lakes. *J. Great Lakes Res.,* 359–364.

Huang, J., Matsumura, T., Yu, G., Deng, S., Yamauchi, M., Yamazaki, N., & Weber, R., (2011). Determination of PCBs, PCDDs and PCDFs in insulating oil samples from stored Chinese electrical capacitors by HRGC/HRMS. *Chemosphere,* 239–246.

Ivanciuc, T., Ivanciuc, O., & Klein, D. J., (2006). Modeling the bio-concentration factors and bioaccumulation factors of polychlorinated biphenyls with poetic quantitative superstructure/activity relationships (Qssar). *Molec. Diversity, 10*, 133–145.

Jacobson, J. I., & Jacobson, S. W., (1996). Intellectual impairment in children exposed to polychlorinated biphenyls in utero. *N. Engl. J. Med., 335*, 783–789.

Javorska, H., Tlustos, P., & Kaliszova, R., (2011). Distribution of polychlorinated biphenyl congeners in root vegetables. *Pol. J. Environ. Stud., 20*, 93–99.

Jensen, T. K., Carlsen, E., Jorgensen, N., Berthelsen, J. G., Keiding, N., Christensen, K., Petersen, J. H., et al., (2002). Poor semen quality may contribute to recent decline in fertility rates. *Hum Reprod., 17*, 1437–1440.

Jirsová, S., Mašata, J., Jech, L., & Zvárová, J., (2010). Effect of polychlorinated biphenyls (PCBs) and 1,1,1trichloro-2,2,-bis (4-chlorophenyl)-ethane (DDT) in follicular fluid on the results of *in vitro* fertilization-embryo transfer (Ivf-Et) programs. *Fertil. Steril., 93,* 1831–1836.

Joensen, U. N., Veyrand, B., Antignac, J. P., Blomberg, J. M., Petersen, J. H., Marchand, P., Skakkebæk, N. E., et al., (2013). PFOs (perfluorooctanesulfonate) in serum is negatively associated with testosterone levels, but not with semen quality, in healthy men. *Human Reprod., 28*, 599–608.

Johansson, M., Nilsson, S., & Lund, B. O., (1998). Interactions between methyl sulfonyl PCB sand the glucocorticoid receptor. *Environ. Health Perspect., 106*, 769–772.

Jorgensen, N., Joensen, U. N., Jensen, T. K., Jensen, M. B., Almstrup, K., Olesen, I. A., Juul, A., et al., (2012). Human semen quality in the new millennium: A prospective cross-sectional population-based study of 4867 men. *BMJ Open, 2*, 000990.

Kandaraki, E., Chatzigeorgiou, A., Livadas, S., Palioura, E., Economou, F., Koutsilieris, M., Palimeri, S., et al., (2011). Endocrine disruptors and polycystic ovary syndrome (PCOs): Elevated serum levels of bisphenolain women with PCOs. *J. Clin. Endocrinol. Metab., 96*, 480–484.

Kania-Korwel, I., & Lehmler, H. J., (2016). Toxicokinetics of chiral polychlorinated biphenyls across different species: A review. *Environ. Sci. Pollut. Res. Int.,* 23, 2058–2080.

Kania-Korwel, I., Wu, X., Wang, K., & Lehmler, H. J., (2017). Identification of lipidomic markers of chronic 3,3,'4,4,'5-pentachlorobiphenyl (PCB 126) exposure in the male rat liver. *Toxicol., 390*, 124–134.

Karman, B. N., Basavarajappa, M. S., Hannon, P., & Flaws, J. A., (2012). Dioxin exposure reduces the steroidogenic capacity of mouse antral follicles mainly at the level of Hsd17b1 without altering atresia. *Toxicol. Appl. Pharmacol., 264*, 1–12.

Katsikantami, I., Sifakis, S., Tzatzarakis, M. N., Vakonaki, E., Kalantzi, O. I., Tsatsakis, A. M., & Rizos, A. K., (2016). A global assessment of phthalates burden and related links to health effects. *Environ. Int., 97,* 212–236.

Khan, M. A., Lichtensteiger, C. A., Faroon, O., Mumtaz, M., Schaeffer, D. J., & Hansen, L. G., (2002). The hypothalamic-pituitary-thyroid (Hpt) axis: A target of nonpersistent ortho-substituted PCB congeners. *Toxicol. Sci., 65,* 52–61.

Kissa, E., (2001). Fluorinated surfactants and repellents. In: Hubbard A. T., (ed), *Surfactant Science Series* (2nd edn., Vol. 97, p. 615). New York: Marcel Dekker Inc.

Kitamura, S., Jinno, N., Suzuki, T., Sugihara, K., Ohta, S., Kuroki, H., & Fujimoto, N., (2005). Thyroid hormone-like and estrogenic activity of hydroxylated PCBs in cell culture. *Toxicol., 65,* 52–61.

Kjaer, M. M., Wu, X., Wang, K., & Lehmler, H. J., (2017). The impact of gastric bypass surgery on sex hormones and menstrual cycles in premenopausal women. *Gynecol. Endocrinol., 33*, 160–163.

Kobayashi, K., Miyagawa, M., Wang, R. S., Suda, M., Sekiguchi, S., & Honma, T., (2008). Effects of in utero exposure to 2, 20,4,40,5,50 -hexachlorobiphenyl (PCB 153) on somatic growth and endocrine status in rat offspring. *Congenital Anomalies, 48*, 151–157.

Koopman-Esseboom, C., Morse, D. C., Weisglas-Kuperus, N., Lutkeschipholt, I. J., Van, D. P. C. G., Tuinstra, L. G. M., Brouwer, A., & Sauer, P. J., (1994). Effects of dioxins and polychlorinated biphenyls on thyroid hormone status of pregnant women and their infants. *Pediatr. Res., 36*, 468–473.

Kuwatsuka, Y., Shimizu, K., Akiyama, Y., Koike, Y., Ogawa, F., Furue, M., & Utani, A., (2014). Yusho patients show increased serum Il-17, Il-23, Il-1β, and Tnfα levels more than 40 years after accidental polychlorinated biphenyl poisoning. *J. Immunotoxicol., 11*, 246–269.

Lasram, M. M., Dhouib, I. B., Annabi, A., Fazaa, S. E., & Gharbi, N., (2014). A review on the molecular mechanisms involved in insulin resistance induced by organophosphorus pesticides. *Toxicol., 322*, 1–13.

Lauby-Secretan, B., Loomis, D., Baan, R., El Ghissassi, F., Bouvard, V., Benbrahim-Tallaa, L., Guha, N., et al., (2016). Use of mechanistic data in the IARC evaluations of the carcinogenicity of polychlorinated biphenyls and related compounds. *Environ Sci. Pollut. Res. Int., 23*, 2220–2229.

Laukers, J. D., (1986). Disposal and destruction of waste PCBs. In: Waid, J. S., (ed.), *PCBs and the Environment* (pp. 83–152). Florida: CRC Press.

Le Fol, V., Ait-Aissa, S., Sonavane, M., Porcher, J. M., Balaguer, P., & Cravedi, J. P., (2017). *In vitro* and *in vivo* estrogenic activity of BPA, BPF and BPS in zebrafish-specific assays. *Ecotoxicol. Environ. Saf., 142*, 150–156.

Le May, C., Chu, K., Hu, M., Ortega, C. S., Simpson, E. R., Korach, K. S., Tsai, M. J., & Mauvais-Jarvis, F., (2006). Estrogens protect pancreatic β-cells from apoptosis and prevent insulin-deficient diabetes mellitus in mice. *Proc. Natl. Acad. Sci., 103*, 9232–9237.

Lee, D. H., Jacobs, D. R. J. R., & Porta, M., (2006). Could low-level background exposure to persistent organic pollutants contribute to the social burden of type 2 diabetes? *Jepidemiol. Community Health., 60*, 1006–1008.

Lee, D. H., Lee, I. K., & Song, K., (2006). *A Strong Dose-Response Relation Between Serum Concentrations of Persistent Organic Pollutants and Diabetes: Results from the National Health and Examination Survey,* 1999–2002.

Lee, D. H., Lee, I. K., Jin, S. H., Steffes, M., & Jacobs, D. R., (2007a). Association between serum concentrations of persistent organic pollutants and insulin resistance among non-diabetic adults: Results from the national health and nutrition examination survey 1999–2002. *Diab. Care., 30*, 622–628.

Lee, D. H., Porta, M., Jacobs, D. R., & Vandenberg, L. N., (2014). Chlorinated persistent organic pollutants, obesity, and type 2 diabetes. *Endocr. Rev., 35*, 557–601.

Lee, D. H., Steffes, M. W., Sjödin, A., Jones, R. S., Needham, L. L., & Jacobs, Jr. D. R., (2011). Low dose organochlorine pesticides and polychlorinated biphenyls predict obesity, dyslipidemia, and insulin resistance among people free of diabetes. *PloS One, 6*, 15977.

Lee, D. H., Steffes, M., & Jacobs, D. R., (2007b). Positive associations of serum concentration of polychlorinated biphenyls or organochlorine pesticides with self-reported arthritis, especially rheumatoid type, in women. *Environ. Health Perspect., 115*, 883–888.

Lee, K. W., (1995). *Practical Management of Chemicals and Hazardous Wastes: An Environmental and Safety Professional Guide.* New Jersey: Prentice Hall.

Lehmler, H. J., & Robertson, L. W., (2001). Atropisomers of PCBs. In: Robertson, L. W., et al., (eds.), *Recent Advances in the Environmental Toxicology and Health Effects of PCBs* (pp. 62–65). University Press of Kentucky, Lexington, KY.

Li, J., Lai, H., Chen, S., Zhu, H., & Lai, S., (2017). Interaction of sex steroid hormones and obesity on insulin resistance and type 2 diabetes in men: The third national health and nutrition examination survey. *J. Diabetes Complicat., 31*, 318–327.

Li, L. A., & Wang, P. W., (2005). PCB126 induces differential changes in androgen, cortisol, and aldosterone biosynthesis in human adrenocortical H295r cells. *Toxicol Sci., 85*, 530–540.

Liu, S., Le May, C., Wong, W. P. S., Ward, R. D., Clegg, D. J., Marco, M., Korach, K. S., Franck, M. J., (2009). Importance of extranuclear estrogen receptor-alpha and mem-brane g protein-coupled estrogen receptor in pancreatic islet survival. *Diabetes, 58*, 2292.

Luthy, R. G., Aiken, G. R., Brusseau, M. L., Cunningham, S. D., Gschwend, P. M., Pignatello, J. J., Reinhard, M., et al., (1997). Sequestration of hydrophobic organic contaminants by geosorbents. *Environ. Sci. Technol., 31*, 3341–3347.

Macek, T., Macková, M., & Káš, J., (2000). Exploitation of plants for the removal of organics in environmental remediation. *Biotechnol. Advance., 18*, 23–34.

Machala, M., Blaha, L., Lehmler, H. J., Pliskova, M., Majkova, Z., Kapplova, P., Sovadinova, I., et al., (2004). Toxicity of hydroxylated and quinoid PCB metabolites: Inhibition of gap junctional intercellular communication and activation of aryl hydrocarbon and estrogen receptors in hepatic and mammary cells. *Chem. Res. Toxicol., 17*, 340–347.

Mahmood, A., Malik, R. N., Li, J., & Zhang, G., (2014). Levels, distribution profile, and risk assessment of polychlorinated biphenyls (PCBs) in water and sediment from two tributaries of the River Chenab, Pakistan. *Environ. Sci. Pollut. Res., 21*, 7847–7855.

Manojlović-Stojanoski, M., Nestorovic, N., Negić, N., Filipovic, B., Šošić-Jurjević, B., Milosević, V., & Sekulić, M., (2006). The pituitary-adrenal axis of fetal rats after maternal dexamethasone treatment. *Anat. Embryology., 211*, 61–69.

Maroni, M., Colombi, A., Arbosti, G., Cantoni, S., & Foa, V., (1981). Occupational exposure to polychlorinated biphenyls in electrical workers. II. Health effects. *Br. J. Ind. Med., 38*, 55–60.

Martin, L., & Klaassen, C. D., (2010). Differential effects of polychlorinated biphenyl congeners on serum thyroid hormone levels in rats. *Toxicol. Sci.,* 36–44.

Meagher, R. B., (2000). Phytoremediation of toxic elemental and organic pollutants. *Curre. Opini. Plant Biol., 3*, 153–162.

Meerts, I. A. T. M., Assink, A., Cenijn, P. H., Van, D. B. J. H. J., Weijers, B. M., Bergman, Å., Koeman, J. H., & Brouwer, A., (2002). Placental transfer of a hydroxylated polychlorinated biphenyl and effects on fetal and maternal thyroid hormone homeostasis in the rat. *Toxicol. Sci., 68*, 361–372.

Meggo, R. E., Schnoor, J. L., & Hu, D., (2013). Dechlorination of PCBs in the rhizosphere of switchgrass and poplar. *Environ. Pollution., 178*, 312–321.

Men, B., He, M., Tan, L., & Lin, C., (2014). Distributions of polychlorinated biphenyls in the Daliao river estuary of Liaodong bay, Bohai Sea (China). *Mar. Environ. Res., 78*, 77–84.

Mendiola, J., Jorgensen, N., Andersson, A., Calafat, A. M., Ye, X., & Redmon, J. B., (2010). Are environmental levels of bisphenol A associated with reproductive function in fertile men? *Environ. Health Perspect., 118*, 1286–1291.

Merino, B., Alonso-Magdalena, P., Lluesma, M., Ñeco, P., Gonzalez, A., Marroquí, L., García-Arévalo, M., et al., (**2015**). Pancreatic alpha-cells from female mice undergo morpho functional changes during compensatory adaptations of the endocrine pancreas to diet-induced obesity. *Sci. Rep., 5*, 11622.

Merkel, T., Graeber, M., & Pagel, L., (1999). A new technology for fluidic microsystems based on PCB technology. *Sens. Actuators Animal Phys., 77*, 98–105.

Moeckel, C., Nizzetto, L., Guardo, A. D., Steinnes, E., Freppaz, M., Filippa, G., Camporini, P., et al., (2008). Persistent organic pollutants in boreal and montane soil profiles: Distribution, evidence of processes and implications for global cycling. *Environ. Sci. Technol., 42*, 8374–8380.

Morris, P. J., Mohn, W. W., Quensen, I. J. F., Tiedje, J. M., & Boyd, S. A., (1992). Establishment of a PCB degrading enrichment culture with predominantly meta-dechlorination. *Appl. Environ. Microbiol., 58*, 3088–3094.
Muller, M. H., Polder, A., Brynildsrud, O. B., Karimi, M., Lie, E., Manyilizu, W. B., Mdegela, R. H., et al., (2017). Organochlorine pesticides (OCPs) and polychlorinated biphenyls (PCBs) in human breast milk and associated health risks to nursing infants in Northern Tanzania. *Environ. Res., 154*, 425–434.
Muller, M. H., Polder, A., Brynildsrud, O. B., Lie, E., Løken, K. B., Manyilizu, W. B., Mdegela, R. H., et al., (2016). Brominated flame retardants (BFRs) in breast milk and associated health risks to nursing infants in Northern Tanzania. *Environ. Int., 89, 90*, 38–47.
Nadal, A., Alonso-Magdalena, P., Soriano, S., Quesada, I., & Ropero, A. B., (2009). The pan-creatic B-cell as a target of estrogens and xenoestrogens: Implications for blood glucose homeostasis and diabetes. *Mol. Cell. Endocrinol., 304*, 63–68.
Navarro, G., Xu, W., Jacobson, D. A., Wicksteed, B., Allard, C., Zhang, G., De Gendt, K., et al., (2016). Extranuclear actions of the androgen receptor enhance glucose-stimulated insulin secretion in the male. *Cell Metab., 23,* 837–851.
Nemerow, N. I., & Agardy, F. J., (1998). *Strategies of Industrial and Hazardous Waste Management* (pp. 562, 563). Van Nostrand, Reinhold.
Niu, J. F., Yang, Z. F., Shen, Z. Y., & Wang, L. L., (2006). QSPRs for the prediction of photodegradation half-life of PCBs in N-hexane. SAR QSAR. *Environ. Res.,* 173–182.
Nordkap, L., Joensen, U. N., Blomberg, J. M., & Jorgensen, N., (2012). Regional differences and temporal trends in male reproductive health disorders: Semen quality may be a sensitive marker of environmental exposures. *Mol. Cell. Endocrinol., 355*, 221–230.
Nussdorfer, G. G., (1996). Paracrine control of adrenal cortical function by medullary chromaffin cells. *Pharmacol. Rev., 48*, 495–530.
Patandin, S., Lanting, G. I., Mulder, P. G., Boersma, E. R., Sauer, P. J., & Weisglas-Kuperus, N., (1999). Effects of environmental exposure to polychlorinated biphenyls and dioxins on cognitive Arbil PCBs Ities in Dutch children at 42 months of age. *J. Pediatr., 134*, 33–41.
Pelletier, C., Doucet, E., Imbeault, P., & Tremblay, A., (2002). Associations between weight loss-induced changes in plasma organochlorine concentrations, serum T(3) concentration, and resting metabolic rate. *Toxicol. Sci., 67*, 46–51.
Pěnčíková, K., Brenerová, P., Svržková, L., Hrubá, E., Pálková, L., Vondráček, J., Lehmler, H. J., & Machala, M., (2018). Atropisomers of 2,2′,3,3′,6,6′-hexachlorobiphenyl (PCB 136) exhibit stereoselective effects on activation of nuclear receptors *in vitro*. *Environ. Sci. Pollut. Res.,* 16411–16419.
Perkins, J. T., Petriello, M. C., Newsome, B. J., & Hennig, B., (2016). polychlorinated biphenyls and links to cardiovascular disease. *Environ. Sci. Pollut. Res. Int., 23*, 2160–2172.
Persson, Y., Hemström, K., Öberg, L., Tysklind, M., & Enell, A., (2008). Use of a column leaching test to study the mobility of chlorinated HOCs from a contaminated soil and the distribution of compounds between soluble and colloid phases. *Chemosphere, 71*, 1035–1042.
Petersen, M. S., Halling, J., Jorgensen, N., Nielsen, F., Grandjean, P., & Jensen, T. K., (2018). Reproductive function in a population of young Faroese men with elevated exposure to polychlorinated biphenyls (PCBS) and perfluorinated alkylate substances (PFAS). *Int. J Environ. Res. Public. Health., 15,* 1880.
Petrakis, D., Vassilopoulou, L., Mamoulakis, C., Psycharakis, C., Anifantaki, A., Sifakis, S., Docea, A. O., et al., (2017). Endocrine disruptors leading to obesity and related diseases. *Int. J. Environ. Res. Public Health., 14,* 1282.

Petriello, M. S., Hoffman, J. B., Vsevolozhskaya, O., Morris, A. J., & Hennig, B., (2018). Dioxin-like PCB 126 increases intestinal inflammation and disrupts gut microbiota and metabolic homeostasis. *Environ. Pollut., 242,* 1022–1032.

Philippat, C., Mortamais, M., Chevrier, C., Petit, C., Calafat, A. M., & Ye, X., (2012). Exposure to phthalates and phenols during pregnancy and offspring size at birth. *Environ. Health. Perspect., 120*, 464–470.

Philippat, C., Nakiwala, D., Calafat, A. M., Botton, J., De Agostini, M., & Heude, B., (2017). Prenatal exposure to nonpersistent endocrine disruptors and behavior in boys at 3 and 5 years. *Environ. Health Perspect., 125*, 097014.

Pocar, P., Fiandanese, N., Secchi, C., Berrini, A., Fischer, B., Schmidt, J. S., Schaedlich, K., et al., (2012). Effects of polychlorinated biphenyls in Cd-1 mice: Reproductive toxicity and intergenerational transmission. *Toxicol. Sci.*, 213–226.

Püttmann, M., Mannschreck, A., Oesch, F., & Robertson, L., (1989). Chiral effects in the induction of drug-metabolizing enzymes using synthetic atropisomers of polychlorinated biphenyls (PCBs). *Biochem. Pharmacol., 38*, 1345–1352.

Randoll, R., Wondrak, W., & Schletz, A., (2014). Dielectric strength and thermal performance of PCB embedded power electronics, microelectron. *Reliab.*, 1872–1876.

Rattan, S., Zhou, C., Chiang, C., Mahalingam, S., Brehm, E., & Flaws, J. A., (2017). Exposure to endocrine disruptors during adult hood: Consequences for female fertility. *J. Endocrinol.*, *233*, 109–R29.

Raymond, R. M., Christensen, Y. K., Thompson, A. B., & Anderson, A. H., (2016). Associations between fish Consumption and contaminant biomarkers with cardiovascular conditions among older male anglers in Wisconsin. *J. Occup. Environ. Med., 58*, 676–682.

Rhodes, C. J., (2005). Type 2 diabetes-a matter of beta-cell life and death? *Sci., 307*, 380–384.

Rignell-Hydbom, A., Rylander, L., & Hagmar, L., (2007). Exposure to persistent organochlorine pollutants and type 2 diabetes mellitus. *Human Exp. Toxicol., 26*, 447–452.

Ruel, M. V. M., Bos, A. F., Soechitram, S. D., Meijer, L., Sauer, P. J. J., & Berghuis, S. A., (2019). Prenatal exposure to organohalogen compounds and children's mental and motor development at 18 and 30 months of age. *Neurotoxicol., 72,* 6–14.

Ruzzin, J., Petersen, R., Meugnier, E., et al., (2010). Persistent organic pollutant exposure leads to insulin resistance syndrome. *Environ. Health Perspect., 118*, 465–471.

Rylander, L., Rignell-Hydborn, A., & Hagmar, L., (2005). A cross-sectional study of the association between persistent organochlorine pollutants and diabetes. *Environ. Health., 4*, 28–33.

Saboor, A. S. A., Kumar, S., & Barber, T. M., (2013). The role of obesity and type 2 diabetes mellitus in the development of male obesity associated secondary hypogonadism. *Clin. Endocrinol., 78*, 330–337.

Safe, S., (2013). Endocrine disruptors and falling sperm counts: Lessons learned or not! *Asian. J. Androl., 15*, 191–194.

Safe, S., Bandiera, S., Sawyer, T., Robertson, L., Safe, L., Parkinson, A., Thomas, P. E., et al., (1985). A. PCBs: Structure-function relationships and mechanism of action. *Environ. Health Perspect., 60*, 47–56.

Schecter, A., Päpke, O., Harris, T. R., Tung, K. C., Musumba, A., Olson, J., & Birnbaum, L., (2006). Polybrominated diphenyl ether (PBDE) levels in an expanded market basket survey of us food and estimated PBDE dietary intake by age and sex. *Environ. Health Perspect., 114*, 1515–1520.

Schell, L. M., Gallo, M. V., & Ravenscroft, J., (2009). Persistent organic pollutants and anti-thyroid peroxidase levels in Akwesasne mohawk young adults. *Environ. Res., 109*, 86–92.

Schreiber, G., Southwell, B. R., & Richardson, S. J., (1995). Hormone delivery systems to the brain: Transthyretin. *Exp. Clin. Endocrinol. Diabetes., 103*, 75–80.

Seegal, R. F., Fitzgerald, E. F., Hills, E. A., Wolff, M. S., Haase, R. F., Todd, A. C., Parsons, P., et al., (2011). Estimating the half-lives of PCB congeners in former capacitor workers measured over A 28-year interval. *J. Expo. Sci. Environ. Epidemiol.*, 234–246.

Shanahan, C. E., Spak, S. N., Martinez, A., & Hornbuckle, K. C., (2015). Inventory of PCB in Chicago and opportunities for reduction in airborne emissions and human exposure. *Environ. Sci. Technol., 49*, 13878–13888.

Shi, H., Jan, J., Hardesty, J. E., Falkner, K. C., Prough, R. A., Balamurugan, A. N., Mokshagundam, S. P., Chari, S. T., & Cave, M. C., (2019). Polychlorinated biphenyl exposures differentially regulate hepatic metabolism and pancreatic function: Implications for nonalcoholic steatohepatitis and diabetes. *Toxicol. Appl. Pharmacol., 363*, 22–33.

Shi, X., Wahlang, B., Wei, X., Yin, X., Falkner, K. C., Prough, R. A., Kim, S. H., et al., (2012). Metabolomic analysis of the effects of polychlorinated biphenyls in nonalcoholic fatty liver disease. *J. Proteome. Res., 11*, 3805–3815.

Sifakis, S., Androutsopoulos, V. P., Tsatsakis, A. M., & Spandidos, D. A., (2017). Human exposure to endocrine disrupting chemicals: Effects on the male and female reproductive systems, *Environ. Toxicol. Pharmacol.*, 56–70.

Sifakis, S., Androutsopoulos, V. P., Tsatsakis, A. M., & Spandidos, D. A., (2017). Human exposure to endocrine disrupting chemicals: Effects on the male and female reproductive systems. *Environ. Toxicol. Pharmacol., 51*, 56–70.

Singh, K., & Chan, H. M., (2017). Persistent organic pollutants and diabetes among Inuit in the Canadian arctic. *Environ. Int., 101*, 183–189.

Sipka, S., Eum, S. Y., Son, K. W., Xu, S., Gavalas, V. G., Hennig, B., & Toborek, M., (2008). Oral administration of PCBs induces proinflammatory and prometastatic responses. *Environ. Toxicol. Pharma., 25*, 251–259.

Slater, H., Gouin, T., & Leigh, M. B., (2011). Assessing the potential for rhizoremediation of PCB contaminated soils in northern regions using native tree species. *Chemosphere, 84*, 199–206.

Souteri, Smith, K. W., Dimitriadis, I., Ehrlich, S., Williams, P. L., & Calafat, A. M., (2013). The association of bisphenol-A urinary concentrations with antral follicle counts and other measures of ovarian reserve in women undergoing infertility treatments. *Reprod. Toxicol., 42*, 224–231.

Southwell, B. R., Duan, W., Alcorn, D., Brack, C., Richardson, S. J., Kohrle, J., & Schreiber, G., (1993). Thyroxine transport to the brain: Role of protein synthesis by the choroid plexus. *Endocrinology., 133*, 2116–2126.

Sowers, K. R., & May, H. D., (2013). *In situ* treatment of PCBs by anaerobic microbial dichlorination in aquatic sediment: Are we there yet? *Curr. Opin. Biotechnol., 24*, 482–488.

Steenland, K., Piacitelli, L., Deddens, J., Fingerhut, M., & Chang, L. I., (1999). Cancer, heart disease, and diabetes in workers exposed to 2,3,7,8-tetrachlorodibenzo-p-dioxin. *J. Natl. Cancer Inst., 91*, 779–786.

Sugantha, P. E., Sathish, K. T., Balaji, S., Bavithra, S., Raja, S. P., Sakthivel, D., Ravi, S. B., & Arunakaran, J., (2017). Lactational exposure effect of polychlorinated biphenyl on rat Sertoli cell markers and functional regulators in prepuberal and puberal F1 offspring. *J. Endocrinol. Invest., 4013*, 91–100.

Sugiura-Ogasawara, M., Ozaki, Y., Sonta, S., Makino, T., & Suzumori, K., (2005). Exposure to bisphenol A is associated with recurrent miscarriage. *Human Reprod., 20*, 2325–2329.

Taylor, S. R., Meadowcraft, L. M., & Williamson, B., (2015). Prevalence, pathophysiology, and management of androgen deficiency in men with metabolic syndrome, type 2 diabetes mellitus, or both. *Pharmacother.*, *35*, 780–792.

Totten, L. A., Panangadan, M., Eisenreich, S. J., Cavallo, G. J., & Fikslin, T. J., (2006). Direct and indirect atmospheric deposition of PCBs to the Delaware River watershed. *Environ. Sci. Technol., 40*, 2171–2176.

Trasande, L., Zoeller, R. T., Hass, U., Kortenkamp, A., Grandjean, P., Myers, J. P., Digangi, J., et al., (2015). Estimating burden and disease costs of exposure to endocrine-disrupting chemicals in the European Union. *J. Clin. Endocrinol. Metab.*

Trudel, D., Scheringer, M., Von, G. N., & Hungerbühler, K., (2011). Total consumer exposure to polybrominated diphenyl ethers in North America and Europe. *Environ. Sci. Technol.*, *45*, 2391–2397.

Tu, C., Teng, Y., Luo, Y., Li, X., Sun, X., Li, Z., Liu, W., & Christie, P., (2011). Potential for biodegradation of polychlorinated biphenyls (Pcbs) By sinorhizobium meliloti. *J. Hazar. Materi., 186*, 1438–1444.

Turyk, M., Anderson, H., Knobeloch, L., Imm, P., & Persky, V., (2009). Organochlorine exposure and incidence of diabetes in a cohort of great lakes sport fish consumers. *Environ. Health Perspec., 117*, 1076–1082.

Tyagi, M., Dafonseca, M. M. R., & De Carvalho, C. C., (2011). Bioaugmentation and biostimulation strategies to improve the effectiveness of bioremediation processes. *Biodegradation, 22*, 231–241.

Valera, B., Ayotte, P., Poirier, P., & Dewailly, E., (2013). associations between plasma persistent organic pollutant levels and blood pressure in Inuit adults from Nunavik. *Environ. Int., 59*, 282–289.

Vandenberg, L. N., Colborn, T., Hayes, T. B., Heindel, J. J., Jacobs, Jr. D. R., Lee, D. H., Shioda, T., et al., (2012). Hormones and endocrine-disrupting chemicals: Low-dose effects and nonmonotonic dose responses. *Endocr. Rev., 33*, 378–455.

Vasiliu, O., Cameron, L., Gardiner, J., Deguire, P., & Karmaus, W., (2006). Polybrominated biphenyls, polychlorinated biphenyls, body weight, and incidence of adult-onset diabetes mellitus. *Epidemiol.,* 352–359.

Vasiliu, O., Cameron, L., Gardiner, J., Deguire, P., & Karmaus, W., (2006). Polybrominated biphenyls, polychlorinated biphenyls, body weight, and incidence of adult-onset diabetes mellitus. *Epidemiology., 17*, 352–359.

Verboven, N., Verreault, J., Letcher, R. J., Gabrielsen, G. W., & Evans, N. P., (2010). Adrenocortical function of arctic-breeding glaucous gulls in relation to persistent organic pollutants. *Gen. Comp. Endocrinol., 166*, 25–32.

Vested, A., Giwercman, A., Bonde, J. P., & Toft, G., (2014). Persistent organic pollutants and male reproductive health. *Asian J. Androl., 16*, 71–80.

Wahlang, B., Beier, J. I., Clair, H. B., Bellis-Jones, H. J., Falkner, K. C., McClain, C. J., & Cave, M. C., (2013). Toxicant-associated steatohepatitis. *Toxicol. Pathol., 41*, 343–360.

Wahlang, B., Falkner, K. C., Gregory, B., Ansert, D., Young, D., Conklin, D. J., Bhatnagar, A., et al., (2013). Polychlorinated biphenyl 153 is a diet dependent obesogen that worsens nonalcoholic fatty liver disease in male C57bl6/J mice. *J. Nutr. Biochem., 24*, 1587–1595.

Wahlang, B., Jin, J., Hardesty, J. E., Head, K. Z., Shi, H., Falkner, K. C., Prough, R. A., et al., (2019). Sex differences arising from polychlorinated biphenyl exposures in toxicant-associated liver disease. *Food Chem. Toxicol., 129*, 64–76.

Wahlang, B., Perkins, J. T., Petriello, M. C., Hoffman, J. B., Stromberg, A. J., & Hennig, B., (2017). A compromised liver alters polychlorinated biphenyl-mediated toxicity. *Toxicol., 380*, 11–22.

Wahlang, B., Song, M., Beier, J. I., Falkner, K. C., Al-Eryani, L., Clair, H. B., Prough, R. A., et al., (2014). Evaluation of Aroclor 1260 exposure in a mouse model of diet-induced obesity and non-alcoholic fatty liver disease. *Toxicol. Appl. Pharmacol., 279*, 380–390.

Walkowiak, J., Wiener, J. A., Fastabend, A., Heinzow, B., Kramer, U., Schmidt, E., Steingruber, H. J., et al., (2001). Environmental exposure to polychlorinated biphenyls and quality of the home environment: Effects on psycho development in early childhood. *Lancet, 358*, 1602–1607.

Wang, S. L., Su, P. H., Jong, S. B., Guo, Y. L., Chou, W. L., & Päpke, O., (2005). In utero exposure to dioxins and polychlorinated biphenyls and its relations to thyroid function and growth hormone in newborns. *Environ. Health Perspect., 113*, 1645–1650.

Watts, R. J., (1998). *Hazardous Wastes: Sources, Pathways, Receptors*. New York: John Wiley and Sons Inc.

Wilson, J., Berntsen, H. F., & Zimmer, K. E., (2016). Do persistent organic pollutants interact with the stress response? Individual compounds, and their mixtures, interaction with the glucocorticoid receptor. *Toxicol. Lett., 241*, 121–132.

Windham, G. C., Pinney, S. M., Sjodin, A., Lum, R., Jones, R. S., Needham, L. L., Biro, F. M., et al., (2010). Body burdens of brominated flame retardants and other persistent organo-halogenated compounds and their descriptors in us girls. *Environ. Res., 110*, 251–257.

Wolska, L., Mechlinska, A., Rogowska, J., & Namiesnik, J., (2012). Sources and fate of PAHs and PCBs in the marine environment. *Crit. Rev. Environ. Sci. Technol.*, 1172–1189.

Wong, P. W., Brackney, W. R., & Pessah, I. N., (1997). Ortho-substituted polychlorinated biphenyls alter microsomal calcium transport by direct interaction with ryanodine receptors of mammalian brain. *J. Biol. Chem.*, 272, 15145–15153.

Wu, H., Yu, W., Meng, F., Mi, J., Peng, J., Liu, J., Zhang, X., Hai, C., & Wang, X., (2017). Polychlorinated biphenyls-153 induces metabolic dysfunction through activation of Ros/Nf-Kappab signaling via downregulation of Hnf1b. *Redox Biol., 12*, 300–310.

Wu, J., Hu, J. C., Wang, S. J., Xi, J. J., Wang, R., Wang, Y., & Jin, J., (2018). Levels, sources, and potential human health risks of PCNs, PCDD/FS and PCBs in an industrial area of Shandong province, China, *Chemosphere*, 382–389.

Wyrwicka, A., Steffani, S., & Urbaniak, M., (2014). The effect of PCB-contaminated sewage sludge and sediment on metabolism of cucumber plants (*Cucumis sativus* L.). *Ecohydr. Hydrobio., 14*, 75–82.

Xi, Z., Fang, L., Xu, J., Li, B., Zuo, Z., Lv, L., & Wang, C., (2019). Exposure to Aroclor 1254 persistently suppresses the functions of pancreatic B-cells and deteriorates glucose homeostasis in male mice. *Environ. Pollut., 249*, 822–830.

Xia, J. W., Tan, S. J., Zhang, X. L., Jun, T., Sun, X. K., & Ling, W., (2015). Correlation of serum testosterone with insulin resistance in elderly male type 2 diabetes mellitus patients with osteoporosis. *J. Diabetes Investig., 6*, 548–552.

Xing, B., & Pignatello, J. J., (1997). Dual mode sorption of low polarity compounds in glassy poly (vinylchloride) and soil organic matter. *Environ. Sci. Technol., 31*, 792–799.

yang, h. y., Zhuo, s. s., xue, b., zhang, c. l., & liu, w. p., (2012). Distribution, historical trends and inventories of polychlorinated biphenyls in sediments from Yangtze River estuary and adjacent East China Sea. *Environ. Pollut., 169*, 20–26.

Yeap, B. B., Chubb, S. P., Hyde, Z., Jamrozik, K., Hankey, G. J., Flicker, L., & Norman, P. E., (2009). Lower serum testosterone is independently associated with insulin resistance in non-diabetic older men: The health in men study. *European J. Endocrinol., 161*, 591.

Yesildaglar, N., Yildirim, G., Attar, R., Karateke, A., Ficicioglu, C., & Yilmaz, B., (2010). Exposure to industrially polluted water resulted in regressed endometriotic lesions and enhanced adhesion formation in a rat endometriosis model: A preliminary study. *Fertil. Steril., 93*, 1722–1724.

Yilmaz, B., Canpolat, S., Sandal, S., Akpolat, N., Kutlu, S., & Ilhan, N., (2006). Paint thinner exposure inhibits testosterone synthesis and secretion in a reversible manner in the rat. *Reprod. Toxicol., 22*, 791–796.

Yilmaz, B., Gilmore, D. P., & Wilson, C. A., (1998). Effects of DPDPE (a specific delta-opioid receptor agonist) and naloxone on hypothalamic monoamine concentrations during the pre-ovulatory Lh surge in the rat. *Eur. J. Endocrinol., 139*, 546–551.

Yilmaz, B., Gilmore, D., & Wilson, C. A., (1996). Inhibition of the preovulatory Lh surge in the rat by central noradrenergic mediation: Involvement of an anaesthetic (urethane) and opioid receptor agonists. *Biogenic Amines., 12*, 423–435.

Yilmaz, B., Kutlu, S., Canpolat, S., Sandal, S., Ayar, A., & Mogulkoc, R., (2001). Effects of paint thinner exposure on serum LH, FSH and testosterone levels and hypothalamic catecholamine contents in the male rat. *Biol. Pharm. Bull., 24*, 163–166.

Yilmaz, B., Sandal, S., & Andcarpenter, D. O., (2012). PCB9 exposure induces endothelial cell death while increasing intracellular calcium and ROS Levels. *Environ. Toxicol., 27*, 185–191.

Younossi, Z. M., Koenig, A. B., Abdelatif, D., Fazel, Y., Henry, L., & Wymer, M., (2016). Global epidemiology of non-alcoholic fatty liver disease-meta-analytic assessment of prevalence, incidence and outcomes. *Hepatology., 64*, 73–84.

Zachow, R., & Uzumcu, M., (2006). The methoxychlor metabolite,2,2-bis-(hydroxyphenyl)-1,1,1-trichloroethane, inhibits steroidogenesis in rat ovarian granulosa cells *In vitro*. *Reprod. Toxicol., 22*, 659–665.

Zeng, S., Gan, N., Mera, R. W., Cao, Y., Li, T., & Sang, W., (2013). Enrichment of polychlorinated biphenyl 28 from aqueous solutions using Fe_3O_4 grafted graphene oxide. *Chem. Eng. J.,* 108–115.

Zhang, Q., Guo, Y., Mengyu, H., Hui, L., & Cheng, G., (2015). Degradation of selected polychlorinated biphenyls by montmorillonite clay-templated Fe^0/Ni^0 bimetallic system. *Chem. Eng. J.*, 122–129.

Zhang, S., Wu, T., Chen, M., Guo, Z., Yang, Z., Zuo, Z., & Andwang, C., (2015). Chronic exposure to Aroclor 1254 disrupts glucose homeostasis in male mice via inhibition of the insulin receptor signal pathway. *Environ. Sci. Technol., 49*, 10084–10092.

Zhang, Y., Guo, H., Mengyu, L., & Hui, G., (2015). Degradation of selected polychlorinated biphenyls by montmorillonite clay-templated Fe^0/Ni^0 bimetallic system. *Chem. Eng. J.,* 122–129.

Zhou, J. L., Maskaoui, K., Qiu, Y. W., Hong, H. S., & Wang, Z. D., (2001). Polychlorinated biphenyl congeners and organochlorine insecticides in the water column and sediments of Daya bay, China. *Environ. Pollut., 113*, 373–384.

Zhu, N. M., Li, Y., & Zhang, F. S., (2011). Catalytic dechlorination of polychlorinated biphenyls in subcritical water by Ni/Fe nanoparticles. *Chem. Eng. J.,* 919–925.

Zhu, N. M., Wang, C. F., & Zhang, F. S., (2012). An integrated two-stage process for effective dechlorination of polychlorinated biphenyls in subcritical water in the presence of hydrogen donors. *Chem. Eng. J.*, 135–142.

Zimmer, K. E., Gutleb, A. C., Lyche, J. L., Dahl, E., Oskam, I. C., Krogenæs, A., Skaare, J. U., & Ropstad, E., (2009). Altered stress-induced cortisol levels in goats exposed to polychlorinated biphenyls (PCB 126 and PCB 153) during fetal and postnatal development. *J. Toxicol. Environ. Health. A.*, *72*, 164–172.

CHAPTER 9

Effect of Bisphenol-A (BPA) on the Endocrine System

BINDUSHREE BHAGHEL

School of Studies in Life Science, Pt. Ravishankar Shukla University, Raipur, Chhattisgarh, India, E-mail: bshri06@gmail.com

ABSTRACT

Bisphenol-A (BPA) is a synthetic organic chemical classified to the group of phenols, which is widely used for manufacturing plastics, paints, lacquers, binding material, and filling material. Recently BPA has been banned from use as BPA is an acute BPA is considered to be an acute endocrine toxicant. The aim of the present review chapter we will discuss about BPA as an endocrine disruptor (ED), properties of BPA, uses of BPA, production of BPA, mechanism of BPA in endocrine disruption, and effects of BPA on animals and humans. BPA is emitted into the environment through diffuse sources or through point sources. Once BPA is absorbed into the body of an organism, it can result in developmental and metabolic disorders, and in specific, it alters endocrine functioning. The alteration in the functioning of the endocrine system alters the physiological, biochemical, genetic, neurological, histopathological, and hematological pathways of the cells/tissues or organs of the body, which in turn interferes with the growth, development, and reproduction of the organisms. It's very difficult to predict the action mechanism of BPA, though it's accepted that BPA interferes with the activity in receptors of estrogen androgen, thyroid, and hydrocarbon. But BPA can also interfere the endocrine functioning through epigenetic mechanisms or oxidative stress. In the current state of research, BPA prime targets and removal techniques from the environment should be extensively studied, and

Environmental Endocrine Toxicants: Biology, Effects, and Management. Younis Ahmad Hajam, Rouf Ahmad Bhat, Khalid Rehman Hakeem, & Rajesh Kumar (Eds.)
© 2023 Apple Academic Press, Inc. Co-published with CRC Press (Taylor & Francis)

prior strategies should be suggested. Furthermore, people should be made aware about the deleterious effects of BPA so that its harmful effects can be minimized.

9.1 INTRODUCTION

Over the last 10 decades, a tremendous increase in synthetic chemicals has been observed. Many chemicals across these synthetic chemicals have been released into the environment by different anthropogenic activities. Such a chemical revolution across the world has changed the structure and functioning of the ecosystem in a way that has impacted animals. Rachel Carson's book entitled "Silent Springs," which was published in 1962, louds the alarm warning regarding the contamination of the environment by different chemicals, especially DDT, which might cause a reduction in the bird population due to failure of reproduction caused by these toxic chemicals. Exposure of these synthetic chemical is lifelong and cause toxicity to different life forms. Animals and humans exposed to these chemicals have to take a lifelong burden; it depends upon the concentration of these chemicals taken and the duration of exposure. Among all these chemicals, some have a direct effect, and others influence the normal functioning of animals and humans through dysfunctioning the endocrine system. Those chemicals which alter the functioning of the endocrine system are called endocrine disruptors (EDs).

EDs are exogenous chemical that mimics and alter the functioning of endocrine hormones, and longtime exposure to EDs adversely affects the structure and functioning of cells/tissues or organs of animals, including human beings (Colborn et al., 1993). Since the endocrine system controls over other systems of the body through the glands that release hormones. These hormones act on target tissue or cells through various transduction pathways like receptors. The signaling of endocrine system usually occurs in a feedback mechanism in which the central nervous system (CNS) initiates the signaling through the hypothalamus (Yorran et al., 2019). Hormone signaling acts by a hormone binding to a specific nuclear receptor transcription factor. Under normal conditions, endocrine system acts normally. In contrast, the interference of EDs disturbs the synthesis, signaling, transport, and gene expression of these vital hormones. The exposure of living organisms to EDs at some critical times in the life span of an organism could have acute detrimental effects on every system, which requires hormonal signal to function normally. The behavioral alteration, histological alterations, reproductive problems, genotoxicity, early puberty, and cancer are some of the health problems that have been correlated with the acute exposure of EDs (Birnbaum and Fenton, 2003; Safe, 2004). EDs don't only mimic the action of a hormone but also inhibit the hormone action when exposed to these exogenous chemicals, thus raising a concern about the severe impacts of EDs in the environment on a variety of life forms (Kloas, 2002).

EDs are persistent and bioaccumulative. On testing for the presence of EDs on different organisms, the results constantly consistently demonstrate a variety of EDs worldwide. We can see and observe the EDs exposure remains high at the places in which industrial chemicals (EDs produced from industries) directly reaches in the water and soil. Then these EDs are assimilated by lower microorganisms, plants, and algae which further move into the animal kingdom through a process in which animals eat the plants and higher animals eat the smaller animals. This is a classic example that shows the EDs movements from one trophic level to another trophic level in an ecosystem. Since animals are placed at the top of the food chain and thus it's reported that these animals including have the highest concentration of EDs in their tissues. General pathways of contact with EDs occur through water, food, skin absorption, and from the atmosphere. Further, the measures of EDs body burdens doesn't reflect only the contemporary contact of animals with EDs, but they also include past exposure, which might have happened decades ago, to persist these chemicals. Once these EDs pass into the human body they start degradation of tissues and alter the functioning of cells. In human the EDs especially make their presence on the prime target organs like liver, kidney, reproductive organs, pancreas, breasts, etc. Furthermore, it's often reported that these exogenous chemicals (EDs) have a desperate response at different life stages, which depend on the physiological hormone concentration, challenging the current risk assessment methodologies, which aren't in consonance with the changes in different life stages. Due to this reason EDs produce adverse effects on growth, development, neurological disorder, genetic alteration, immune stress, and abnormal growth pattern. In the current context, it has been well-recognized that there are five major classes of EDs like:

- Estrogens (anti);
- Androgens (anti);
- Progestins (anti);
- Aryl hydrocarbon receptor agonists; and
- Thyroid hormone disruptor.

Across these EDs; the estrogenic activity, androgenic activity, and thyroid activity is mediated by their specific nuclear receptors. However, previous reports also suggest other action mechanisms like steroidogenic enzymes inhibition, interaction with globulins, binding to membrane receptors, and/or other nuclear receptors can't be excluded (Zacharewski et al., 2002). EDs can act directly by interfering with some stages of action mechanism or normal pathway of hormonal action. EDs can also affect the endocrine system by alteration of synthesis, secretion, transport, uptake, metabolism of normal hormones (Bunce et al., 2000). Across, all of the EDs, the EDs which act directly have received great attention, due to the influence of these EDs in reproduction, growth, behavior, physiology, histology, hematology, etc. The severe concern of EDs is that they might be harmful below the lethal level and the exposure to EDs at the critical period of growth and development may cause irreversible adverse effects in adults (Zacharewski et al., 2002; Kime, 1999). It has been reported that there are lacks of manufactured chemicals, of which thousands are regarded as EDs. A short list representative EDs and with their pathway of production is shown in Table 9.1. There are many other EDs whose influence is reported as EDs but they have very low effects. Across the thousands of EDs, a large body of data has revealed that bisphenol (BP) is the most harmful one, and human beings have to face this ED on a daily basis.

Bisphenols are different components of epoxy resins which are enormously used in thermal paper, filling, paints, and lining of containers as well as water pipes. Chemically Bisphenols are a group of different chemicals with two hydroxyphenyl functionalities. There are a number of derivatives of Bisphenols as EDs. Most of them are based on diphenylmethane. In these chemicals, Bisphenols is a common name, but the letter following to the Bisphenol refers to one of the reactants. A wide array of Bisphenols regarded as environmental contaminants are given in Table 9.2.

Across all these types of bisphenols, bisphenol-A (BPA) is the most popular and toxic ED. That is why BPA is often simply called "Bisphenol" only, which represents the whole group of Bisphenols. In the present review, we will discuss about Bisphenol-A (BPA) and its effects as an ED.

9.2 BISPHENOL-A (BPA)

Bisphenol-A (BPA) is regarded as a chemical of high production volume, which is used for manufacturing of polycarbonate plastics, epoxy resin,

TABLE 9.1 Some Common Endocrine Disruptors (EDs)

Sl. No.	Endocrine Disruptors (EDs)	EDs Pathway of Production
1.	Bisphenol-A (BPA)	BPA is used to make polycarbonate plastics and epoxy resin which is found in different plastic products and storage containers.
2.	Dioxins	These are used as byproducts of different manufacturing process like herbicide production, bleaching of paper, waste burning and wildfires.
3.	PFAS (per- and poly-fluoroalkyl substances)	These are basically large group of chemicals used in industrial applications like freighting foam, nonstick pans, and textile coating.
4.	Phthalates	These chemicals are used to make plastics more flexible. They are found in food packing cosmetics, toys, medical devices, and perfumes.
5.	Phytoestrogens	These are naturally occurring chemicals found in plants. They have hormone-like activities like genistein and daidzein in soy products.
6.	Polybrominated diphenyl ether (PBDE)	These chemicals are used to make flame retardants for different products like furniture foam and carpet.
7.	Polychlorinated biphenyls (PCBs)	These chemicals are used to make electrical equipment like transformers. They are also present in hydraulic fluids, heat transfer fluids, lubricants, and plasticizers.
8.	Pesticides	Majority of fungicides, insecticides, herbicides, rodenticides, plant growth hormones and then bio-pesticides
9.	Triclosan	Used as an antibacterial

TABLE 9.2 Different Types of Bisphenols with Their Structural Formula, Chemical Abstract Service (CAS) and Reactants

Name	Structural Formula	CAS	Reactant	
Bisphenol-A (BPA)	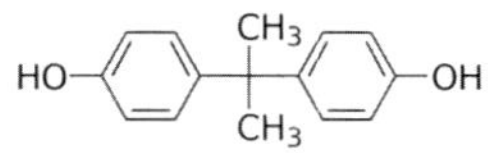	80-05-7	Phenol	Acetone
Bisphenol-AP (BPAP)	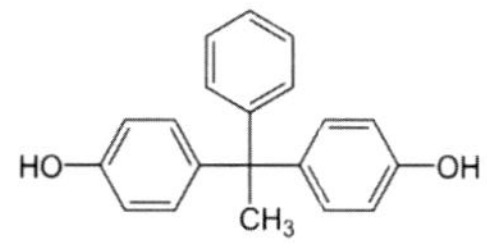	1571-75-1	Phenol	Acetophenone
Bisphenol-AF (BPAF)	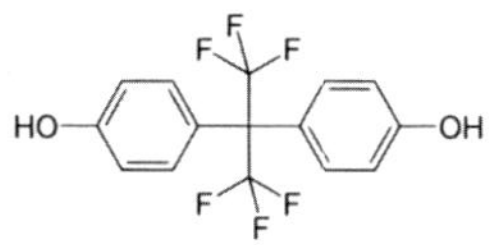	1478-61-1	Phenol	Hexafluoroacetone
Bisphenol-B (BPB)	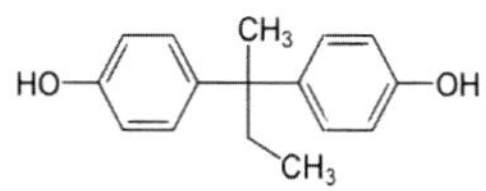	77-40-7	Phenol	Butanone
Bisphenol-BP (BPBP)	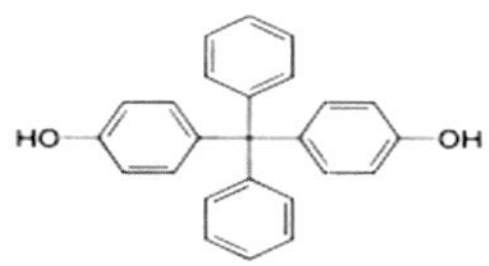	1844-01-5	Phenol	Benzophenone
Bisphenol-C (BPC)	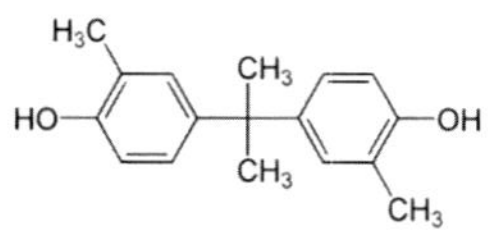	79-97-0	o-cresol	Acetone

TABLE 9.2 *(Continued)*

Name	Structural Formula	CAS	Reactant	
Bisphenol-C2 (BPC)	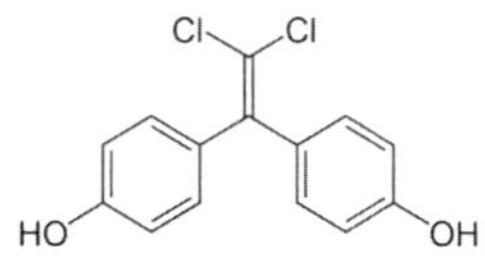	14868-03-02	Phenol	Chloral
Bisphenol-E (BPE)	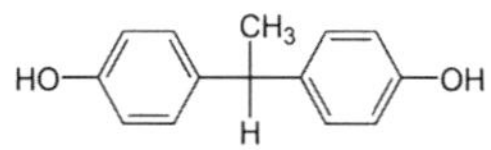	2081-08-5	Phenol	Ethanal
Bisphenol-F (BPF)	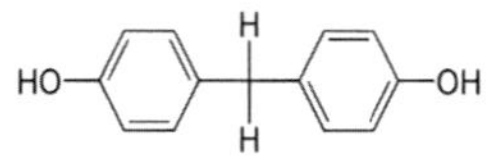	620-92-8	Phenol	Formaldehyde
Bisphenol-G (BPG)	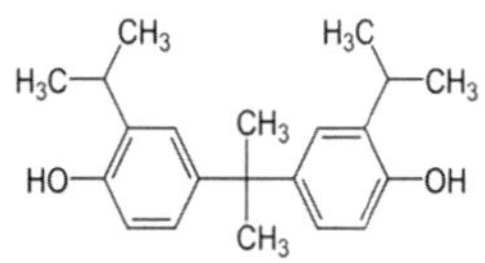	127-54-8	2-isopropyl phenol	Acetone
Bisphenol-M (BPM)	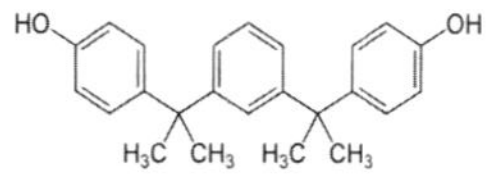	13595-25-0	NA	NA
Bisphenol-S (BPS)	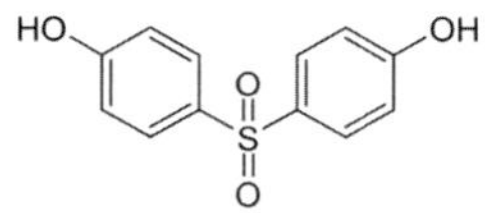	80-09-1	Phenol	Sulfur trioxide
Bisphenol-P (BPP)	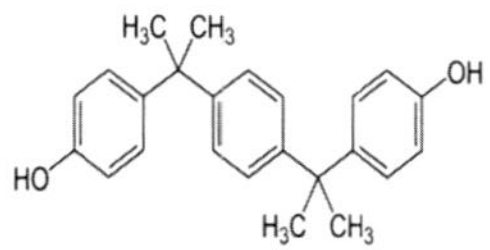	2167-51-3	NA	NA

TABLE 9.2 *(Continued)*

Name	Structural Formula	CAS	Reactant	
Bisphenol-PH (BPPH)	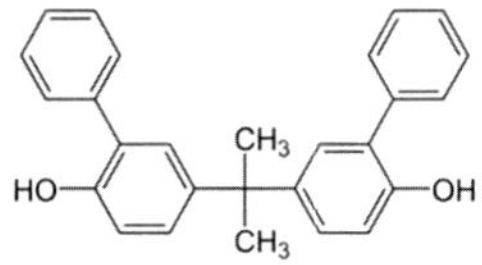	24038-68-4	2-Phenylphenol	Acetone
Bisphenol-TMC (BPTMC)	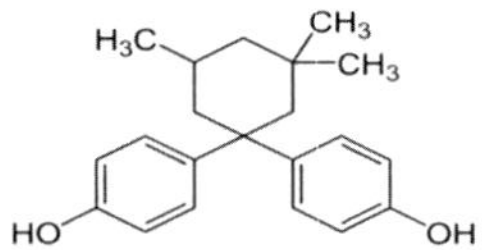	129188-99-4	Phenol	3,3,5-Trimethylcyclohexanone
Bisphenol-Z (BPZ)	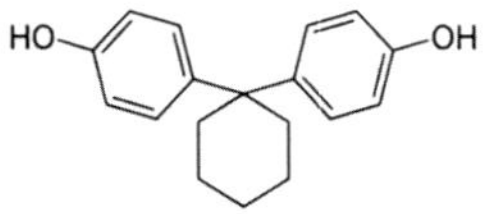	843-55-0	Phenol	Cyclohexanone
Dinitobisphenol-A (DBPA)	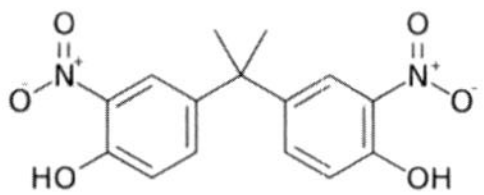	5329-21-5	Bisphenol-A	Nitric acid
Tetrabromobisphenol-A (TBPA)	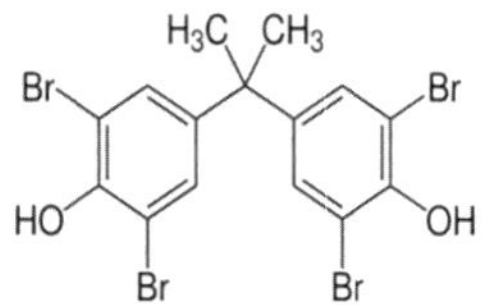	79-94-7	Bisphenol-A	Bromine

and other products such as thermal papers, plastics, and flame retardant (GOC, 2010; NTP, 2008, 2017; Eladak, 2014). Across all bisphenols, BPA is regarded as the most predominant. BPA was initially manufactured in the year 1891 by a Russian chemist Alexander P Dianin "through a condensation reaction of two phenol molecule and one acetone molecule with hydrogen chloride and ion exchange resin acting as a catalyst." This BPA basically remains white solid at room temperature and having a molecular weight of 228.29, and it has a melting point varying from 150°C to 155°C. Due to low vapor pressure of BPA (5×10^{-6} Pa at 20°C), it also vaporizes slightly at room temperature. The use of BPA as a synthetic estrogen was abandoned, as diethylstilbestrol (DES) was determined to have more severe effects on the endocrine system than BPA. Till the year 1940, BPA was potentially used in the industries for the manufacturing plastic containers, painting, furnishing, DVDs, lining, filling, and thermal papers. It was reported that about 8 million tons of BPA was manufactured every year, thus making BPA as a chemical produced at highest volumes across the globe. Till 1940, BPA was extensively utilized in the industrial production of plastic containers, linings of cans, DVDs, and thermal papers. It was estimated that about 8 million tons of BPA was produced, thus making BPA one of the highest volumes of chemicals produced worldwide. Further, it was reported by Grand View Research (2015) that the global demand for BPA was 7 million tons in year 2013 and reached to 9.6 million tons by 2020 (Table 9.3).

9.2.1 USES OF BISPHENOL-A (BPA)

During 1940's BPA started to be used for the production of plastics. The wide use of BPA has made it as one of the most produced chemicals in the world. In this context Grand View Research (2015) that the global demand for BPA was 7 million tons in year 2013 and reached to 9.6 million tons by 2020. The production of BPA was stimulated due to more utilization of this chemical in the following major ways:

- BPA is used as a plasticizer;
- BPA is used for manufacturing polycarbonate plastics, drinking containers, epoxy resins, water pipes, toys, bottles of kids, etc.;
- BPA is used as a hardening agent;
- BPA is used in dental sealant and filling, household electronics, and CDs/DVDs;
- BPA is used in packaging of food, beverage, and storage containers;

- The Polycarbonate plastic and epoxy resins manufactured from BPA are also used in different products like cell phones, household appliances, computers, bicycle helmets, and flooring;
- BPA is used for unsaturated polyester, polysulfone, and polyetherimide.

TABLE 9.3 Properties of Bisphenol-A (BPA)

SL. No.	Bisphenol-A (BPA): A Potential Endocrine Disruptor (EDs)	
1.	Company name	Aldrich (2001)
2.	Synonyms	4,4'-(l-methylethylidene) bisphenol.
		4,4'-isopropylidene diphenol.
		2,2'-bis (4-hydroxyphenyl) propane.
3.	Molecular formula	$C_{15}H_{16}O_2$
4.	CAS No	80-05-7
5.	Purity	99%
6.	Molecular weight	228.29
7.	Chemical constituents	C (78.92%)
		H (7.07%)
		O (14.02%)
8.	Melting point	150°C to 155°C
9.	Boiling point	220°C/4 mm
10.	Physical property	Irritant
11.	Decay	Above 8 mm pressure when heated above 220
12.	Solubility	Insoluble in water. Soluble in aqueous alkaline soluble in alcohol and acetone. But, slightly soluble in CCL_4.
13.	Vapor pressure (Pa)	5×10^{-6} Pa at 20°C
14.	Density (25C/25°C)	1.195

9.2.2 PRODUCTION OF BISPHENOL-A (BPA)

The wide use of BPA has led it to become one of the potential chemicals to be produced across the globe, at approximately 2 million tons per year (Lang et al., 2008). BPA is regarded as one of the highest chemicals used across the world. The increased production and the common use have made BPA a ubiquitous chemical in terms of environmental aspects. The production capacity of BPA in the world was 1 million tons in year 1980's. Grand View Research (2015) that the global demand for BPA was 7 million tons in year 2013 and reached to 9.6 million tons by 2020. In the year 2003, US consumption was 856,000 tons, 72% of which was used to make polycarbonate plastic and 21% going into epoxy resins. In addition, the exponential growth in the construction industries and automobile industry has already increased the market growth of BPA. The reason for increased production and market size of BPA are as follows:

- The increasing demand for BPA in different end-user industries like paints and coating, electronics, adhesive, packaging, and other industries is regarded as one of the major factors responsible for the growth of the BPA market and production.
- Epoxy resins are created by mixing BPA and epichlorohydrin. The properties of BPA like versatility, durability, adhesion, high resistance, toughness, and ease they offer while pouring on cast without forming any bubble, thus BPA are becoming an integral part of various commercial and industrial sectors, which simultaneously increase the market size and production of BPA.
- One of the major applications of BPA is in paint, filling, and coating industry. They are commonly used for construction, shipbuilding, water, and sewage treatments, as well as home appliances, thus increasing the demand for BPA.
- Furthermore, the growing application of epoxy resins in different paint, lubricant, and coating industries has exponentially increased the demand of BPA, and that is, in turn, driving the market and production of this chemical.

9.2.3 BPA MECHANISM OF ACTION: AS AN ENDOCRINE DISRUPTOR (EDS)

The endogenous hormone binds its specific hormone receptor to mediate its action and signaling pathway. EDs have structural similarities with those of endogenous hormones. Thus they interfere with the nuclear receptors of hormones which result in disruption of the functioning of normal hormones. Since these EDs leads to the inhibition in the signaling, or it might increase the receptor activity than those of endogenous hormones leading to increased signaling. The endogenous hormones are regulated by different ways in a body of an organism (Zoeller, 2005; Ghisari and Bonefield-Jorgensen, 2005). BPA is regarded as a potential ED, due to its rate of consumption and production throughout the globe.

BPA has the ability to interfere with the nuclear receptor and affect the optimal functioning of the nuclear receptors, which makes the BPA as most potential and controversial ED. Furthermore, the structural analysis suggests that BA can interfere with the nuclear receptors by copying the action of normal hormone and thus keeping the nuclear receptor in an active conformation. BPA has the ability to affect the endocrine system of an organism in different ways (Figure 9.1). Basically, the action of BPA as an ED can be divided into five categories (Figure 9.1), i.e.:

1. **Mimicking:** BPA that mimic hormones bind to hormone receptor and tricks cell into action. Thus alters the functioning of the Endocrine system. The pathway of mimicking is given in Figure 9.1.

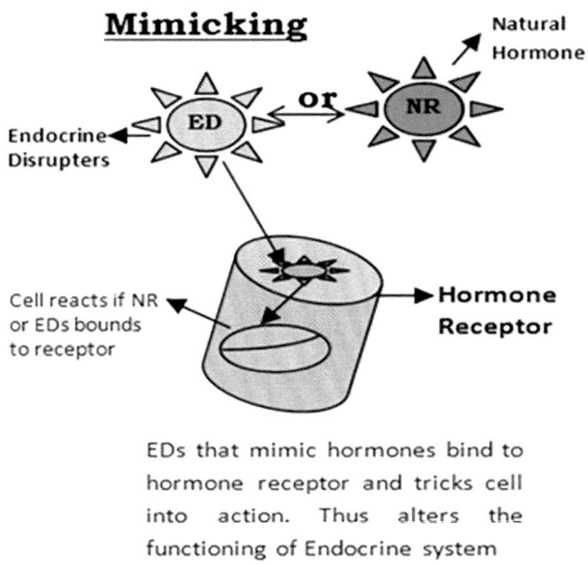

FIGURE 9.1 Pathway showing the mimicking action of BPA.

2. **Blocking:** When BPA is bounded to hormone receptor, the signal pathway isn't turned on. Thus, the working of hormone signaling is stopped, which blocks the functioning of the endocrine system. This pathway is given in Figure 9.2.
3. **Triggering:** Under these conditions, BPA elicit the unusual or abnormal reactions in cells.
4. **Epigenetic:** In these conditions BPA alters the genes, regulating the synthesis, secretion, and transport of hormones. This pathway is explained in Figure 9.3.

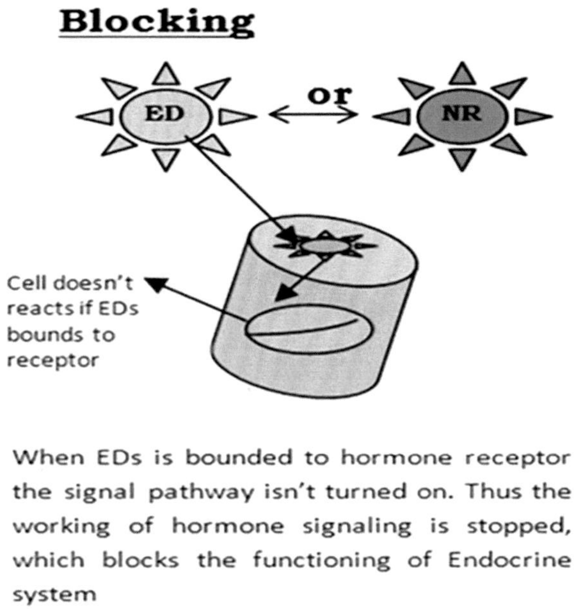

FIGURE 9.2 Figure showing the blocking action of BPA.

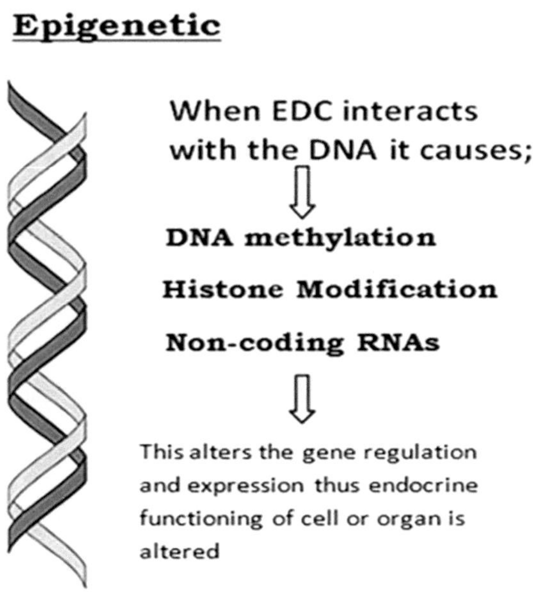

FIGURE 9.3 Figure showing the epigenetic modification caused by BPA.

Further, these actions of BPA disrupt the functioning of the endocrine system in some ways which are discussed in subsections.

9.2.4.1 ALTERING HORMONE METABOLISM

BPA upset the hormone balance by interfering with their metabolism like; synthesis, secretion, or transport from the body (Figure 9.2). The main reason behind alteration of hormone metabolism is that, BPA disrupts endocrine pathway and it also accumulate in different cells, tissues organs.

9.2.3.2 HORMONAL CONTROL OF BPA

The pituitary gland in the brain is the main control center of the hormone system. The pituitary gland receives signal from the hypothalamus, which constantly monitor the level of hormone in the blood of an organism. BPA appear to affect the control of brain on hormone system, either directly through interaction with steroid hormone level, or directly by affecting the activities of brain in other ways (Figure 9.2).

9.2.3.3 AH RECEPTOR

Ah receptor isn't regarded as a hormone receptor, but is known as a sort of receptor in a cell controlling some activities of the endocrine system. BPA has the ability to bind these receptors and triggers many different biological effects, which can disrupt the normal hormone functioning of the endocrine system. It is reported that EDs (especially BPA) can cause anti-estrogenic effects and alter levels of thyroid hormone (Safe and Krishnan, 1995).

9.2.3.4 BPA CAN BIND AND ACTIVATE THE ESTROGEN RECEPTOR WHICH FURTHER ACTS AS AN ESTROGEN

It has been observed that the stimulation of female hormone 17f3-estradiol occurs due to exposure of BPA. BPA has the ability to bind and activate the estrogen receptor (Sumpter and Jobling, 2005). BPA conduct this activity by mimicking the estrogen. Thus, they behave like estrogen and regulate different functions of the endocrine system (Figure 9.4).

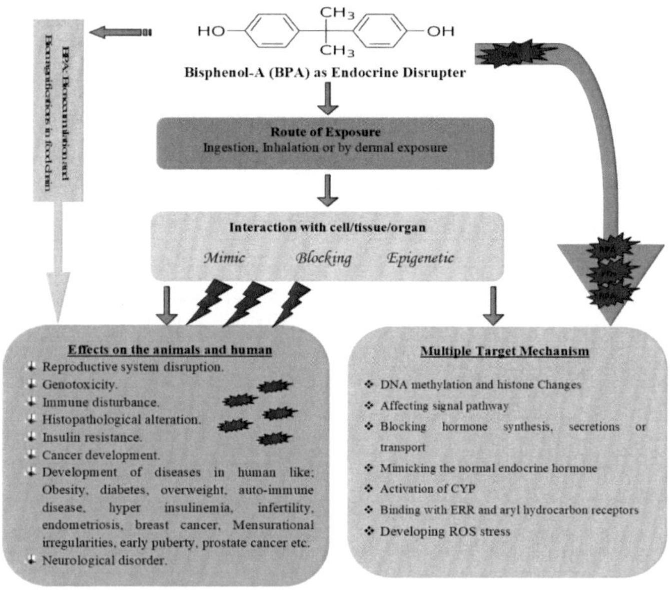

FIGURE 9.4 The route of EDs exposure and their targets.

9.3 BISPHENOL-A (BPA): MULTIPLE ACTION MECHANISM AND EFFECTS ON ANIMAL AND HUMAN BEINGS

9.3.1 BPA HAVE THE PROPERTY OF BINDING BUT NOT ACTIVATING THE ESTROGEN (LIKE AN ANTI-ESTROGEN)

BPA has the ability to bind with the hormone receptor and block them. This prevents the hormones from binding to the receptors and exerting their normal biological affects (Mclachlan, 1993). A few chemicals including BPA, are known by the ability of blocking the male sex hormone "androgen" receptor. Thus, these chemicals are known as "anti-androgenic chemicals."

9.3.2 BINDING WITH OTHER RECEPTORS

There are many other receptors which are involved in the functioning of hormones. The EDs like BPA can block the androgen receptor (AR), i.e., the receptor for the male hormone testosterone. This binding of EDs can either activate or deactivate the receptors (Greim, 2004).

9.3.3 BPA CAN MODIFY THE METABOLISM OF NATURAL HORMONES

BPA can affect the metabolic pathway of estradiol directly or indirectly; thus, more estrogenic metabolites are produced. Thus, producing more estrogenic metabolites. BPA also activate enzymes which in turn speed up the functioning of the natural hormones, so disrupting their natural state. It has been revealed that there are some specific sets of enzymes which metabolize the estrogen. These enzymes break down the estrogen rapidly to form where they can no longer bind their receptor. However, if these enzymes are affected by xenoestrogen this metabolism will be decreased (Toppari et al., 1996).

9.3.4 INHIBITING SYNTHESIS, SECRETION, AND TRANSPORT OF HORMONE

BPA can directly or indirectly affect the natural hormone production and transport by simply interfering with the functioning of the hormone system (Figure 9.2) (Hotckiss et al., 2008).

9.3.5 EPIGENETIC MECHANISM OF BPA AS AN ENDOCRINE DISRUPTOR (ED)

When BPA interacts with DNA of a cell of an organism. It causes DNA methylation, modification of histones and non-coding RNAs. This alters the gene regulation and expression of hormone production, secretion, and transport, thus endocrine functioning of cell or organ is altered.

9.3.6 DEVELOPMENT OF OXIDATIVE STRESS

When BPA enters in cell through the cell membrane, it interacts with the different compounds of the cell and cell membrane. This result in the formation of reactive oxygen species (ROS). The excessive ROS alters the physiological functioning of cell. Such a deviation from normal functioning due to ROS is known as oxidative stress. Once this mechanism happens in the tissues of the endocrine system, it alters the functioning of the endocrine process (Figure 9.5).

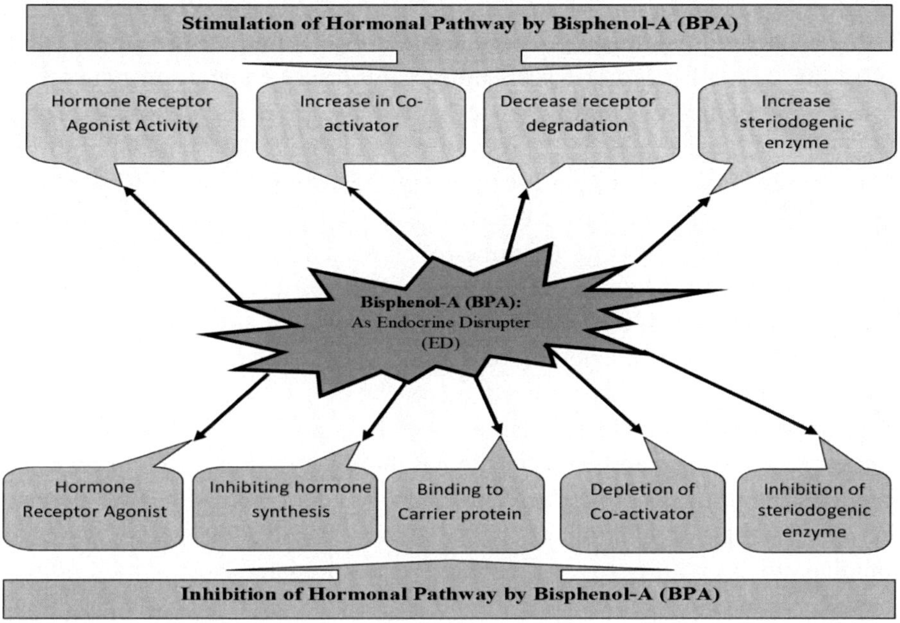

FIGURE 9.5 Bisphenol-A: (BPA): Mechanism of BPA as endocrine disruptor (ED) and alteration of hormonal pathway.

9.4 EFFECTS OF BISPHENOL-A (BPA)

Bisphenol-A (BPA) is a phenolic, synthetic organic and high productive volume (HPV) chemical used to make polycarbonate plastics or epoxy resins. BPA is regarded as a reproductive, developmental, or a systemic toxicant, due to which it's regarded as a potential ED. Due to severe adverse health effects of BPA, it has been banned in some products. Therefore, presence of BPA and its analogs in the environment has recently attracted global attention. Reports on exposure of BPA to human beings revealed that human are exposed to different concentrations of BPA (0.27–10.6 ng/ml), with global estimated intake of BPA was 30.76 ng/kg per body weight per day by dermal exposure, dust inhalation, and ingestion. Further, it was revealed that the major route of BPA exposure to different animals and humans is via ingestion, these accounts for about 90% of BPA exposure. BPA usually leach into the food products and further is ingested orally. Once BPA reaches to the target cell, tissue, or organ of the endocrine system, it alters the normal functioning of specific nuclear receptors even in very low doses. Thus, making BPA as controversial EDs. In addition, structural analysis of BPA suggested that BPA can also mimic the action of natural hormone. Furthermore, BPA act by blocking the cell receptors, preventing naturally occurring hormones from affecting cells in the usual way and lastly BPA alters the genes, regulating the synthesis, secretion, and transport of hormones. BPA has the ability to alter the density of estrogen receptors in respective tissues; this alters the sensitivity of that tissue towards natural estrogen. Since estrogen plays a key role in the development of a number of tissues. Thus, the interference with the estrogen activity during development can result in permanent changes that affect the reproductive process later in the life of an organism.

Since BPA affects the normal functioning of the body of animals including human beings. But in specific, it alters the functioning of the endocrine system, which in turn affects the animals/humans in different ways. It has been revealed by researchers that BPA has the ability to bind with the receptors of estrogen, receptors of aryl hydrocarbon, and peroxisome proliferators activated receptors, and induce an alternative estrogen signaling pathway. The binding of BPA with estrogen receptors has resulted in distortion of hormonal balance. Thus, the mechanism of mimicking, blocking, triggering of hormonal signaling/pathway by BPA results in the inhibition of reproductive activity through modulations in the transcript of steroid hormone and steroidogenic mediating genes.

The BPA exposure disrupts the action of another steroid hormone like testosterone and thyroid hormone. This alteration of reproductive activity is also associated with the epigenetic mechanism induced by BPA caused the alteration in the transcription genes exhibiting sex difference during neonatal periods, which results in decreased expression of sex genes (Eckstrum et al., 2016). The reproductive disorder and impaired behavior in animals and human beings by the exposure of BPA has been proven (Dianin, 1891; vom Saal et al., 2012; Rezg et al., 2014; Sabanayagam et al., 2013). BPA exposure to animals reduced egg quality and other aspects of egg viability in patients seeking fertility treatments (Han and Hong, 2016; Bae et al., 2017). The exposure of BPA also causes sperm-oocyte fusion in human beings.

BPA binds with estrogen Y (ERRY), this receptor is expressed strongly in mammalian brain development and promotes differentiation and maturation of fetal brain disorders (Liu et al., 2014). BPA blocks the androgen-dependent enhancement of dendritic spines in the hippocampus, an effect which suggests BPA may interfere with neural plasticity (Leranth et al., 2008). Reports show that BPA results in inhibitory effects on hormonal adrenocortical function (Feng et al., 2016). In addition, with that BPA can affect the thyroid development and the development of the hypothalamic-pituitary testicular axis by modulating the synthesis of hormones (Chan et al., 2012). Furthermore, BPA cause disruption of thalamus and spleen. By inducing a conformational change in the tertiary ability to recruit coactivator proteins and the inhibition of transcriptional changes in the tertiary ability to recruit coactivator proteins and the inhibition of transcriptional activity induced by triiodothyronine (T3) (Peyre et al., 2014; Monke et al., 2010).

It was reported that BPA increased tumor necrosis factor (TNF α), which was associated with the changes in estrogen-related receptors, further it has been observed that DNA methylation and histone changes might be key events in the mechanism of BPA toxicity (Peyre et al., 2014; Monje et al., 2010). Since, BPA causes DNA methylation changes in neuroendocrine pathway fundamental to reproductive health, energy balance, and behavior, including the estrogen receptor pathway (Dolinoy et al., 2007; Kundakovic et al., 2013; Yeo et al., 2013). Moreover, histopathological alterations were recorded on exposure of BPA depending on the route and dose (Khan and Ahmad, 2015).

BPA exposure to living organisms causes oxidative stress and disruption of the β-cell function (Ozaydın et al., 2018), which result in complication in glucose regulation (Mullainadhan et al., 2017; Wei et al., 2017; Sivashanmugam et al., 2017). BPA exposure discovered that the insulin signaling include insulin phosphorylated extracellular signal-regulated protein kinase (p-ERK) and glucose transporter (GLUT) were significantly decreased also glucose intolerance, hyperinsulinemia, and insulin resistance (IR) were reported to be associated with BPA exposure (Khalil et al., 2014). It was further revealed that BPA exposure to animals including human cause hypertension (Han and Hong, 2016; Bae et al., 2017; Cypher et al., 2015) and it severely affects the cardiovascular system during a hypoxic event (Cypher et al., 2015). BPA affects angiogenesis and causes uncontrolled neovascularization (Basini et al., 2017) and increase in interventricular septal thickness (Kumar et al., 2017). A number of diseases are associated with the BPA exposure (Figure 9.1).

9.5 CONCLUSION

Considerable attention has been attracted by the chemicals that disrupt the endocrine activities. Across all of these chemicals, BPA has received much attention due to its potential of affecting endocrine process. BPA are phenolic, synthetic organic, and HPV chemical used to make polycarbonated plastics and epoxy resins. BPA is a reported to be a reproductive, developmental, and systematic toxicant, due to which its regarded as a potential ED. Due to potential adverse health effects of BPA, it has been banned in some products. Once BPA is absorbed into the body of an organism, it can result in developmental and metabolic disorders, and in specific, it alters the functioning of the endocrine system. The findings of different research suggested that BPA paradigm and alternative signaling of estrogen are dominant contributors to estrogen activity, affecting hormonal regulation. Further elite studies are necessary to substantiate the molecular mechanism of BPA action. A better understanding concerning the presence and mobilization of BPA during different phases of the life cycle of animals, and also development of modern technologies to remove the BPA from the environment. BPA is considered to an acute endocrine toxicant as it interferes with any respect of endocrine hormone action, and now BPA has received attention at the global level as its estrogen effect, persistence, and biomagnifications. BPA is basically a synthetic chemical classified to the group of phenols, which is widely used for manufacturing plastics, paints, lacquers, binding material, and filling material. Once BPA is absorbed into the body of an organism, it alters the reproductive and growth process, behavior, physiology, biochemical activities, genetics, neurological, histopathological, and hematological pathways of the cells/tissues or organs of the body. In the current state of research BPA prime targets and removal techniques from the environment should be extensively studied and prior strategies should be suggested.

KEYWORDS

- **bisphenol-A**
- **chemical abstract service**
- **diethylstilbestrol**
- **endocrine disrupters**
- **metabolic disorders**
- **polychlorinated biphenyls**

REFERENCES

Bae, S. Y., Lim, H., Lee, Y. A., Shin, C. H., Oh, S. Y., & Hong, Y. C., (2017). Maternal urinary bisphenol a concentration during midterm pregnancy and children's blood pressure at age 4Novelty and significance. *Hypertension, 69,* 367–374.

Basini, G., Bussolati, S., Grolli, S., Ramoni, R., & Grasselli, F., (2017). Bisphenol a interferes with swine vascular endothelial cell functions. *Can. J. Physiol. Pharmacol., 95,* 365–371.

Birnbaum, L. S., & Fenton, S. E., (2003). Cancer and developmental exposure to endocrine disruptors. *Environ. Health Perspect., 111,* 389–394.

Cao, J., Mickens, J. A., McCaffrey, K. A., Leyrer, S. M., & Patisaul, H. B., (2012). Neonatal bisphenol A exposure alters sexually dimorphic gene expression in the postnatal rat hypothalamus. *Neurotoxicology., 33,* 23–36.

Chan, W. K., & Chan, K. M., (2012). Disruption of the hypothalamic-pituitary-thyroid axis in zebrafish embryo-larvae following waterborne exposure to BDE-47, TBBPA and BPA, *Aquat. Toxicol., 108,* 106–111.

Colborn, T., Vom, S. F. S., & Soto, A. M., (1993). Developmental effects of endocrine-disrupting chemicals in wildlife and humans. *Environ. Health Perspect., 101,* 378–384.

Cypher, A. D., Ickes, J. R., & Bagatto, B., (2015). Bisphenol A alters the cardiovascular response to hypoxia in Danio rerio embryos. *Comp. Biochem. Physiol. Part C Toxicol. Pharmacol., 174*, 39–45.

Dianin, A. P., (1891). Kondensation von ketonen mit phenolen (Übers. aus d. russ.), zhurnal rus. fiz. obs. *J. Russ. Phys. Chem. Soc., 23,* 601–611.

Dolinoy, D. C., Huang, D., & Jirtle, R. L., (2007). Maternal nutrient supplementation counteracts bisphenol induced DNA hypomethylation in early development. *Proc. Natl. Acad. Sci. USA, 104,* 13056–13061.

Eckstrum, K. S., Weis, K. E., Baur, N. G., Yoshihara, Y., & Raetzman, L. T., (2016). Icam5 expression exhibits sex differences in the neonatal pituitary and is regulated by estradiol and bisphenol A. *Endocrinology, 157*, 1408–1420.

Eladak, S., Grisin, T., Moison, D., Guerquin, M., Tumba, B. T. N., Pozzi-Gaudin, S., Benachi, A., et al., (2014). A new chapter in the bisphenol A story: Bisphenol S and bisphenol F are not safe alternatives to this compound. *Fertil. Steril., 103,* 11–21.

Feng, Y., Jiao, Z., Shi, J., Li, M., Guo, Q., & Shao, B., (2016). *Effects of Bisphenol Analogues on Steroidogenic Gene Expression and Hormone Synthesis in H295R Cells*, *147,* 9–19.

Ghisari, M., & Bonefeld-Jorgensen, E. C., (2009). Effects of plasticizers and their mixtures on estrogen receptor and thyroid hormone functions. *Toxicol. Lett., 189,* 67–77.

Government of Canada, (2010). *Bisphenol A - Order Adding a Toxic Substance to Schedule 1 to the Canadian Environmental Protection Act.* https://www.canada.ca/en/health-canada/services/home-garden-safety/bisphenol-bpa (accessed on 30 August 2022).

Grand View Research, (2015). *Bisphenol A Market Size Worth $20.03 Billion by 2020.* https://www.grandviewresearch.com/press-release/global-bisphenol-a-bpa-market (accessed on 26 October 2022).

Greim, H. A., (2004). The endocrine and reproductive system: Adverse effects of hormonally active substances? *Pediatrics*, *113,* 1070.

Han, C., & Hong, C., (2016). Bisphenol A, hypertension, and cardiovascular diseases: Epidemiological, laboratory, and clinical trial evidence. *Curr. Hypertens. Rep., 18,* 11.

Hotchkiss, A. K., Rider, C. V., Blystone, C. R., Wilson, V. S., Hartig, P. C., Ankley, G. T., Foster, P. M., et al., (2008). Fifteen years after "wingspread"—Environmental endocrine disruptors and human and wildlife health: Where we are today and where we need to go. *Toxicol. Sci., 105,* 235–259.

Khalil, N., Ebert, J. R., Wang, L., Belcher, S., Lee, M., Czerwinski, S. A., & Kannan, K., (2014). Bisphenol a and cardiometabolic risk factors in obese children. *Sci. Total Environ., 470,* 726–732.

Khan, D., & Ahmed, S. A., (2015). Epigenetic regulation of non-lymphoid cells by bisphenol A, a model endocrine disruptor: Potential implications for immunoregulation. *Front. Endocrinol. (Lausanne), 6,* 91.

Kloas, W., (2015). Amphibians as a model for the study of endocrine disruptors. *Int. Rev. Cytol., 216,* 1–57.

Kundakovic, M., Gudsnuk, K., Franks, B., Madrid, J., Miller, R. L., Perera, F. P., & Champagne, F. A., (2013). Sex-specific epigenetic disruption and behavioral changes following low-dose in utero bisphenol A exposure. *Proceedings of the National Academy of Sciences of the United States of America, 110,* 9956–9961.

Leranth, C., Hajszan, T., Szigeti-Buck, K., Bober, J., & MacLusky, N. J., (2008). Bisphenol A prevents the synaptogenic response to estradiol in hippocampus and prefrontal cortex of ovariectomized nonhuman primates. *Proc. Nat. Acad. Sci., 105,* 14187–14191.

Liu, X., Matsushima, A., Shimohigashi, M., & Shimohigashi, Y., (2014). A characteristic back support structure in the bisphenol A-binding pocket in the human nuclear receptor ERR γ. *PLoS One, 9,* 101252.

MacLatchy, D., Peters, L., Nickle, J., & Van, D. K. G., (1997). Exposure to sitosterol alters the endocrine status of goldfish differently than 17p-estradiol. *Environ. Toxicol. Chem., 16,* 1895–1904.

Mohan, K. S. M. J., Rajendran, T. D., Vyas, A. K., Hoang, V., Asirvatham-Jeyaraj, N., Veiga-Lopez, A., Olivier, N. B., et al., (2017). Effects of prenatal bisphenol-A exposure and postnatal overfeeding on cardiovascular function in female sheep. *J. Dev. Orig. Health Dis., 8,* 65–74.

Monje, L., Varayoud, J., Munoz-de-Toro, M., Luque, E. H., & Ramos, J. G., (2010). Exposure of neonatal female rats to bisphenol a disrupts hypothalamic LHRH pre- mRNA processing and estrogen receptor alpha expression in nuclei controlling estrous cyclicity. *Reprod. Toxicol., 30,* 625–634.

Mullainadhan, V., Viswanathan, M. P., & Karundevi, B., (2017). Effect of bisphenol-a (BPA) on insulin signal transduction and GLUT4 translocation in gastrocnemius muscle of adult male albino rat. *Int. J. Biochem. Cell Biol., 90,* 38–47.

NTP, NTP Report on Biological Activity of Bisphenol A (BPA) Structural Analogues and Functional Alternatives, (2017). https://ntp.niehs.nih.gov/ntp/results/pubs/rr/reports/rr04_508.pdf?utm_source¼direct&utm_medium¼prod&utm_campaign¼ntpgolinks&utm_term¼rr04 (accessed on 30 August 2022).

NTP, NTP-CERHR., (2008). *Monograph on the Potential Human Reproductive and Developmental Effects of Bisphenol A.* https://ntp.niehs.nih.gov/ntp/ohat/bisphenol/bisphenol.pdf (accessed on 30 August 2022).

Ozaydın, T., Oznurlu, Y., Sur, E., Celik, I., Uluısık, D., & Dayan, M., (2018). Effects of bisphenol a on antioxidant system and lipid profile in rats. *Biotech. Histochem.,* 1–8.

Peyre, L., Rouimi, P., De Sousa, G., Helies-Toussaint, C., Carre, B., Barcellini, S., Chagnon, R., & Rahmani, M. C., (2014). Comparative study of bisphenol a and its analogue bisphenol s on human hepatic cells: A focus on their potential involvement in nonalcoholic fatty liver disease. *Food Chem. Toxicol., 70,* 9–18.

Rezg, R., El-Fazaa, S., Gharbi, N., & Mornagui, B., (2014). Bisphenol and human chronic diseases: Current evidence, possible mechanisms, and future perspectives. *Environ. Int., 64,* 83–90.

Sabanayagam, C., Teppala, S., & Shankar, A., (2013). Relationship between urinary bisphenol a levels and prediabetes among subjects free of diabetes. *Acta. Diabetol., 50,* 625–631.

Safe, S., & Krishnan, V., (1995). Chlorinated hydrocarbons: Estrogens and antiestrogens. *Toxicol. Lett., 82,* 731–736.

Safe, S., (2004). Endocrine disruptors and human health: Is there a problem. *Toxicology, 205,* 3–10.

Sivashanmugam, P., Mullainadhan, V., & Karundevi, B., (2017). Dose-dependent effect of bisphenol-a on insulin signaling molecules in cardiac muscle of adult male rat. *Chem. Biol. Interact., 266,* 10–16.

Sumpter, J. P., & Jobling, S., (1995). Vitellogenesis as a biomarker for estrogenic contamination of the aquatic environment. *Environ. Health Perspect., 103,* 173–178.

Toppari, J., Larsen, J. C., Christiansen, P., Giwercman, A., Grandjean, P., Guilllette, L. J. Jr., Jegou, B., Jensen, T. K., et al., (1995). Male reproductive health and environmental chemicals with estrogenic effects. *Milijoprojekt., 290,* 166.

Vom, S. F. S., Nagel, S. C., Coe, B. L., Angle, B. M., & Taylor, J. A., (2012). The estrogenic endocrine disrupting chemical bisphenol A (BPA) and obesity, *Mol. Cell. Endocrinol., 354,* 74–84.

Wei, J., Ding, D., Wang, T., Liu, Q., & Lin, Y., (2017). MiR-338 controls BPA-triggered pancreatic islet insulin secretory dysfunction from compensation to decompensation by targeting Pdx-1. *The FASEB Journal, 31*, 5184–5195.

Yeo, M., Berglund, K., Hanna, M., Guo, J. U., Kittur, J., Torres, M. D., Abramowitz, J., et al., (2013). Bisphenol A delays the perinatal chloride shift in cortical neurons by epigenetic effects on the Kcc2 promoter. *Proc. Natl. Acad. Sci. U.S.A, 110*, 4315–4320.

Zoeller, R. T., (2007). Environmental chemicals impacting the thyroid: Targets and consequences. *Thyroid, 17,* 811–817.

CHAPTER 10

Effects of Endocrine Toxicants on the Growth and Metabolism of Thyroid Glands

YOGESH SINGH[1], PREETI KUMARI[1], and ARUP GIRI[2]

[1]*DRDO–Defense Institute of High-Altitude Research (DIHAR), Leh, Ladakh UT–194101, Jammu and Kashmir, India*

[2]*Baba Mastnath University, Asthal Bohar, Rohtak, Haryana–124021, India, E-mail: arupsatadal@gmail.com*

ABSTRACT

Global human health is changing. This changing pattern is caused by a changing lifestyle, contaminated food, altered drinking water quality, etc. All these agents are severe enough to damage normal human physiology after disrupting most of the endocrine glands. The thyroid gland (TG) is responsible for the synthesis of tri-iodothyronine (T3) and tetra-iodothyronine (T4) in humans. These hormones mostly regulate metabolism. This study found that a lot of substances act as thyro-endocrine disruptors (EDs), like bisphenol-A (BPA), flame retardants (polybrominated diphenyl ethers (PBDEs), polychlorinated biphenyl (PCB) (dioxins, furans, polycarbonate plastics, and epoxy resins), tetrabromobisphenol-A (TBBPA), perfluorinated compounds (PFCs) (perfluorooctanoic acid (PFOA) perfluorooctane sulfonate (PFOS) and perfluorononanoate (PFNA)), pharmaceuticals (Phenothiazines, methimazole, propylthiouracil, amiodarone, alemtuzumab, and imatinib), palatable (di-(2-ethylhexyl) phthalate (DEHP)), household materials (benzophenone, benzhydrol, PFOA, octyl methoxycinnamate (OMC), homosalate (HMS), 4-aminobenzoic acid (PABA), nonylphenol

Environmental Endocrine Toxicants: Biology, Effects, and Management. Younis Ahmad Hajam, Rouf Ahmad Bhat, Khalid Rehman Hakeem, & Rajesh Kumar (Eds.)
© 2023 Apple Academic Press, Inc. Co-published with CRC Press (Taylor & Francis)

(NP), triclosan (TCS), and heavy metals (arsenic, lead, cadmium, etc.), are disrupting normal function of TG.

10.1 INTRODUCTION

A large number of substances that are suspected of causing disruptions in endocrine systems are exposed to humans (Street et al., 2018). Most of the research is focused on finding the mechanistic approach after the entry of different endocrine disruptive agents. The hypothalamic-pituitary-thyroid (HPT) axis (T3) is responsible for the regulation and synthesis of TRH, TSH, and T4. This gland is situated in the tracheal region of the neck. It is responsible for the synthesis of T3 and T4 hormone. At first, primarily, T4 hormone secreted from the glands and the most bio-active form is T3 (Koibuchi, 2012).

The concentration of T3 determines the biological activity. This is biologically active and binds to nuclear receptors (TR). The majority of circulated T3 is produced by extra-thyroidal T4 deiodination, which occurs primarily in the liver (Harvey and Williams, 2002). Thyroid disorders are the endocrine system's widespread glandular disorders (Bajaj et al., 2016), including hypothyroidism, hyperthyroidism, goiter, etc. A lot of endocrine disruptors (EDs) like bisphenol-A (BPA), flame retardants (polybrominated diphenyl ethers (PBDEs), tetrabromobisphenol-A (TBBPA)), polychlorinated biphenyl (PCB) (dioxins, furans, polycarbonate plastics, and epoxy resins), perfluorinated compounds (PFCs) (perfluorooctane sulfonate (PFOS), perfluorooctanoic acid (PFOA) and perfluorononanoate (PFNA), pharmaceuticals (Phenothiazines, methimazole, propylthiouracil, amiodarone, alemtuzumab, and imatinib), palatable (di-(2-ethylhexyl) phthalate (DEHP), household materials (Benzophenone, benzhydrol, PFOA, octylmethoxycinnamate (OMC), homosalate (HMS), 4-aminobenzoic acid (PABA), nonylphenol (NP), triclosan (TCS)), heavy metals (arsenic, lead, cadmium, etc.), hampered the thyroid regulation (Zoeller et al., 2007). There is still little understanding about the mechanistic approach of all these thyroid disruptors. This chapter reveals the findings of the mechanistic approach which relies on thyroid disruption.

10.2 BRIEF ABOUT THYROID GLAND (TG)

The highly developed butterfly-shaped bi-lobed glandular tissue evolved from the primitive endostyle of the lower protochordate, affecting almost every cell type in the body. This is situated under the Adams' apple (Figure 10.1).

The lobes are surrounded by the windpipe and they are connected to each other by an isthmus, which is mainly connective tissue. By origin, the thyroid gland (TG) endo-dermally originated. The most abundant follicular cells with varying shapes from round to oval are the primary unit of this gland. They are mainly responsible for hormone synthesis. The gland has another type of cells, i.e., C-cells. Calcitonin hormone is secreted from this type of cells, but their number is very less. The TG produces hormones that regulate energy metabolism through generating downstream effects that control basal metabolic rate (BMR). Therefore, disruptions of these hormones will lead to disturbed homeostasis of the body.

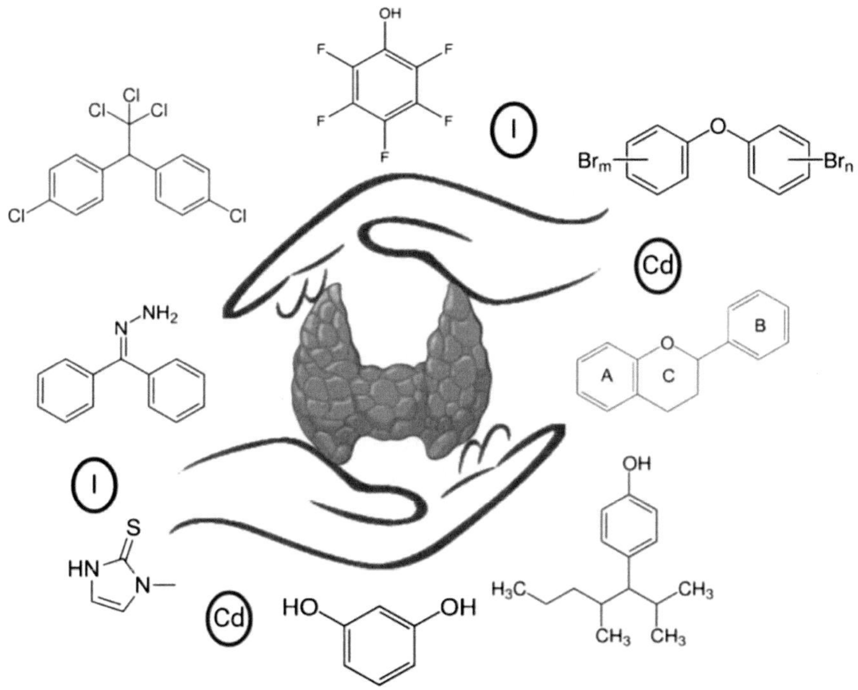

FIGURE 10.1 Thyroid gland and its hormones structure.

10.3 SYNTHESIS OF THYROID HORMONES (THS) AND THEIR METABOLISM

This gland is situated in the tracheal region of the neck. It is responsible for the synthesis of T3 and T4 hormones. At first, primarily, T4 hormone is secreted from the glands, and the most bio-active form is T3 (Koibuchi, 2012).

The synthesis of the thyroid hormones (THs) is revealed by the four stages. At first, the iodine trapping process was completed in the follicular cell. Through diet, we are taking iodine, and that is getting absorbed by the gastrointestinal tract (GI). Furthermore, they are absorbed by the calcium-iodine co-transporter pump. This is located on the basal membrane of the thyro-follicular cells. In the second stage, thyroglobulin will be synthesized in the follicular cells after glycosylation of the highly tyrosine-rich protein in the Golgi complex of follicular cells. Then, in the next stage, absorbed inactive iodine will be oxidized by the thyroperoxidase enzyme (TPO) on the tip of the follicular cells. The oxidized iodine molecules then adjoin with the rings of the tyrosine molecules. The presence of one iodine molecule with the tyrosine ring at 3' position is called mono-iodo-tyrosine (MIT). Meanwhile, the adjoining of two iodine molecules with tyrosine molecules at 3' and 5' is called di-iodo-tyrosine (DIT). Finally, the coupling process will occur between MIT and DIT. T3 will be formed after the coupling of one MIT and one DIT molecule. In the meantime, T4 will be formed after the coupling of two DIT molecules. Finally, during requirements in the body, through the exocytosis process, the synthesized TH will be secreted from the TG. They are normally stored in the TG for at least 3–4 months. During secretion; thyroglobulin will be separated from thyroxine (T4) and T3 hormone. Through the pinocytosis process, the separated thyroglobulin molecules are taken up by the follicular cells for the further synthesis of T4 and T3 hormones. It has been found that T3 is the most active derivative of these hormones (Figure 10.2) (Koibuchi, 2012).

In animal body, most of the cellular metabolic function has been mainly regulated by the THs. This is mainly responsible for the carbohydrate, lipid, and protein metabolism (Mullur et al., 2014).

10.3.1 BROWN ADIPOSE TISSUE (BAT)

The characteristics of brown adipose tissue (BAT) are highly enriched with mitochondria and lipid droplets. Due to its high mitochondrial population, BAT is the most responsible for heat production in the body. The regulation of heat production in BAT is mainly dependent upon the environmental temperature, body weight, and requirements of body energy. It has been found that in hypothermic conditions, this BAT has been induced to produce heat after being up-regulated by T3 hormone. Generally, it has been found that during hypothermic conditions or low environmental temperature conditions, they upregulate the expression of type 2 iodothyronine deiodinase (D2). This is responsible to interchange T4 to T3. In the meantime, this T3 hormone is also responsible for lipid oxidation and lipogenesis (Oppenheimer et al., 1991; de Jesus et al., 2001; Obregon, 2014; Shu et al., 2017).

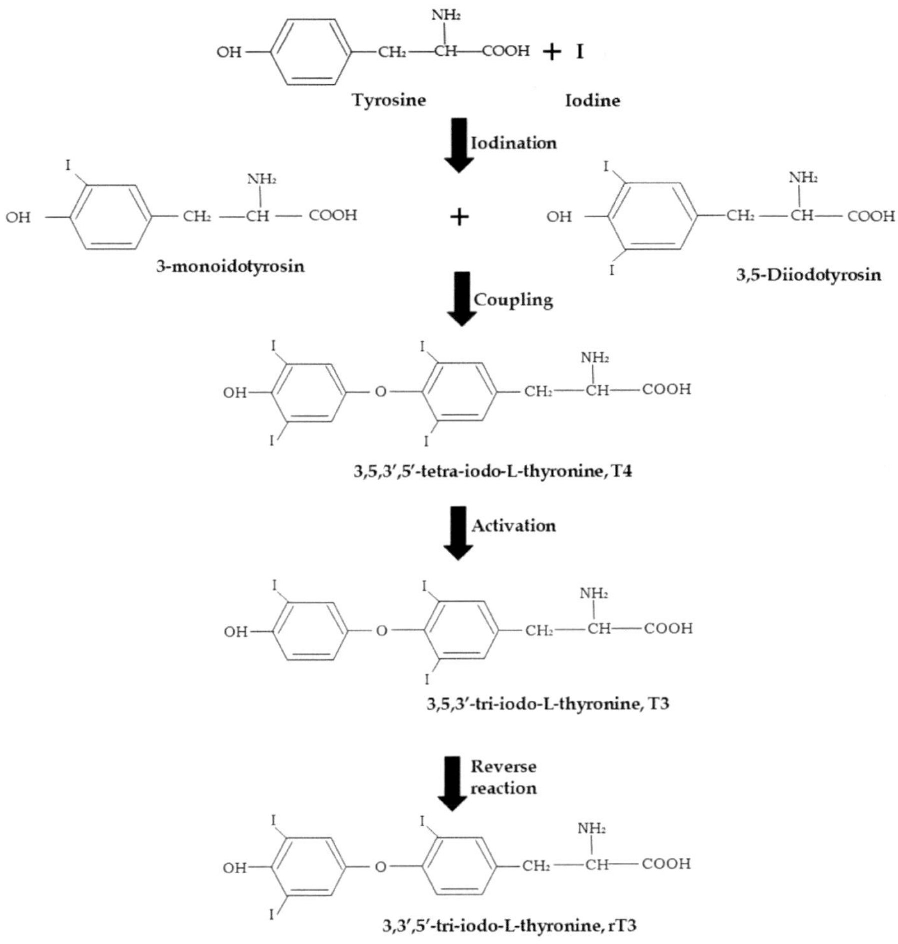

FIGURE 10.2 Synthesis of thyroid hormone.

10.3.2 WHITE ADIPOSE TISSUE (WAT)

As with BAT, white adipose tissue (WAT) is also characterized by a higher number of lipid droplets. These lipid droplets are primarily in charge of storing energy during a critical condition. In WAT, D2 and D3 are the two key players in energy regulation. D2 is responsible for the genes expression which is related with lipogenesis and adipocyte differentiation while D3 acts for the adipocyte proliferation related genes (Young et al., 1984; Obregon, 2014).

10.3.3 SKELETAL MUSCLES

The architecture of the human body is orchestrated by skeletal muscle. Most of the muscles in the human body are skeletal muscles. They are in charge of movement, body posture, energy production, and so on. Most of their functions are regulated by the thyroid hormone. During the requirements of energy, D2-D3 regulates the expression of T3 and energy produced. However, thyroid hormone is also responsible for the regulation of stem cell differentiation and regeneration (Salvatore et al., 2014; Dentice et al., 2014; Bloise et al., 2018; Simonides and Van Hardeveld, 2008).

10.3.4 LIVER

The hepatocytic BMR is greatly influenced by the THs. T3-TRβ (TH receptor β) plays the main role for the completion of the mechanistic approach of lipid metabolism and cholesterol homeostasis (Malik and Hodgson, 2002; Mullur et al., 2014). All three types of deiodinases (D1, D2, and D3) signify the T3 level in the liver and act as biomarkers to some extent (Schneider et al., 2006; Jakobs et al., 1997; Christoffolete et al., 2010; Fonseca et al., 2015; Rodriguez-Perez et al., 2008).

10.3.5 PANCREAS

The pancreas is the most crucial organ for the regulation of glucose in the human body as this organ is responsible for the synthesis of insulin and glucagon hormone from β and α cells, respectively. The T3 influences directly the activation and deactivation of β cell (Mastracci and Evans-Molina, 2014). T3 hormone over activation of cells causes glucose intolerance in the body (Falzacappa et al., 2006; Tomita, 2017). Hyperthyroidism then indirectly influences pancreatic activity (Cicatiello et al., 2018).

10.3.6 HYPOTHALAMUS

The hypothalamus is responsible for coordinating the function of T3 hormonal action through melanocortin neurons. The hypothalamus controls feeding and appetite by this mechanistic approach. The D2-related induction of T3 causes the hyperphagic condition. In the case of food deprivation, T3 is highly secreted. It has been found that the hypothalamus mainly reduces the secretion of T3 to store energy in the body under normal conditions (Elmquist et al., 1999; Mullur et al., 2014; Van Haasteren et al., 1995).

10.4 SOURCES OF THYROID DISRUPTING CONTAMINANTS

The TG may be hampered by different natural contaminants and uncontrolled drug use. Some of those chemicals, heavy metals, drugs, natural contaminants are tabulated in Table 10.1 and Figure 10.3.

10.5 EFFECTS OF SYNTHETIC CHEMICALS, DRUGS, PHYTOCHEMICALS ON THYROID

Many insecticides, pesticides, and industrial by-products after application in the environment remain for years or decades in the ecosystem. These chemical compounds are called POPs and include dioxins, furans, organohalogen contaminants (OHCs), chlordanes (CHLs), organochlorine pesticides (OCPs), and PCBs (Tables 10.2 and 10.3). Although most of these contaminants are either banned or their usage greatly restricted by many countries, they are still found in the environment, and thus, their uses are for limited purposes (Sparling, 2016). Many harmful and hazardous chemicals present in the environment may interfere with the thyroid's normal functioning and homeostasis. Several environmental chemicals resemble T4 and T3 structures, therefore replacing them with harmful chemicals leads to subclinical hypothyroidism, hyperthyroidism, which affect child growth and development (Boas et al., 2006) (Figures 10.7–10.10).

10.5.1 ORGANOHALOGEN CONTAMINANTS (OHCS)

Organohalogens are environmental contaminants, and these are organic compounds that contain chlorine, bromine, and fluorine atoms. They are

TABLE 10.1 Sources of Different Thyroid Disrupting Molecules

Category of Chemicals	Main Chemical Agents	Sources of Chemicals	References
BPA (Bisphenol-A)	4,40-Isopropylidenediphenol	Polycarbonate plastics, epoxy resins, thermal paper, water bottles, canned foods, certain construction materials. CDs, DVDs, safety equipment, and medical devices.	Boas et al. (2006); Gentilcore et al. (2013); Meeker (2012); Mughal et al. (2018)
Metallic	Cadmium, mercury, and lead	Tobacco smoking, nickel-cadmium batteries, nonferrous metal mining, fossil fuel combustion waste incineration and disposal, smelting, refining, battery manufacturing, and soldering procedures.	Buha et al. (2018); Pearce and Braverman (2009)
Flame retardants	Polybrominated diphenyl ethers (PBDEs), tetrabromobisphenol-A (TBBPA), polybrominated biphenyls (PBBs)	Textiles, wire insulation, foams, plastics, carpets, and upholstery.	Boas et al. (2006); Jarque and Piña (2014); Jugan et al. (2010); Mughal et al. (2018); Terrien et al. (2011)
PCB (Polychlorinated biphenyl)	Dioxins, furans, polycarbonate plastics, and epoxy resins	Electrical cables, transformers, switches, hydraulic fluids, capacitors, heat transfer fluids, plasticizers, electrical cables and thermostats, and lubricants.	Boas et al. (2006); Jarque and Piña (2014); Pinson et al. (2016)
PFC (Perfluorinated compounds)	Perfluorooctane sulfonate (PFOS), perfluorooctanoic acid (PFOA), perfluorononanoate (PFNA)	Surfactants, oil-resistant coatings, floor polishes, emulsifiers, Teflon, fire-fighting foams, and fabrics.	Boas et al. (2006); Jarque and Piña (2014); Mughal et al. (2018)
Pharmaceuticals	Phenothiazines, methimazole, propylthiouracil, amiodarone, alemtuzumab, and imatinib	Pharma industry, hospital, and home waste.	Eskes and Wiersinga (2009); Hamnvik et al. (2011); Nakashima et al. (1978); Sauvage et al. (1998)

TABLE 10.1 *(Continued)*

Category of Chemicals	Main Chemical Agents	Sources of Chemicals	References
Natural	Quercetin, genistein, daidzein, catechins, thiocyanate, chalcones, iodine, and various flavonoids	Soy, green tea, cyanogenic plant foods, tomato, potato, seaweed, millet, and green leafy vegetables.	Chandra et al. (2009); Gaitan et al. (1989); Radović et al. (2006); de Souza dos Santos et al. (2011)
Palatable	Di-(2-ethylhexyl) phthalate (DEHP)	Toys, medical tubing, plastic bottles, packing cases.	Boas et al. (2006); Jugan et al. (2010); Mughal et al. (2018); Terrien et al. (2011)
Household	Benzophenone, benzhydrol, perfluorooctanoic acid (PFOA), octylmethoxycinnamate (OMC), homosalate (HMS), 4-aminobenzoic acid (PABA), nonylphenol (NP), triclosan (TCS)	Ultraviolet (UV) filters, various cosmetics, soaps, toothpaste, floor cleaners, and disinfectants.	Boas et al. (2006); Duntas (2015); Pinson et al. (2016); Terrien et al. (2011)
Perchlorates	Ammonium perchlorate	Solid propellants for rockets and missiles, fireworks, matches, airbag inflation systems, rocket fuel, paints, fertilizers, and lubricants.	Jugan et al. (2010); Stoker et al. (2006)
Agricultural	Triadimefon, DDT, lindane, parathion, carbamate, toxaphene, amitrole, mancozeb, and nitrous compounds	Pesticides, herbicides, fungicides, and nitrogenous fertilizers.	Carr and Patiño (2011); Duntas (2015); Terrien et al. (2011)

TABLE 10.2 Drugs Affecting the Thyroid Gland Function

Drug type	Agent	Mechanism of Action	Possible Target Molecule/Cell	Clinical Manifestations	References
Antipsychotic and antidepressants	Phenothiazines	MHC-Ag genetic activity induced on thyroid cells, monoaminergic systems	Thyrocytes, adrenergic nerve receptors	Graves disease Hypothyroidism	Sauvage et al. (1998)
	Lithium	Increase tyrosine kinase activity induce cell proliferation, compete for iodine, thyroiditis	Follicular cells, iodine	Goiter, hyperthyroidism, hypothyroidism	Ramschak-Schwarzer et al. (2000)
	Chlorpromazines, imipramine, desipramine	Iodine complexation	TPO, I	Hypothyroidism	Raby et al. (1990); Sauvage et al. (1998)
Strong electron donors Known antithyroid	Methimazole, propylthiouracil, amiodarone	Disrupt iodination and/or coupling	TPO, I	Hypothyroidism	Eskes and Wiersinga (2009); Roy et al. (2007)
Others	Levamisole, promethazine, TMTU	Iodine complexation	Iodine	Hypothyroidism, increase thyroid size	Raby et al. (1990)
Iodide drugs	Amiodarone, iopanoic acid, iodoquinol	The appearance of the Jod-Basedow phenomenon, Hashimoto's thyroiditis, autoimmunity	Follicular cells, thyrocytes, immune cells	Graves disease, hypothyroidism, hyperthyroidism	Papanastasiou et al. (2007)
Immune modulators drugs	Alemtuzumab	Anti-CD52 antibody-induced cross-reactivity	Thyroid cells	Graves disease	Barbesino (2010); Hamnvik et al. (2011)
	Ipilimumab and tremelimumab	Anti-CTLA-4 mAbs	CTLA-4 gene	Hyper and hypothyroidism	Torino et al. (2012)

TABLE 10.2 *(Continued)*

Drug type	Agent	Mechanism of Action	Possible Target Molecule/Cell	Clinical Manifestations	References
Radiotherapy	CD-20 antibody with I^{131}	Uptake of I^{131} into thyrocytes	Thyrocytes damaged by radiation	Hypothyroidism	Torino et al. (2012)
	Imatinib, sunitinib, sorafenib	Tyrosine kinases activity inhibited	–	Hypothyroidism	Kamba et al. (2006); Torino et al. (2012)
	Tamoxifen (TAM)	Disrupt iodination and/or coupling	VEGF factor	Hypothyroidism	Beyssen et al. (2000)
Other drugs	–	Thyrocyte destruction	TPO, I	Thyroiditis, hypothyroidism	Ghori et al. (2006)

TABLE 10.3 Phytochemicals Involved in Thyroid Gland Disruption

Key	Metabolic Step	Inhibition Target	Inhibitory Compound	Plant/Natural Source	Possible Mechanism of Action	References
A	TG synthesis	Proteins involved in the pathway	Quercetin	Apple, onion, grapes	Decrease expression of thyroid-restricted genes involved	Giuliani et al. (2014)
B	Iodine transport into thyroid cells	NIS Na, K^+ ATPase	Genistein Quercetin Kaempferol Apigenin, Luteolin Arachinoside and anacardioside Kojic acid Thiocyanate Cardiac glycosides like Ouabain	Soy Apple Onion Grapes Green leafy vegetables Peanut Fungal metabolites from fermentation Millet Cyanogenic plants *A. schimperi* and *S. gratus*	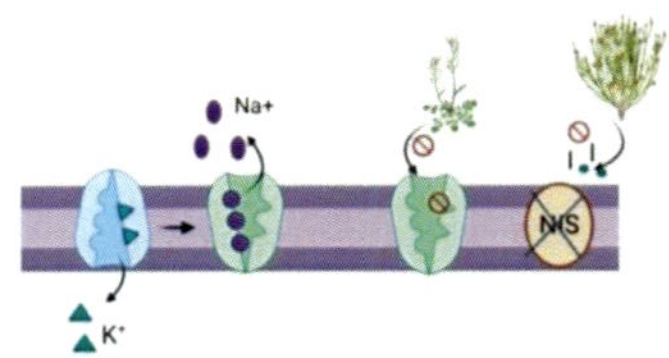	Burdock et al. (2001); Chandra et al. (2009); Radović et al. (2006); Moudgal et al. (1958); Pearce and Braverman (2009)
B	Iodine transport	NIS, thyrocyte, immune cells	Iodine	Seaweed, iodinated salt, preservatives, packaged foods	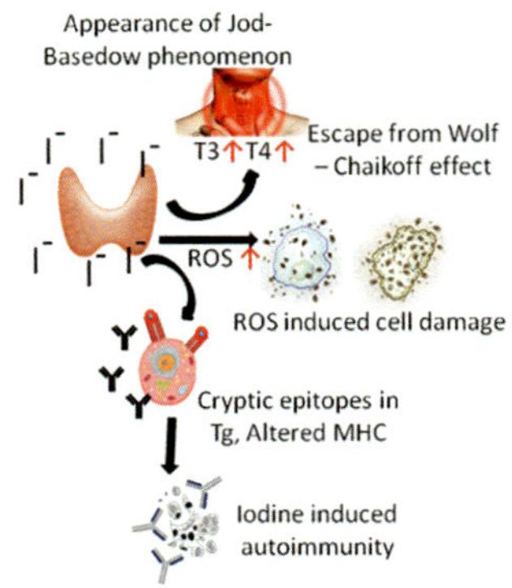	Papanastasiou et al. (2007)

TABLE 10.3 *(Continued)*

Key	Metabolic Step	Inhibition Target	Inhibitory Compound	Plant/Natural Source	Possible Mechanism of Action	References
C	Iodination (2I⁻ I_2)	H_2O_2 TPO	Genistein and daidzein Various flavonoids Various flavonoids Phytoestrogen (genistein and daidzein) Catechins Various flavonoids Various flavonoids	Soy *M. uniflora* *Kalanchoe brasiliensis* Soy and soy products Green tea, red wine, chocolate *Kalanchoe brasiliensis M. uniflora*	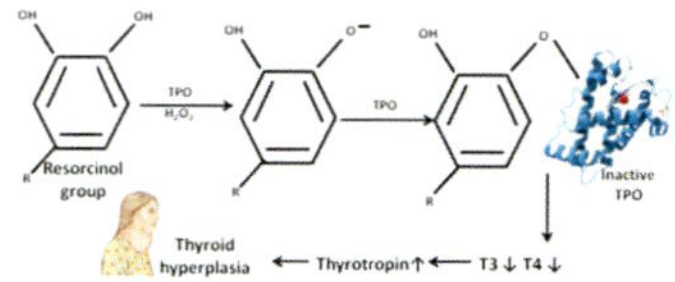	de Souza dos Santos et al. (2011)
D	Organification and coupling	H_2O_2, TPO, TG, and I_2	Genistein, quercetin, kaempferol, and naringenin Thiocyanate C-Glycosylflavones, glucosylvitexin, myricetin, luteolin biochanin A	Apples, grapes, green tea, and berries Cyanogenic plant foods like cabbage, mustard, turnip, radish, bamboo shoot, and cassava Pearl millet, green leafy vegetables Peanut, chickpea, and soya	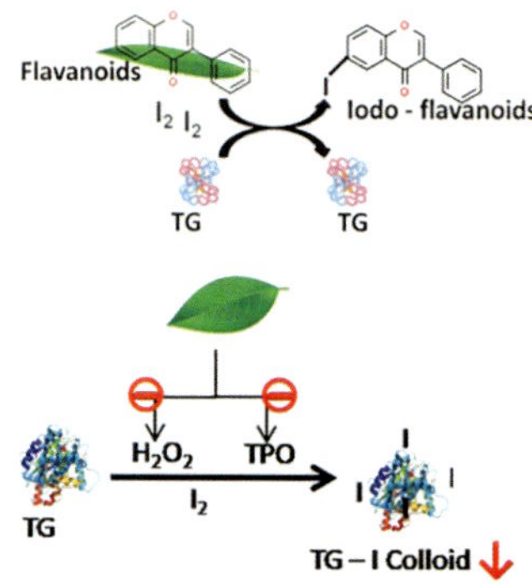	de Souza dos Santos et al. (2011) Gaitan et al. (1989)
E	Hormone transport	TTR	Genistein, phytoestrogen (glycitein) chalcones	Soy and soy products Tomato, potato, apple	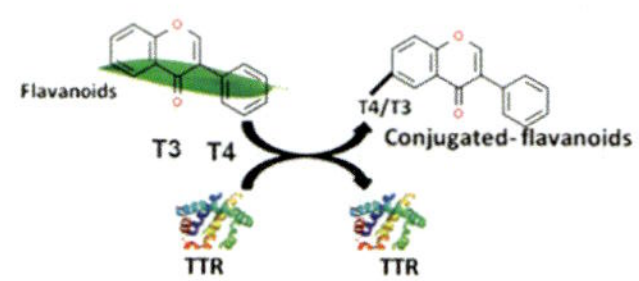	de Souza dos Santos et al. (2011); Radović et al. (2006)

TABLE 10.3 *(Continued)*

Key	Metabolic Step	Inhibition Target	Inhibitory Compound	Plant/Natural Source	Possible Mechanism of Action	References
F	Deiodination	sulfotransferase enzymes 5'-monodeiodinase I (5'-DI) Deionisases (D1 and D2)	Genistein and daidzein Catechin, soy phytoestrogens Mostly flavonoids	Soy and soy products Soy, green tea, red wine, dark chocolate Green vegetables	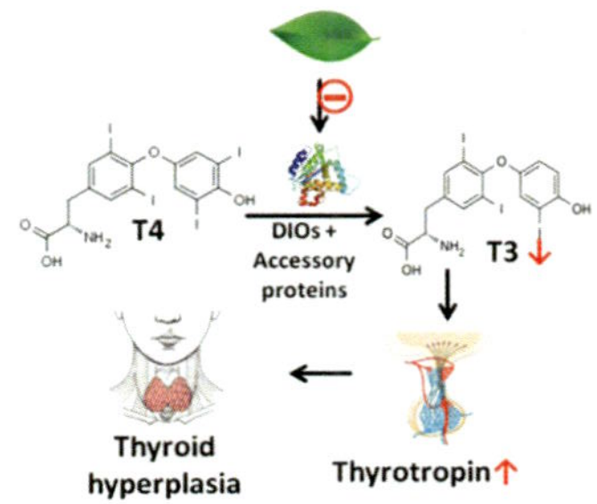	Chandra and De (2013); Radović et al. (2006)

present in insecticides such as PCBs, CHLs, dichlorodiphenyltrichloroethane (DDTs), hexachlorocyclohexane (HCHs), etc. (Kodavanti and Loganathan, 2016). Many OHCs are banned or restricted in most countries, but still, OHCs are found in almost all global ecosystem components that cause biomagnification and bioaccumulation in living organisms.

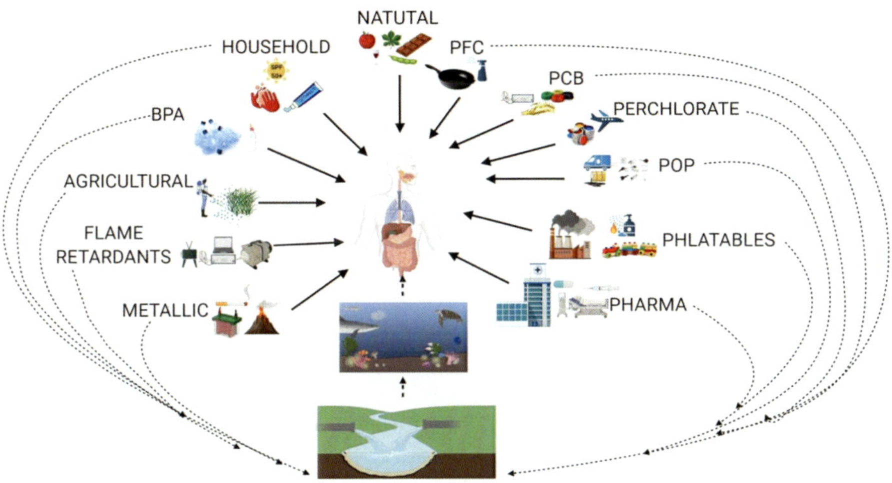

FIGURE 10.3 Different sources of thyroid-disrupting chemicals.

OHCs harm the endocrine system. Their main target is the hypothalamic-pituitary-thyroid (HPT) axis. After the production of THs, they moved to the distant part of the body after using different carrier proteins like albumin, globulin, transthyretin (TTR), etc. (Jugan et al., 2010; Brouwer et al., 1998; Zoeller et al., 2007). Some OHCs have structural similarities with the T3 and T4. They control the negative feedback loop of the hypothalamo-thyroidal axis and control the release of thyroid hormone. The invasion of OHCs in a pregnant mother causes the development of an impaired child with a defective brain. Some OHCs have also been shown to impair the mechanistic approach for neurotransmitter release in developing children (Brouwer et al., 1998; Boas et al., 2006; Langer et al., 2007; Zoeller et al., 2007; Darras, 2008).

10.5.2 POLYCHLORINATED BIPHENYLS (PCBS)

PCBs are mainly secret from industrial sources. A lot of different types of PCBs are in the environment. These are not easily degraded in the environment. They can enter the cell easily due to their lipophilic nature. In the meantime, there is the chance of bioaccumulation of PCBs in higher-order animals. Studies discovered that these PCBs have an analogic nature to T3 and T4. Therefore, they are the most potent chemicals to disrupt the thyro-endocrine axis (Figure 10.4). Figure 10.4 shows a PCB with chlorine attached at the 2, 4, 5, and 2, 3, 4, and 5 positions on one ring and the 2, 3, 4, and 5 locations on the second ring (Johnson et al., 1964).

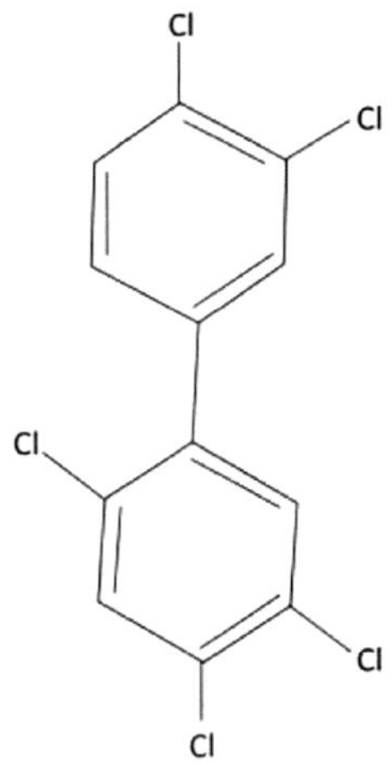

FIGURE 10.4 2,3,'4,4,'5-pentachlorobiphenyl.

In three different ways, these PCBs disrupt the T3 and T4 synthesis. They mainly affect the binding capacity of T3/T4 with the carrier protein, they reduce the synthesis of T3/T4 hormone, and they reduce the thyroid hormone level in the body after higher secretion by bile (Brouwer et al., 1998; Yang et al., 2010; Meerts et al., 2002; Schreiber et al., 1995; Southwellp et al., 1993; Grandjean et al., 2001; Walkowiak et al., 2001).

10.5.3 ORGANOCHLORINE PESTICIDES (OCPS)

Organochlorine pesticides (OCPs) are molecules that have at least one ring structure and multiple chlorine atoms, thus the name organochlorine. They include many different compounds, including dichlorodiphenyltrichloroethane (DDT) and its relatives, γ-hexachlorocyclohexane (γ-HCH), methoxychlor, endrin, endosulfan, heptachlor, CHL, toxaphene, and mirex (Sparling, 2016).

Most OCPs are lipophilic, which means that they readily dissolve in fats and oils, causing bioaccumulation through bioconcentration and biomagnification in living organisms (Sparling, 2016). OCPs resemble T4 and T3, and they disrupt thyroidal synthesis and function (Langer et al., 2007; lvarez-Pedrerol et al., 2008; Zoeller, 2007; Meeker et al., 2009; Chevrier et al., 2008). Chronic exposure to certain OC pesticides in humans may interfere with circulating levels of free T4, total T4, total T3, and TSH in cohorts characterized by low or moderate OC exposure (Bloom et al., 2003; Langer, 2010; Meeker et al., 2009; Sala et al., 2001; Schell et al., 2004, 2008). The developing fetus is more toxic to OCPs than the adult. They can cross the placenta and hamper the growing neurological circuit in the fetus (Sala et al., 2001; Maervoet et al., 2007).

10.5.4 CHLORDANES (CHLS)

Chlordane (CHL) (Figure 10.5) is mainly used as a pesticide and sometimes as an insulator on electrical cables. They are toxic in nature and cause skin irritation. For a long time, CHL may persist in the environment (PubChem, 2021a).

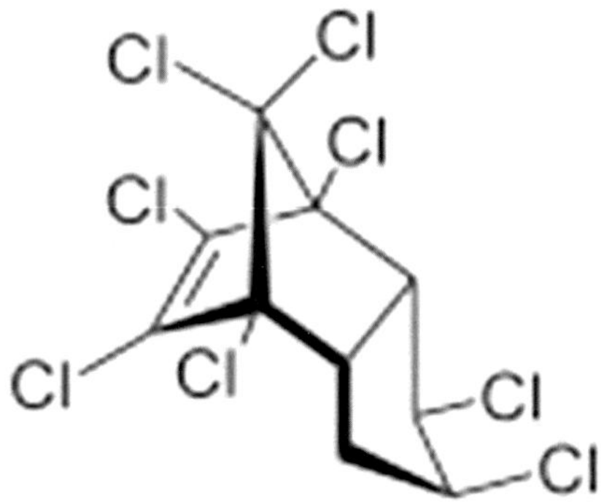

FIGURE 10.5 Chlordane.

CHL, like other pesticides, disrupts thyroid synthesis and function. In mothers, CHL was negatively associated with free T4 and total T4 levels and positively associated with TSH in the blood (Kim et al., 2005; Takser et al., 2005). CHL reaches the mother's milk, and high levels of CHL in breast milk have been associated with higher levels of congenital hypothyroidism in offspring (Nagayama et al., 2007).

10.5.5 PETROLEUM AND THYROID TOXICITY

Petroleum is a fossil fuel; to use it for vehicles and other purposes, it undergoes downstream refining, during which various other substances are also produced as a by-product of petroleum. These by-products share many standard physical, chemical, and toxicological properties. Gasoline, motor oils, kerosines, heavy gas oil, highly refined base oils, heavy fuel oil, naphthenic solvents, aromatic extracts, cracked gas oils, bitumen, etc., are the products of petroleum. Every human being is getting exposed to these products directly or indirectly. They have very less toxic level but in long term exposure that cause lethal condition in the body (McMillen et al., 2001). PAHs are found in some petroleum streams; they can cause mutations, carcinogenicity, and other hazards in humans and animals (Gray et al., 2013). PAHs have a significant effect on human thyroid peroxidase (TPO) enzyme activity, which affects the thyroid synthesis in the TG (Song et al., 2012). However, the mechanism behind such effects is unknown, but reports claim that benzo(k) fluoranthene, benzo(e) pyrene and pyrene are PAHs that inhibit TPO activity (Song et al., 2012) (Figure 10.7). The function of TPO is easily affected by dibenzo(a,h)anthracene but still the mechanism is not well clear (Song et al., 2012).

10.5.6 THYROID HORMONE EXPRESSION AND BISPHENOL-A (BPA) ACTIVITY

Bisphenol-A (BPA) (Figure 10.6) is used to make epoxy, polycarbonate, phenoxy, polysulfone, and polyester resins. It is used in flame retardants and rubber chemicals. BPA is also used in thermal papers (ex., register receipts) and food packaging products (can liners, plastic bottles). Although its use is banned for making baby bottles, sippy cups for babies, and packaging for baby food products, animals, and humans are exposed to BPA (Figure 10.6) by its leaching from articles prepared from it. The general population may be exposed by eating food in contact with certain plastics, coated cans, and bottle tops (PubChem, 2021b).

BPA can disturb thyroid hormone action (Figure 10.7). Due to the analogic nature with thyroid hormone, they can resemble the normal function of TH. Mainly BPA disturb the binding of receptor and reduce the transcriptional activity. It enhances the co-repressor activity against the nuclear receptor of TH (Moriyama et al., 2002).

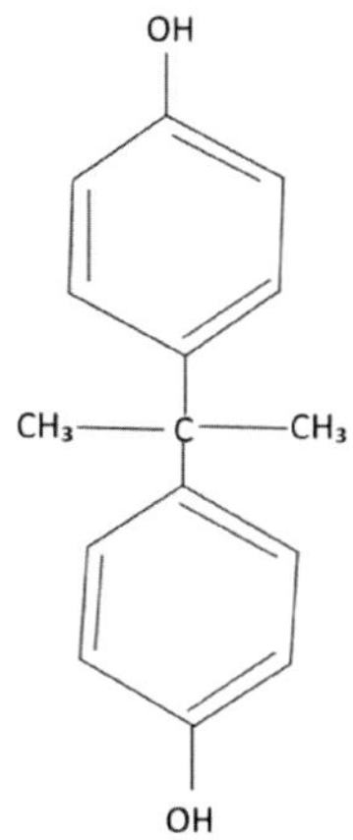

FIGURE 10.6 4-[2-(4-hydroxyphenyl) propan-2-yl] phenol.

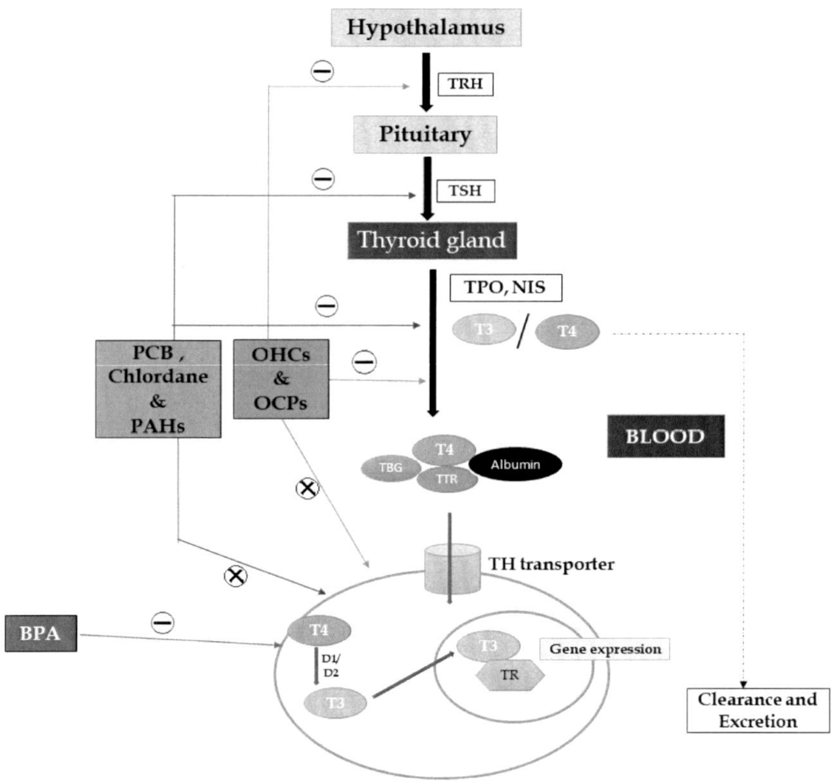

FIGURE 10.7 Inhibition by chemical toxins on thyroid hormone synthesis and metabolism. ⊖: Inhibition of process; ⊗: Prohibit the gene expression.

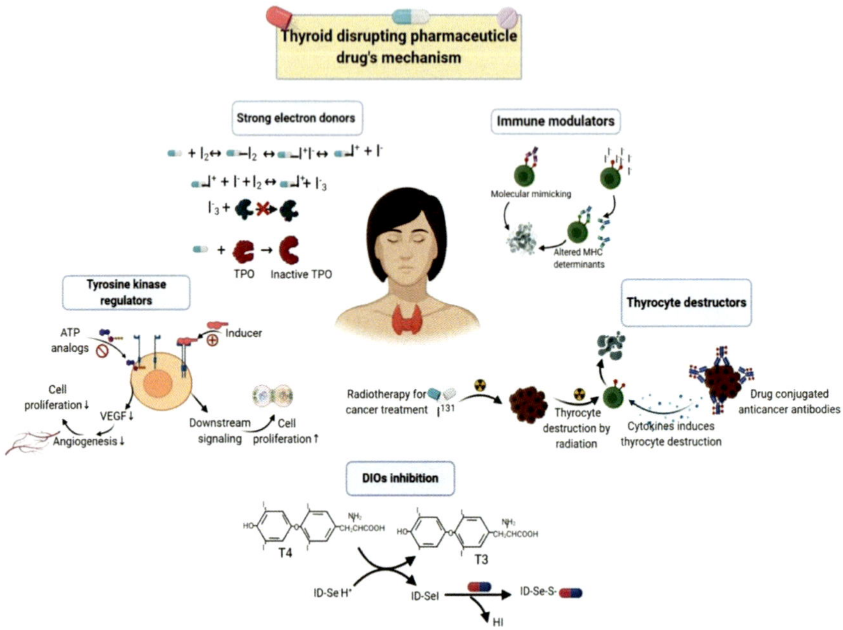

FIGURE 10.8 Drugs affecting the thyroid gland function.

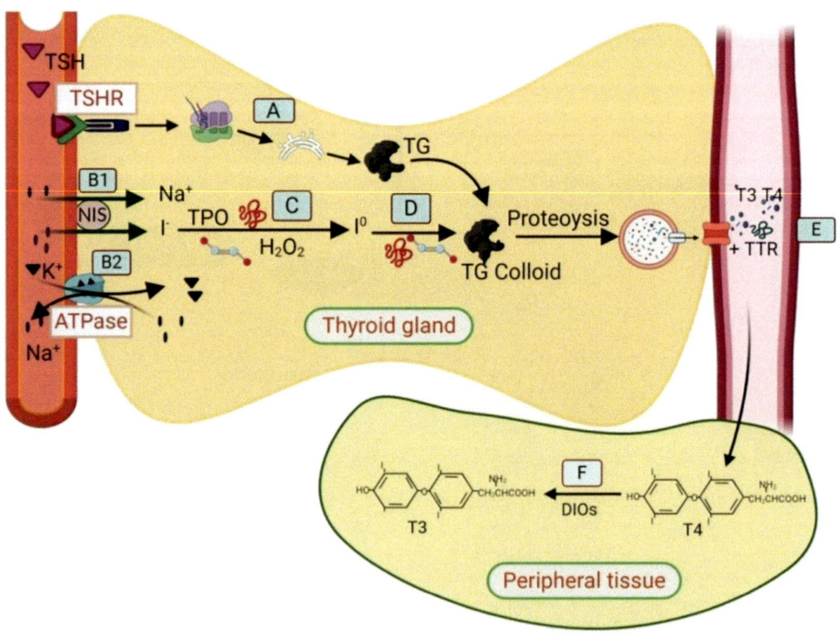

FIGURE 10.9 Pathways involved in thyroid hormone disruption by different phytochemicals.

10.6 EFFECTS OF HEAVY METALS ON THYROID

Many studies have found that arsenic (As), lead (Pb), cadmium (Cd), and other heavy metals have a negative impact on the TG (Table 10.4). In the case of As, thyroid follicular cells and colloidal regions are mostly affected in zebrafish (Liu et al., 2006). In the mouse model, it was found that As exposure caused the down-regulation of thyroperoxidase enzymes, which are responsible for thyroid synthesis (Palazzolo and Jansen, 2008). In another study, it was reported that As also inhibited the activity of hormone response element-binding protein (Davey et al., 2008). Studies found that Cd also disrupts the thyroidal cell activity and inhibits peripheral deiodinase enzyme activity (Liu et al., 2007; Yoshida et al., 1987) (Figure 10.10). A lot of other findings have been listed in Table 10.4.

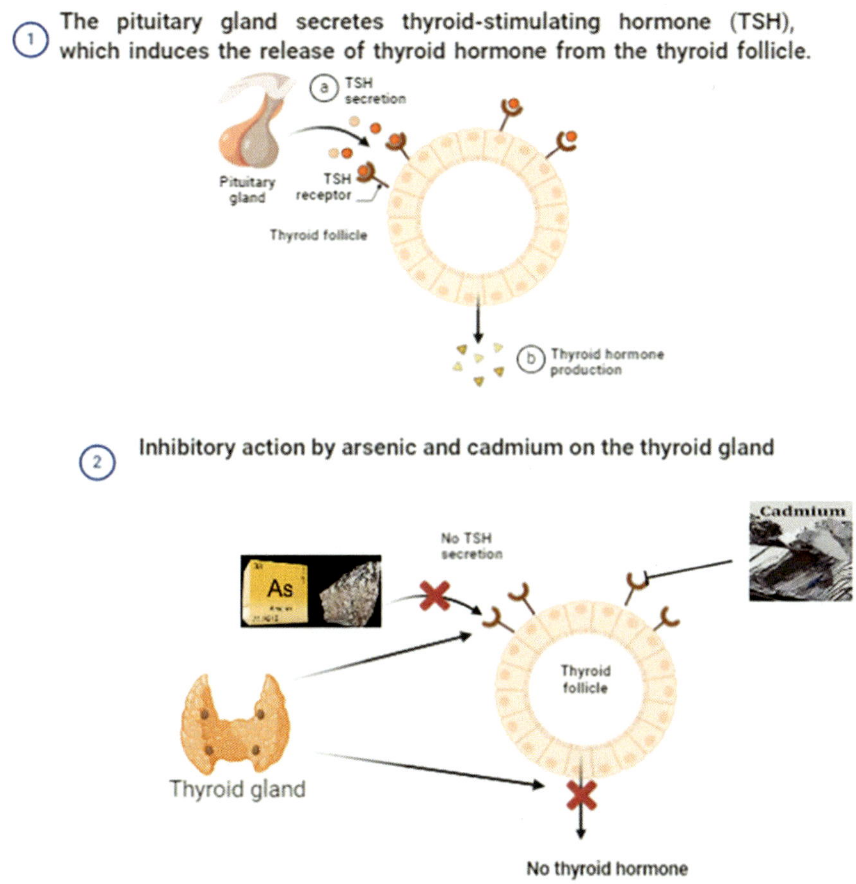

FIGURE 10.10 Effects of heavy metals on thyroid hormone synthesis.

TABLE 10.4 Effects of As and Pb on Thyroid Hormone Synthesis in Different Animals

Animal Model	Time/Duration	Metals/Concentration	Effects	References
Oncorhynchus mykiss (Juvenile)	2, 4, 24, or 96 h	0.4, 0.8, and 2.4 mg Cd/L	T4↓ (2–4 h exposure)	Hontela et al. (1996)
Oncorhynchus mykiss (Juvenile)	1 week	0.4 and 0.8 mg Cd/L	T4↓	Hontela et al. (1996)
Oreochromis niloticus	24, 48, or 96 h	25 mg $CdCl_2$/L	Plasma T3↓	Garcia-Santos et al. (2013)
Gobiocyprisrarus (Larvae)	96 h	0, 0.5, and 2.5 mg Cd/L	Thyroid hormones↓, tireoglobuline↑	Li et al. (2014)
Oncorhynchus mykiss (Adults)	30 days	10 and 25 mg Cd/L	Plasma T3↓ T4↓	Ricard et al. (1998)
Oncorhynchus mykiss (Juvenile)	30 days	1 and 5 mg Cd/L	No effects plasma T4 and T3	Ricard et al. (1998)
Rats (Adults)	35 days	200 ppm Cd	Serum T4↓, serum TSH↑, thyroid weight↑,	Hammouda et al. (2008)
Rats (Adults)	4 weeks	50 mg Cd/L	Serum T4↓, T3↓, serum TSH↑	Mohamed et al. (2015)
Rats (Adults)	1 month	5, 10, 25, 50, or 100 ppm	Plasma TSH↑	Lafuente et al. (2003)
Rats (Adults)	30 days	25 and 50 mg/L	TSH ↑	Caride et al. (2010)
Rats (Adults)	12 weeks	0.55 and 2.19 mg/L	T4↓, T3↓, TSH↓	Zienab and Heba (2009)
Rats (Adults)	12 months	5 and 50 mg Cd/L	Serum T4↓; no effects serum T3 and TSH	Piłat-Marcinkiewicz et al. (2003)
Rats (Adults)	28 days	0.3, 0.6, 1.25, 2.5, 5, and 10 mg Cd/kg	T3↓, FT3↓, T4↓, FT4↓	Buha et al. (2013)
Rats (Adults)	28 days	2.5, 7.5, and 15 mg Cd/kg	T4↓, FT4↓	Ćurčić et al. (2012)

TABLE 10.4 *(Continued)*

Animal Model	Time/Duration	Metals/Concentration	Effects	References
Bighead carp	78 h	150 μg L^{-1}AsV	T4↑	Sun et al. (2016)
Guinea pigs	11 weeks	50 mg kg^{-1}AsIII	T3↓, T4↓	Mohanta et al. (2014)
Rat pituitary tumor cells	24 h	37.5–150 μg L^{-1}AsIII	THRE↓, TR-regulated type I deiodinase gene↓	Davey et al. (2008)
Zebrafish	10–30 days	1 mg L^{-1}AsV	Thyroid histopathology disrupted	Liu et al. (2016)
Zebrafish	48 h	0.1–4.2 mg L^{-1}AsIII	T4↑	Sun et al. (2015)
Human	–	AsIII exposure in air	TSH↑, Thyroglobulin↑, T3↓, T4↓	Ciarrocca et al. (2012)

10.7 CONCLUSION

This chapter described the brief of the TG, the synthesis of THs and their metabolism, and the sources of thyro-contaminants with special emphasis on their mechanistic pathway. It has been found that the regulatory mechanism of thyroid synthesis in different animals may be altered due to different phytochemicals, heavy metals, drugs, etc. Their consequent effects may impair the other metabolic pathways, and the animal's health will be in doom.

KEYWORDS

- **endocrine toxicants**
- **hormone**
- **perfluorononanoate**
- **perfluorooctane sulfonate**
- **perfluorooctanoic acid**
- **polybrominated diphenyl ethers**
- **thyroid gland**

REFERENCES

Álvarez-Pedrerol, M., Ribas-Fitó, N., Torrent, M., Carrizo, D., Garcia-Esteban, R., Grimalt, J. O., & Sunyer, J., (2008). Thyroid disruption at birth due to prenatal exposure to β-hexachlorocyclohexane. *Environ. Int., 34*, 737–740. https://doi.org/10.1016/j.envint.2007.12.001.

Bajaj, J. K., Salwan, P., & Salwan, S., (2016). Various possible toxicants involved in thyroid dysfunction: A review. *J. Clin. Diagnostic Res., 10*, FE01–FE03. https://doi.org/10.7860/JCDR/2016/15195.7092.

Barbesino, G., (2010). Drugs affecting thyroid function. *Thyroid: Official Journal of the American Thyroid Asso., 20,* 763–770. https://doi.org/10.1089/thy.2010.1635.

Beyssen, M. L., Lagorce, J. F., Clédat, D., Jambut, A. C., & Buxeraud, J., (2000). Antithyroid action of tamoxifen in the rat: *In vivo* and *in vitro* studies. *Pharmacol., 61*, 22–30. https://doi.org/10.1159/000028376.

Bloise, F. F., Cordeiro, A., & Ortiga-Carvalho, T. M., (2018). Role of thyroid hormone in skeletal muscle physiology. *J. Endocrinol., 236*, R57–R68. https://doi.org/10.1530/JOE-16-0611.

Bloom, M. S., Weiner, J. M., Vena, J. E., & Beehler, G. P., (2003). Exploring associations between serum levels of select organochlorines and thyroxine in a sample of New York state sportsmen: The New York State Angler Cohort Study. *Environ. Res., 93,* 52–66. https://doi.org/10.1016/S0013-9351(02)00085-3.

Boas, M., Feldt-Rasmussen, U., Skakkebæk, N. E., & Main, K. M., (2006). Environmental chemicals and thyroid function. *European J Endocrinol., 154*, 599–611. https://doi.org/10.1530/eje.1.02128.

Brouwer, A., Morse, D. C., Lans, M. C., Gerlienke, S. A., Murk, A. J., Klasson-Wehler, E., Bergman, Å., & Visser, T. J., (1998). Interactions of persistent environmental organohalogens with the thyroid hormone system: Mechanisms and possible consequences for animal and human health. *Toxicol. Ind. Health, 14*, 59–84. https://doi.org/10.1177/074823379801400107.

Buha, A., Antonijević, B., Bulat, Z., Jaćević, V., Milovanović, V., & Matović, V., (2013). The impact of prolonged cadmium exposure and co-exposure with polychlorinated biphenyls on thyroid function in rats. *Toxicol. Lett., 221*, 83–90.https://doi.org/10.1016/j.toxlet.2013.06.216.

Buha, A., Matovic, V., Antonijevic, B., Bulat, Z., Curcic, M., Renieri, E. A., Tsatsakis, A. M., et al., (2018). Overview of cadmium thyroid disrupting effects and mechanisms. *Int. J. Mol. Sci., 19*, 1501. https://doi.org/10.3390/ijms19051501.

Burdock, G. A., Soni, M. G., & Carabin, I. G., (2001). Evaluation of health aspects of kojic acid in food. *Regul. Toxicol. Pharmacol., 33*, 80–101. https://doi.org/10.1006/RTPH.2000.1442.

Caride, A., Fernández-Pérez, B., Cabaleiro, T., Tarasco, M., Esquifino, A. I., & Lafuente, A., (2010). Cadmium chronotoxicity at pituitary level: Effects on plasma ACTH, GH, and TSH daily pattern. *J. Physiol. Biochem., 66*, 213–220.https://doi.org/10.1007/s13105-010-0027-5.

Carr, J. A., & Patiño, R., (2011). The hypothalamus-pituitary-thyroid axis in teleosts and amphibians: Endocrine disruption and its consequences to natural populations. *Gen. Comp. Endocrinol., 170*, 299–312. https://doi.org/10.1016/j.ygcen.2010.06.001.

Chandra, A. K., & De, N., (2013). Catechin induced modulation in the activities of thyroid hormone synthesizing enzymes leading to hypothyroidism. *Mol. Cell. Biochem., 374*, 37–48. https://doi.org/10.1007/s11010-012-1503-8.

Chandra, R., Bharagava, R. N., Yadav, S., & Mohan, D., (2009). Accumulation and distribution of toxic metals in wheat (*Triticum aestivum* L.) and Indian mustard (*Brassica campestris* L.) irrigated with distillery and tannery effluents. *J. Hazard. Mater., 162*, 1514–1521. https://doi.org/10.1016/j.jhazmat.2008.06.040.

Chevrier, J., Eskenazi, B., Holland, N., Bradman, A., & Barr, D. B., (2008). Effects of exposure to polychlorinated biphenyls and organochlorine pesticides on thyroid function during pregnancy. *Am. J. Epidemiol., 168*, 298–310. https://doi.org/10.1093/aje/kwn136.

Christoffolete, M. A., Doleschall, M., Egri, P., Liposits, Z., Zavacki, A. M., Bianco, A. C., & Gereben, B., (2010). Regulation of thyroid hormone activation via the liver x-receptor/retinoid X-receptor pathway. *J. Endocrinol., 205*, 179–186. https://doi.org/10.1677/JOE-09-0448.

Ciarrocca, M., Tomei, F., Caciari, T., Cetica, C., Andrè, J. C., Fiaschetti, M., Schifano, M. P., et al., (2012). Exposure to arsenic in urban and rural areas and effects on thyroid hormones. *Inhal. Toxicol., 24*, 589–598. https://doi.org/10.3109/08958378.2012.703251.

Cicatiello, A. G., Girolamo, D. D., & Dentice, M., (2018). Metabolic effects of the intracellular regulation of thyroid hormone: Old players, new concepts. *Front. Endocrinol., 9*, 474. https://doi.org/10.3389/fendo.2018.00474.

Ćurčić, M., Janković, S., Jaćević, V., Stanković, S., Vučinić, S., Durgo, K., Bulat, Z., & Antonijević, B., (2012). Combined effects of cadmium and decabrominated diphenyl ether on thyroid hormones in rats. *Arh. Hig. Rada Toksikol., 63*, 255–262. https://doi.org/10.2478/10004-1254-63-2012-2179.

Darras, V. M., (2008). Endocrine disrupting polyhalogenated organic pollutants interfere with thyroid hormone signaling in the developing brain. *Cerebellum, 7*, 26–37. https://doi.org/10.1007/s12311-008-0004-5.

Davey, J. C., Nomikos, A. P., Wungjiranirun, M., Sherman, J. R., Ingram, L., Batki, C., Lariviere, J. P., & Hamilton, J. W., (2008). Arsenic as an endocrine disruptor: Arsenic disrupts retinoic acid receptor- and thyroid hormone receptor-mediated gene regulation and thyroid hormone-mediated amphibian tail metamorphosis. *Environ. Health Perspect., 116*, 165–172. https://doi.org/10.1289/ehp.10131.

Dentice, M., Ambrosio, R., Damiano, V., Sibilio, A., Luongo, C., Guardiola, O., Yennek, S., et al., (2014). Intracellular inactivation of thyroid hormone is a survival mechanism for muscle stem cell proliferation and lineage progression. *Cell Metab., 20*, 1038–1048. https://doi.org/10.1016/j.cmet.2014.10.009.

Duntas, L. H., (2015). Chemical contamination and the thyroid. *Endocrine., 48*, 53–64. https://doi.org/10.1007/s12020-014-0442-4.

Elmquist, J. K., Elias, C. F., & Saper, C. B., (1999). From lesions to leptin: Hypothalamic control of food intake and body weight. *Neuron., 22*, 221–232. https://doi.org/10.1016/S0896-6273(00)81084-3.

Eskes, S. A., & Wiersinga, W. M., (2009). Amiodarone and thyroid. *Best Pract. Res. Clin. Endocrinol. Metab., 23*, 735–751. https://doi.org/10.1016/j.beem.2009.07.001.

Falzacappa, C. V., Panacchia, L., Bucci, B., Stigliano, A., Cavallo, M. G., Brunetti, E., Toscano, V., & Misiti, S., (2006). 3,5,3′-triiodothyronine (T3) is a survival factor for pancreatic β-cells undergoing apoptosis. *J. Cell. Physiol., 206*, 309–321. https://doi.org/10.1002/jcp.20460.

Fonseca, T. L., Fernandes, G. W., McAninch, E. A., Bocco, B. M. L. C., Abdalla, S. M., Ribeiro, M. O., Mohácsik, P., et al., (2015). Perinatal deiodinase 2 expression in hepatocytes defines epigenetic susceptibility to liver steatosis and obesity. *Proc. Natl. Acad. Sci. U. S. A., 112*, 14018–14023. https://doi.org/10.1073/pnas.1508943112.

Gaitan, E., Lindsay, R. H., Reichert, R. D., Ingbarf, S. H., Cooksey, R. C., Legan, J., Meydrech, E. F., et al., (1989). Antithyroid and goitrogenic effects of millet: Role of C-glycosylflavones. *J. Clin. Endocrinol. Metab., 68*, 707–714. https://doi.org/10.1210/jcem-68-4-707.

Garcia-Santos, S., Fontaínhas-Fernandes, A., Monteiro, S. M., & Wilson, J. M., (2013). Effects of exposure to cadmium on some endocrine parameters in tilapia, Oreochromis niloticus. *Bull. Environ. Contam. Toxicol., 90*, 55–59. https://doi.org/10.1007/s00128-012-0905-3.

Gentilcore, D., Porreca, I., Rizzo, F., Ganbaatar, E., Carchia, E., Mallardo, M., Felice De, M., & Ambrosino, C., (2013). Bisphenol A interferes with thyroid specific gene expression. *Toxicol., 304*, 21–31. https://doi.org/10.1016/j.tox.2012.12.001.

Ghori, F., Polder, K. D., Pinter-Brown, L. C., Hoff, A. O., Gagel, R. F., Sherman, S. I., & Duvic, M., (2006). Thyrotoxicosis after denileukin diftitox therapy in patients with mycosis fungoides. *J. Clin. Endocrinol. Metab., 91*, 2205–2208. https://doi.org/10.1210/jc.2005-2839.

Giuliani, C., Bucci, I., Santo, S. D., Rossi, C., Grassadonia, A., Piantelli, M., Monaco, F., & Napolitano, G., (2014). The flavonoid quercetin inhibits thyroid-restricted genes expression and thyroid function. *Food Chem. Toxicol., 66*, 23–29. https://doi.org/10.1016/j.fct.2014.01.016.

Grandjean, P., Weihe, P., Burse, V. W., Needham, L. L., Storr-Hansen, E., Heinzow, B., Debes, F., et al., (2001). Neurobehavioral deficits associated with PCB in 7-year-old children prenatally exposed to seafood neurotoxicants. *Neurotoxicol. Teratol., 23*, 305–317. https://doi.org/10.1016/S0892-0362(01)00155-6.

Gray, T. M., Simpson, B. J., Nicolich, M. J., Murray, F. J., Verstuyft, A. W., Roth, R. N., & McKee, R. H., (2013). Assessing the mammalian toxicity of high-boiling petroleum

substances under the rubric of the HPV program. *Regul. Toxicol. Pharmacol., 67*, S4–S9. https://doi.org/10.1016/j.yrtph.2012.11.014.
Haasteren, G. A. C. V., Linkels, E., Klootwijk, W., Toor, H. V., Rondeel, J. M. M., Themmen, A. P. N., Jong, F. H. D., et al., (1995). Starvation-induced changes in the hypothalamic content of prothyrotrophin-releasing hormone (proTRH) mRNA and the hypothalamic release of proTRH-derived peptides: Role of the adrenal gland. *J. Endocrinol., 145*, 143–153.
Hammouda, F., Messaoudi, I., Hani, J. E., Baati, T., Saïd, K., & Kerkeni, A., (2008). Reversal of cadmium-induced thyroid dysfunction by selenium, zinc, or their combination in rat. *Biol. Trace Elem. Res., 126*, 194–203. https://doi.org/10.1007/s12011-008-8194-8.
Hamnvik, O. P. R., Larsen, P. R., & Marqusee, E., (2011). Thyroid dysfunction from antineoplastic agents. *J. Natl. Cancer Inst., 103*(21), 1572–1587. https://doi.org/10.1093/jnci/djr373.
Harvey, C. B., & Williams, G. R., (2002). Mechanism of thyroid hormone action. *Thyroid., 12*, 441–446. https://doi.org/10.1089/105072502760143791.
Hontela, A., Daniel, C., & Ricard, A. C., (1996). Effects of acute and subacute exposures to cadmium on the interrenal and thyroid function in rainbow trout, *Oncorhynchus mykiss*. *Aquat. Toxicol., 35*, 171–182. https://doi.org/10.1016/0166-445X(96)00012-4.
Jakobs, T. C., Schmutzler, C., Meissner, J., & Köhrle, J., (1997). The promoter of the human type I 5'-deiodinase gene: Mapping of the transcription start site and identification of a DR+4 thyroid-hormone- responsive element. *Eur. J. Biochem., 247*, 288–297. https://doi.org/10.1111/j.1432-1033.1997.00288.x.
Jarque, S., & Piña, B., (2014). Deiodinases and thyroid metabolism disruption in teleost fish. *Environ. Res., 135*, 361–375. https://doi.org/10.1016/j.envres.2014.09.022.
Jesus, L. A. D., Carvalho, S. D., Ribeiro, M. O., Schneider, M., Kim, S. W., Harney, J. W., Larsen, P. R., & Bianco, A. C., (2001). The type 2 iodothyronine deiodinase is essential for adaptive thermogenesis in brown adipose tissue. *J. Clin. Invest., 108*, 1379–1385. https://doi.org/10.1172/jci13803.
Johnson, G. W., Quensen, J. F., Chiarenzelli, J. R., & Coreen, H. M., (1964). Polychlorinated biphenyls. In: *Environmental Forensics: Contaminant Specific Guide* (pp. 187–225). Academic Press. https://doi.org/10.1016/B978-012507751-4/50032-X.
Jugan, M. L., Levi, Y., & Blondeau, J. P., (2010). Endocrine disruptors and thyroid hormone physiology. *Biochem. Pharmacol., 79*, 939–947. https://doi.org/10.1016/j.bcp.2009.11.006.
Kamba, T., Tam, B. Y. Y., Hashizume, H., Haskell, A., Sennino, B., Mancuso, M. R., Norberg, S. M., et al., (2006). VEGF-dependent plasticity of fenestrated capillaries in the normal adult microvasculature. *Am. J. Physiol. - Hear. Circ. Physiol., 290*, 560–576. https://doi.org/10.1152/ajpheart.00133.2005.
Kim, H., Hardy, J., Novak, G., Ramet, J., & Weber, F., (2005). Off-tastes in raw and reconstituted milk, food & agriculture organization of the United Nations, Rome. *Anim. Prod. Heal., 35*, 2.
Kodavanti, P. R. S., & Loganathan, B. G., (2016). Organohalogen pollutants and human health. In: *International Encyclopedia of Public Health* (pp. 359–366). Academic Press. https://doi.org/10.1016/B978-0-12-803678-5.00318-0.
Köhrle, J., Lehmphul, I., Pietzner, M., Renko, K., Rijntjes, E., Richards, K., Anselmo, J., et al., (2020). 3,5-T2—A Janus-faced thyroid hormone metabolite exerts both canonical T3-mimetic endocrine and intracrine hepatic action. *Front. Endocrinol. (Lausanne), 10*, 787. https://doi.org/10.3389/fendo.2019.00787.
Koibuchi, N., (2012). [Molecular mechanisms of thyroid hormone synthesis and secretion]. *Nihon rinsho. Japanese J. Clinic. Med., 70*, 1844–1848. https://doi.org/10.1007/978-3-319-44675-2_5.

Lafuente, A., Cano, P., & Esquifino, A. I., (2003). Are cadmium effects on plasma gonadotropins, prolactin, ACTH, GH and TSH levels, dose-dependent? *BioMetals., 16*, 243–250. https://doi.org/10.1023/A:1020658128413.

Langer, P., (2010). The impacts of organochlorines and other persistent pollutants on thyroid and metabolic health. *Front. Neuroendocrinol., 31*, 497–518. https://doi.org/10.1016/j.yfrne.2010.08.001.

Langer, P., Kočan, A., Tajtáková, M., Rádiková, Ž., Petrík, J., Koška, J., Kšinantová, L., et al., (2007). Possible effects of persistent organochlorinated pollutants cocktail on thyroid hormone levels and pituitary-thyroid interrelations. *Chemosphere, 70*, 110–118. https://doi.org/10.1016/j.chemosphere.2007.06.046.

Li, C., Cheng, Y., Tang, Q., Lin, S., Li, Y., Hu, X., Nian, J., et al., (2014). The association between prenatal exposure to organochlorine pesticides and thyroid hormone levels in newborns in Yancheng, China. *Environ. Res., 129*, 47–51. https://doi.org/10.1016/j.envres.2013.12.009.

Liu, C., Zhang, D., Zhao, L., Lu, X., Zhang, P., He, S., Hu, G., & Tang, X., (2016). Synthesis of a thiacalix[4]arenetetrasulfonate-functionalized reduced graphene oxide adsorbent for the removal of lead(ii) and cadmium(ii) from aqueous solutions. *RSC Adv., 6*, 113352–113365. https://doi.org/10.1039/c6ra24353c.

Liu, F. J., Wang, J. S., & Theodorakis, C. W., (2006). Thyrotoxicity of sodium arsenate, sodium perchlorate, and their mixture in zebrafish Danio rerio. *Environ. Sci. Technol., 40*, 3429–3436. https://doi.org/10.1021/es052538g.

Liu, Z. M., Chen, G. G., Shum, C. K. Y., Vlantis, A. C., George, C. M., Koropatnick, J., & Andrew, V. H. C., (2007). Induction of functional MT1 and MT2 isoforms by calcium in anaplastic thyroid carcinoma cells. *FEBS Lett., 581*, 2465–2472. https://doi.org/10.1016/j.febslet.2007.04.049.

Maervoet, J., Vermeir, G., Covaci, A., Larebeke, N. V., Koppen, G., Schoeters, G., Nelen, V., et al., (2007). Association of thyroid hormone concentrations with levels of organochlorine compounds in cord blood of neonates. *Environ. Health Perspect., 115*, 1780–1786. https://doi.org/10.1289/ehp.10486.

Malik, R., & Hodgson, H., (2002). The relationship between the thyroid gland and the liver. *QJM - Monthly Journal of the Association of Physicians., 95*, 559–569.https://doi.org/10.1093/qjmed/95.9.559.

Mastracci, T. L., & Evans-Molina, C., (2014). Pancreatic and islet development and function: The role of thyroid hormone. *J. Endocrinol. Diabetes Obes., 2*.

McMillen, S. J., Magaw, R. I., & Carovillano, R. L. (2001). *Risk-Based Decision-Making for Assessing Petroleum*. Google Scholar. https://nla.gov.au/nla.cat-vn4162609 (accessed on 26 October 2022).

Meeker, J. D., (2012). Exposure to environmental endocrine disruptors and child development. *Arch. Pediatr. Adolesc. Med., 166*, 952–958.

Meeker, J. D., Rossano, M. G., Protas, B., Diamond, M. P., Puscheck, E., Daly, D., Paneth, N., & Wirth, J. J., (2009). Multiple metals predict prolactin and thyrotropin (TSH) levels in men. *Environ. Res., 109*, 869–873. https://doi.org/10.1001/archpediatrics.2012.241.

Meerts, I. A. T. M., Assink, Y., Cenijn, P. H., Berg, J. H. J. V. D., Weijers, B. M., Bergman, Å., Koeman, J. H., & Brouwer, A., (2002). Placental transfer of a hydroxylated polychlorinated biphenyl and effects on fetal and maternal thyroid hormone homeostasis in the rat. *Toxicol. Sci., 68*, 361–371. https://doi.org/10.1093/toxsci/68.2.361.

Mohamed, T. M., Salama, A. F., Nimr, T. M. E., & Gamal, D. M. E., (2015). Effects of phytate on thyroid gland of rats intoxicated with cadmium. *Toxicol. Ind. Health, 31*, 1258–1268. https://doi.org/10.1177/0748233713485887.

Mohanta, R. K., Garg, A. K., Dass, R. S., & Behera, S. K., (2014). Blood biochemistry, thyroid hormones, and oxidant/antioxidant status of guinea pigs challenged with sodium arsenite or arsenic trioxide. Biol. *Trace Elem. Res., 160*, 238–244. https://doi.org/10.1007/s12011-014-0041-5.

Moriyama, K., Tagami, T., Akamizu, T., Usui, T., Saijo, M., Kanamoto, N., Hataya, Y., et al., (2002). Thyroid hormone action is disrupted by bisphenol A as an antagonist. *J. Clin. Endocrinol. Metab., 87*, 5185–5190. https://doi.org/10.1210/jc.2002-020209.

Moudgal, N. R., Raghupathy, E., & Sarma, P. S., (1958). Studies on goitrogenic agents in food. III. Goitrogenic action of some glycosides isolated from edible nuts. *J. Nutr., 66*, 291–303.https://doi.org/10.1093/jn/66.2.291.

Mughal, B. B., Fini, J. B., & Demeneix, B. A., (2018). Thyroid-disrupting chemicals and brain development: An update. *Endocr. Connect., 7*, R160–R186. https://doi.org/10.1530/EC-18-0029.

Mullur, R., Liu, Y. Y., & Brent, G. A., (2014). Thyroid hormone regulation of metabolism. *Physiol. Rev., 94*, 355–382. https://doi.org/10.1152/physrev.00030.2013.

Nagayama, J., Kohno, H., Kunisue, T., Kataoka, K., Shimomura, H., Tanabe, S., & Konishi, S., (2007). Concentrations of organochlorine pollutants in mothers who gave birth to neonates with congenital hypothyroidism. *Chemosphere, 68*, 972–976. https://doi.org/10.1016/j.chemosphere.2007.01.010.

Nakashima, T., Taurog, A., & Riesco, G., (1978). Mechanism of action of thioureylene antithyroid drugs: Factors affecting intrathyroidal metabolism of propylthiouracil and methimazole in rats. *Endocrinol., 103*, 2187–2197. https://doi.org/10.1210/endo-103-6-2187.

Obregon, M. J., (2014). Adipose tissues and thyroid hormones. *Front. Physiol., 5, 479*. https://doi.org/10.3389/fphys.2014.00479.

Oppenheimer, J. H., Schwartz, H. L., Lane, J. T., & Thompson, M. P., (1991). Functional relationship of thyroid hormone-induced lipogenesis, lipolysis, and thermogenesis in the rat. *J. Clin. Invest., 87*, 125–132. https://doi.org/10.1172/JCI114961.

Palazzolo, D. L., & Jansen, K. P., (2008). The minimal arsenic concentration required to inhibit the activity of thyroid peroxidase activity *in vitro*. *Biol. Trace Elem. Res., 126*, 49–55. https://doi.org/10.1172/JCI114961.

Papanastasiou, L., Vatalas, I. A., Koutras, D. A., & Mastorakos, G., (2007). Thyroid autoimmunity in the current iodine environment. *Thyroid, 17*, 729–739. https://doi.org/10.1089/thy.2006.0299.

Pearce, E. N., & Braverman, L. E., (2009). Environmental pollutants and the thyroid. *Best Pract. Res. Clin. Endocrinol. Metab., 23*, 801–813. https://doi.org/10.1016/j.beem.2009.06.003.

Piłat-Marcinkiewicz, B., Brzóska, M. M., Sawicki, B., & Moniuszko-Jakoniuk, J., (2003). Structure and function of thyroid follicular cells in female rats chronically exposed to cadmium. *Bull. Vet. Inst. Pulawy, 47*, 157–163.

Pinson, A., Bourguignon, J. P., & Parent, A. S., (2016). Exposure to endocrine disrupting chemicals and neurodevelopmental alterations. *Andrology, 4*, 706-722. https://doi.org/10.1111/andr.12211.

PubChem. *Bisphenol A* $C_{15}H_{16}O_2$ -PubChem. https://pubchem.ncbi.nlm.nih.gov/compound/Bisphenol-A (accessed on 30 August 2022).

PubChem. *Chlordecone $C_{10}Cl_{10}O$* -PubChem. https://pubchem.ncbi.nlm.nih.gov/compound/Chlordecone (accessed on 30 August 2022).

Raby, C., Lagorce, J. F., Jambut-Absil, A. C., Buxeraud, J., & Catanzano, G., (1990). The mechanism of action of synthetic antithyroid drugs: Iodine complexation during oxidation of iodide. *Endocrinol., 126*, 1683–1691. https://doi.org/10.1210/endo-126-3-1683.

Radović, B., Mentrup, B., & Köhrle, J., (2006). Genistein and other soya isoflavones are potent ligands for transthyretin in serum and cerebrospinal fluid. *Br. J. Nutr., 95*, 1171–1176. https://doi.org/10.1079/bjn20061779.

Ramschak-Schwarzer, S., Radkohl, W., Stiegler, C., Dimai, H. P., & Leb, G., (2000). Interaction between psychotropic drugs and thyroid hormone metabolism an overview. *Acta Med. Austriaca, 27*, 8–10. https://doi.org/10.1046/j.1563-2571.2000.200102.x.

Ricard, A. C., Daniel, C., Anderson, P., & Hontela, A., (1998). Effects of subchronic exposure to cadmium chloride on endocrine and metabolic functions in rainbow trout Oncorhynchus mykiss. *Arch. Environ. Contam. Toxicol., 34*, 377-381. https://doi.org/10.1007/s002449900333.

Rodriguez-Perez, A., Palos-Paz, F., Kaptein, E., Visser, T. J., Dominguez-Gerpe, L., Alvarez-Escudero, J., & Lado-Abeal, J., (2008). Identification of molecular mechanisms related to nonthyroidal illness syndrome in skeletal muscle and adipose tissue from patients with septic shock. *Clin. Endocrinol. (Oxf).*, *68*, 821–827. https://doi.org/10.1111/j.1365-2265.2007.03102.x.

Roy, G., Das, D., & Mugesh, G., (2007). Bioinorganic chemistry aspects of the inhibition of thyroid hormone biosynthesis by anti-hyperthyroid drugs. *Inorganica Chim. Acta, 360*, 303–316. https://doi.org/10.1016/j.ica.2006.07.052.

Sala, M., Sunyer, J., Herrero, C., To-Figueras, J., & Grimalt, J., (2001). Association between serum concentrations of hexachlorobenzene and polychlorobiphenyls with thyroid hormone and liver enzymes in a sample of the general population. *Occup. Environ. Med., 58*, 172–177. https://doi.org/10.1136/oem.58.3.172.

Salvatore, D., Simonides, W. S., Dentice, M., Zavacki, A. M., & Larsen, P. R., (2014). Thyroid hormones and skeletal muscle - New insights and potential implications. *Nat. Rev. Endocrinol., 10*, 206–214. https://doi.org/10.1038/nrendo.2013.238.

Sauvage, M. F., Marquet, P., Rousseau, A., Raby, C., Buxeraud, J., & Lachâtre, G., (1998). Relationship between psychotropic drugs and thyroid function: A review. *Toxicol. Appl. Pharmacol., 149*(2), 127–135. https://doi.org/10.1006/taap.1998.8367.

Schell, L. M., Gallo, M. V., DeCaprio, A. P., Hubicki, L., Denham, M., & Ravenscroft, J., (2004). Thyroid function in relation to burden of PCBs, p,p′-DDE, HCB, mirex and lead among Akwesasne mohawk youth: A preliminary study. *Environ. Toxicol. Pharmacol., 18*(2), 91–99. https://doi.org/10.1016/j.etap.2004.01.010.

Schell, L. M., Gallo, M. V., Denham, M., Ravenscroft, J., DeCaprio, A. P., & Carpenter, D. O., (2008). Relationship of thyroid hormone levels to levels of polychlorinated biphenyls, lead, p,p'-DDE, and other toxicants in Akwesasne Mohawk Youth. *Environ. Health Perspect., 116*(6), 806–813.https://doi.org/10.1289/ehp.10490.

Schneider, M. J., Fiering, S. N., Thai, B., Wu, S. Y., Germain, St. E., Parlow, A. F., Germain, St. D. L., & Galton, V. A., (2006). Targeted disruption of the type 1 selenodeiodinase gene (Dio1) results in marked changes in thyroid hormone economy in mice. *Endocrinol., 147*(1), 580–589. https://doi.org/10.1210/en.2005-0739.

Schreiber, G., Southwell, B. R., & Richardson, S. J., (1995). Hormone delivery systems to the brain - transthyretin. *Exp. Clin. Endocrinol. Diabetes, 103*(2), 75–80. https://doi.org/10.1055/s-0029-1211332.

Shu, L., Hoo, R. L. C., Wu, X., Pan, Y., Lee, I. P. C., Cheong, L. Y., Bornstein, S. R., et al., (2017). A-FABP mediates adaptive thermogenesis by promoting intracellular activation of thyroid hormones in brown adipocytes. *Nat. Commun., 8*(1), 1–16. https://doi.org/10.1038/ncomms14147.

Simonides, W. S., & Hardeveld, C. V., (2008). Thyroid hormone as a determinant of metabolic and contractile phenotype of skeletal muscle. *Thyroid, 18*(2), 205–216. https://doi.org/10.1089/thy.2007.0256.

Song, M., Kim, Y. J., Park, Y. K., & Ryu, J. C., (2012). Changes in thyroid peroxidase activity in response to various chemicals. *J. Environ. Monit., 14*(8), 2121–2126. https://doi.org/10.1039/c2em30106g.

Southwellp, B. R., Köhrle, J., Brack, C., Richardson, S. J., & Schreiber, G., (1993). Thyroxine transport to the brain: Role of protein synthesis by the choroid plexus. *Endocrinology, 133*(5), 2116–2126. https://doi.org/10.1210/endo.133.5.8404661.

Souza, D. S. D. M. C., Gonçalves, C. F. L., Vaisman, M., Ferreira, A. C. F., & Carvalho De, D. P., (2011). *Food Chem. Toxicol., 49*(10), 2495–2502. https://doi.org/10.1016/j.fct.2011.06.074.

Sparling, D. W., (2016). Organochlorine pesticides. *Ecotoxicol. Essentials*, 69–107. https://doi.org/10.1016/B978-0-12-801947-4.00004-4.

Stoker, T. E., Ferrell, J. M., Laws, S. C., Cooper, R. L., & Buckalew, A., (2006). Evaluation of ammonium perchlorate in the endocrine disruptor screening and testing program's male pubertal protocol: Ability to detect effects on thyroid endpoints. *Toxicol., 228*(1), 58–65. https://doi.org/10.1016/j.tox.2006.08.026.

Street, M. E., Angelini, S., Bernasconi, S., Burgio, E., Cassio, A., Catellani, C., Cirillo, F., et al., (2018). Current knowledge on endocrine disrupting chemicals (EDCs) from animal biology to humans, from pregnancy to adulthood: Highlights from a national Italian meeting. *Int. J. Mol. Sci., 19*(6), 1647. https://doi.org/10.3390/ijms19061647.

Sun, H. J., Li, H. B., Xiang, P., Zhang, X., & Ma, L. Q., (2015). Short-term exposure of arsenite disrupted thyroid endocrine system and altered gene transcription in the HPT axis in zebrafish. *Environ. Pollut., 205*, 145–152. https://doi.org/10.1016/j.envpol.2015.05.037.

Sun, H. J., Xiang, P., Tang, M. H., Sun, L., & Ma, L. Q., (2016). Arsenic impacted the development, thyroid hormone and gene transcription of thyroid hormone receptors in bighead carp larvae (*Hypophthalmichthys nobilis*). *J. Hazard. Mater., 303*, 76–82. https://doi.org/10.1016/j.jhazmat.2015.10.032.

Takser, L., Mergler, D., Baldwin, M., Grosbois, S. D., Smargiassi, A., & Lafond, J., (2005). Thyroid hormones in pregnancy in relation to environmental exposure to organochlorine compounds and mercury. *Environ. Health Perspect., 113*(8), 1039–1045. https://doi.org/10.1289/ehp.7685.

Terrien, X., Fini, J. B., Demeneix, B. A., Schramm, K. W., & Prunet, P., (2011). Generation of fluorescent zebrafish to study endocrine disruption and potential crosstalk between thyroid hormone and corticosteroids. *Aquat. Toxicol., 105*(1-2), 13-20. https://doi.org/10.1016/j.aquatox.2011.04.007.

Tomita, T., (2017). Apoptosis of pancreatic β-cells in type 1 diabetes. *Bosn. J. Basic Med. Sci., 17*(3), 183–193. https://doi.org/10.17305/bjbms.2017.1961.

Torino, F., Barnabei, A., Baldelli, R., & Appetecchi, M., (2012). Thyroid function abnormalities in patients receiving anticancer agents. In: *Thyroid Hormone* (p. 311). https://doi.org/10.5772/50375.

Walkowiak, J., Wiener, J. A., Fastabend, A., Heinzow, B., Krämer, U., Schmidt, E., Steingrüber, H. J., et al., (2001). Environmental exposure to polychlorinated biphenyls and quality of the home environment: Effects on psycho development in early childhood. *Lancet, 358*(9293), 1602–1607. https://doi.org/10.1016/S0140-6736(01)06654-5.

Yang, J. M., Salmon, A. G., & Marty, M. A., (2010). Development of TEFs for PCB congeners by using an alternative biomarker - thyroid hormone levels. *Regul. Toxicol. Pharmacol., 56*(2), 225–236.https://doi.org/10.1016/j.yrtph.2009.12.008.

Yoshida, K., Sugihira, N., Suzuki, M., Sakurada, T., Saito, S., Yoshinaga, K., & Saito, H., (1987). Effect of cadmium on T4 outer ring monodeiodination by rat liver. *Environ. Res., 42*(2), 400–405. https://doi.org/10.1016/S0013-9351(87)80206-2.

Young, P., Arch, J. R. S., & Ashwell, M., (1984). Brown adipose tissue in the parametrial fat pad of the mouse. *FEBS Lett., 167*(1), 10–14. https://doi.org/10.1016/0014-5793(84)80822-4.

Zienab, Y. M., & Heba, H. E. G., (2009). Effect of cadmium chloride on function of thyroid gland in rats. *Egyptian J. Comp. Pathol. Clinical Pathol., 22*(3), 10–23.

Zoeller, R. T., (2007). Environmental chemicals impacting the thyroid: Targets and consequences. *Thyroid, 17*(9), 811–817. https://doi.org/10.1089/thy.2007.0107.

Zoeller, R. T., Tan, S. W., & Tyl, R. W., (2007). General background on the hypothalamic-pituitary-thyroid (HPT) axis. *Crit. Rev. Toxicol., 37*(1, 2), 11–53. https://doi.org/10.1080/10408440601123446.

CHAPTER 11

Endocrine Toxicants and Reproductive Health of Males and Females

PREM LATA[1], SONIKA[1], and YOUNIS AHMAD HAJAM[2]

[1]*Department of Biosciences, Division of Zoology, Career Point University, Hamirpur, Himachal Pradesh–176041, India*

[2]*Department of Life Sciences and Allied Health Sciences, Sant Baba Bhag Singh University, Padhiana, Jalandhar, Punjab, India, E-mail: younismajeed64@gmail.com*

ABSTRACT

Endocrine toxicants (ETs) are chemicals having undesirable effects on individuals due to their intrusion into the endocrine system. Endocrine-disrupting chemicals (EDCs) lead to endocrinal hormonal imbalance. Other endocrine processes, such as adrenal axes, thyroid, and reproductive, have proof that endocrine-disrupting cells (EDCs) might act together with numerous endocrine systems. In the endocrine system, balanced functioning is essential for the conservation of the human species by providing them with normal growth, reproduction, and normal functioning of all organs. EDCs can be grouped into: pharmaceutical, industrial (alkyl phenols, polychlorinated biphenyls (PCBs), and dioxins) and agricultural toxicants include as insecticides, pesticides, and herbicides. Therefore, the present chapter summarizes the effect of ETs on male and female reproductive systems.

Environmental Endocrine Toxicants: Biology, Effects, and Management. Younis Ahmad Hajam, Rouf Ahmad Bhat, Khalid Rehman Hakeem, & Rajesh Kumar (Eds.)
© 2023 Apple Academic Press, Inc. Co-published with CRC Press (Taylor & Francis)

11.1 INTRODUCTION

In the endocrine system, balanced functioning is essential for the conservation of the human species by providing them normal growth, reproduction, and normal functioning of all organs. Endocrine disrupting chemicals (EDCs) can be grouped into: pharmaceutical, industrial (alkyl phenols, polychlorinated biphenyls (PCBs) and dioxins), and agricultural toxicants include as insecticides, pesticides, and herbicides (Coster, 2012). Environmental pollutants are endocrine toxicants (ETs), having undesirable effects on individuals due to their intrusion into the endocrine system. EDCs lead to endocrinal hormonal imbalance. Other endocrine processes, such as adrenal axes, thyroid, and reproductive, have proof that endocrine-disrupting cells (EDCs) might act together with numerous endocrine systems (Amaral, 2002; Bauer et al., 2002). Moreover, many organs, including brain development are controlled by hormonal signals; hence, endocrine-disrupting cells EDCs can change growth processes by obstructing these signaling (Figure 11.1) (Panzica et al., 2005). Gonadal hormones 17b-estradiol (E2) and androgens have more effect on brain growth and also on the growth of primary and secondary sex as well as numerous steroid behaviors. Surprisingly in low concentration, the estrogenic effect enormously affects the growth and role of estrogen-dependent neural networks and their behavior (Ottinger et al., 2002). The endocrine role in animals and human are changed by numerous chemicals which are present in the environment and vulnerability to such endocrine-disrupting chemicals EDCs might cause harmful outcomes on the neurological role, fetal/child growth, metabolism, and reproduction. The human species or its offspring badly affected by ETs (Marcoccia et al., 2017). These ETs could be present in numerous products and goods, such as, in deodorant, cosmetic products foodstuff, drinking water, synthetic materials, shampoos, clothes, kinds of toothpaste, body soaps, soil fertilizers, paper, textiles, carpets, utensils, bedding, toys, etc. (Joensen et al., 2009). Human species are exposed to various ETs by using these products; they can be entered into the human body via ingestion, germination, and inhalation and through skin contact. Therefore, ETs are highlighted for the reason of their rising usefulness in everyday self-care products and likely linked to compromised male reproductive health. Testosterone raises the formation of male secondary characteristics and also the process of sperm formation, hence, the endocrine system is primarily essential for male reproductive growth (Nordkap et al., 2012). Due to the increase in different ETs, there is a decline in male testosterone and sperm count (Joensen et al., 2009).

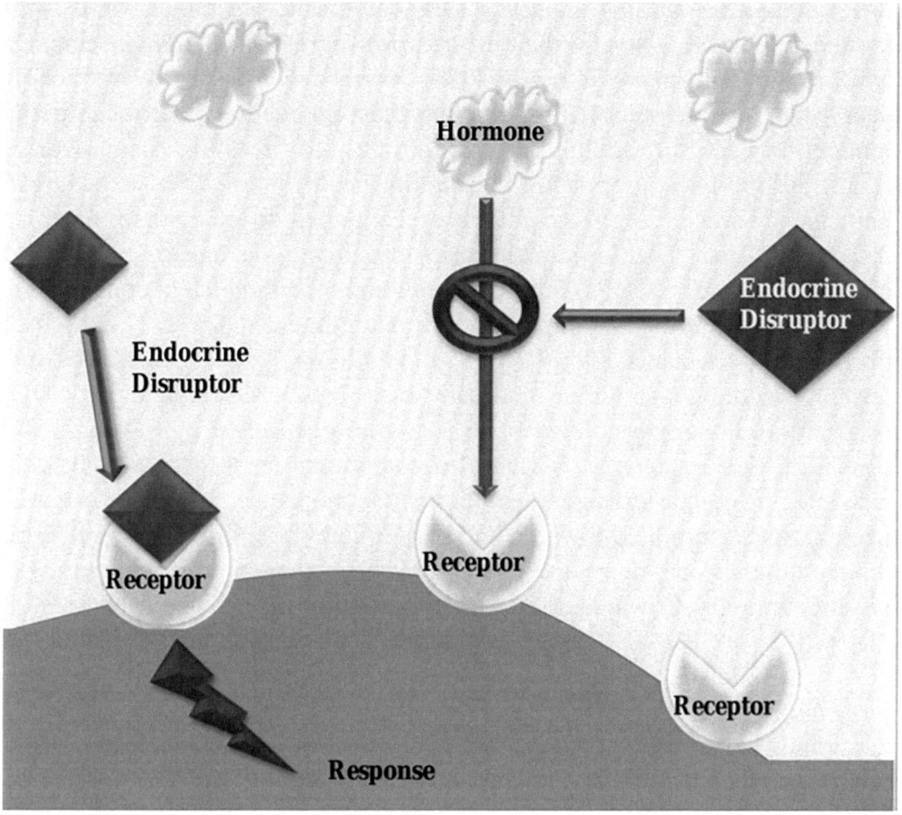

FIGURE 11.1 In a cell, instead of correct hormone binding with receptor, endocrine disruptor binds with the receptors. Thus, blocking the normal signaling and leading to the low response from the body.

11.2 IMPACT OF ENDOCRINE-DISRUPTING CHEMICALS ON MALE REPRODUCTION HEALTH

11.2.1 ORGANOCHLORINES PESTICIDES AND PCBS

For decades PCBs, i.e., PCBs were used for the manufacturing of industrial and consumer products. These compounds belong to the class of synthetic, lipophilic, and tenacious complexes. PCBs are included in the class of non-natural chlorinated compounds, which are also known as EDCs or ethylene dichloride. PCBs were utilized in industrialized and end-user products, for instance, hydraulic fluids and transformers, and as a chemical addition in oils, paints, and building materials. Organochlorines pesticides were familiarized in the year 1940s for their efficacy in contrast to different insects. The consumption and manufacture of PCBs have been debarred ever since the year 1970s, on the other hand, there has been no decline or an insignificant reduction in the surroundings from the time mid of 1990s (Kania-Korwel et al., 2009). There is a static substantial health threat from exposure of humans towards PCBs from intake of contained foods (Norström et al., 2010). DDT or "Dichlorodiphenyltrichloroethane" is one of the organochlorine pesticides (OCPs) but is the maximum examined and consequently was nominated to be encompassed in this argument. DDT or "Dichlorodiphenyltrichloroethane" and its chief "metabolite *p,p′*-DDE" is an extensive, tenacious ecological pollutant (Turusov et al., 2002). Mechanical grade DDT or Dichlorodiphenyltrichloroethane is a combination of *o,p′*-DDT (15%), *p,p′*-DDT (85%), and "*o,o′*-DDT (trace amounts)." Both *o*, *p′*-DDT, and *p,p′*-DDT stimulate estrogenic property (Bhatia et al., 2005). Propagative irregularities accredited towards Dichlorodiphenyltrichloroethane/Dichlorodiphenyldichloroethylene disclosure have been recounted in a diversity of wildlife animals (Edwards et al., 2006; Guillette et al., 2006) and laboratory rats (Kilian et al., 2007). In recent times, epidemiological investigations have inscribed the connection concerning revelation to DDT or Dichlorodiphenyltrichloroethane and male reproductive health (Rignell-Hydbom et al., 2004; Hauser et al., 2003). Consequently, of their existence and ubiquity into the surroundings, assessable intensities of serum PCBs, DDT/DDE, and many other organochlorines are instituted in the common general population (Patterson et al., 2009). Because of their lipophilicity nature and upper trophic level, these chemicals incorporate into the food chain. Numerous composites, including plastic contaminants, dioxins, triclosan (TCS), pesticides, and herbicides (e.g., diphenyl-dichloro-trichloroethane (DDT), metals, and others are known EDCs (Frye et al., 2012).

11.2.2 BISPHENOL-A (BPA)

Bisphenol-A (BPA) belongs to diphenylmethane derivatives and isophenols. Since 1957, it is commercially use and poorly soluble in water. In 2010, US Food and Drug Administration (USFDA) reported the risk in mother womb and infants. Human consume food contaminated with BPA which act as endocrine disruptors (EDs). BPA is used in the epoxy resin, plastic, food industry, e.g., lacquers for food cans and food packaging, etc., and dentistry.

BPA is prevailing in our surroundings and assessable concentrations have been spotted in the maximum of persons. Human beings are chiefly exposed to BPA through nutritive assimilation of leaching from the inside covering of sacks and microwave vessels throughout warming of foodstuffs and by the use of liquid refreshment in polycarbonate thermoses because of repetitive procedure or connection through any acid or alkaline (Biles et al., 1997). BPA is identified to have a scrawny estrogenic function because of its little empathy instead of the ERa (Steinmetz et al., 1997). Despite this fragile requisite property, very little amounts of BPA overseen throughout the prenatal age have significances on male mice, exactly amplified the prostate weight and declined production of the sperm (Nagel et al., 1997; Saal et al., 1998).

11.2.3 PHTHALATES

These are prevalent trade elements that are recounted to poorly disturb human reproductive consequences. Human disclosure towards phthalates is extensive accordingly of their increased use in consumer products and personal care products (PCPs) (Silva et al., 2004). Phthalate's exposure inhibits the response of testosterone hormone in the testis. It follows PPAR-alpha and p53 signaling pathways (Yang et al., 2015). They are classified into two separate classes on the basis of their mass of molecule, exact dissimilar properties, toxicological properties, and classification, i.e., high molecular weight complexes (di-2-Ethylhexyl phthalate (DEHP) with alkyl chain lengths from 8 to 13 carbons) have high weight of molecule and diethyl phthalate (DEP), dibutyl phthalate (DBP) having less weight of molecule. DHEP are produced 3 billion/kg annually all over the world and cause environmental pollution. It can be absorbed from water and food and known as EDs. It affects development and cause obesity and insulin resistance (IR) (Akash et al., 2021). It is principally used as a substance added to resin to promote flexibility and reduce brittleness. "Flexible vinyl," which consecutively, is used covering of wall, consumer products, and in devices which are used for medical purpose (ATSDR, 2002). Phthalates diethyl phthalate, DBP are used in daily used products as plasticizers and solvents for cellulose acetate. Disclosure over inhalation, "ingestion, and dermal contact" is deliberated imperative ways of disclosure for the common population. Upon disclosure, DHEP are promptly absorbed and defecated in urine (ATSDR, 2002). Human exposure to phthalates can be examined by the measuring the phthalate metabolites concentrations in urine samples, which is the most collective bio-monitoring method (Figure 11.2).

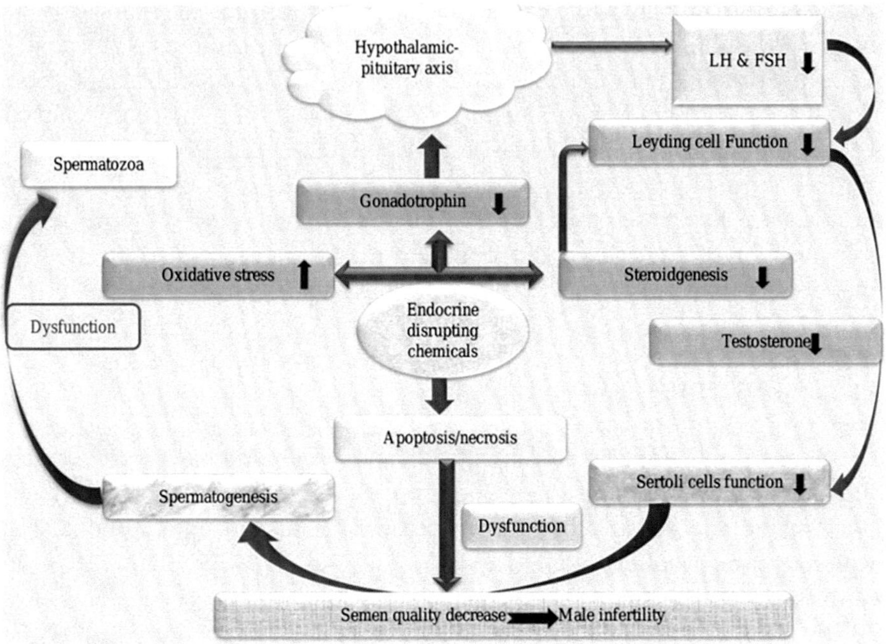

FIGURE 11.2 Various endocrine toxicants and their side effects on male reproductive health.

11.3 BPA CONSEQUENCES ON HUMAN HEALTH

11.3.1 REDUCED SPERM QUALITY

In the male reproductive system, the concentration and motility of sperm are lowered, and higher BPA in the male urine. It affects by chemicals or other metals. BPA was showing a negative correlation with sperm concentration and total sperm count in the control group (Li et al., 2017). According to USEPA, the total BPA concentration is lower (38.7 g/L) than normal daily intake (2678.5 g/L) 0.5 mg/kg per day. At low doses, BPA may be damaging sperm quality and cause males' infertility (Meeker et al., 2010). Due to some differences in the number and urine sample timing was determined by numerous statistical analyzes. There was substantially interconnection with BPA in the urine sample and lower sperm count, when analyzing a urine sample. It decreases/disruption the motion of sperm, DNA damage and sperm motion. Moreover, non-monotonic doses are significant responses in sperm motility, concentration, and percentage of the tail. Though, BPA exposure has not affected the sperm collected on that day, and also urine samples are a good measure of BPA during days (Mahalingaiah et al., 2008), and thus during sperm development may be an equally precise measure of contact with BPA in urine samples (Heller et al., 1964). Several studies show that in early morning concentration of BPA is less whereas in concentration of BPA was higher in the evening. The authors "suggested that men with sub-fertility may be more liable to BPA-related effects than men with normal fertility" (Meeker et al., 2010). Indeed, Mendiola et al. (2010) considered that the research of fertile men and higher BPA with decreased seminal volume ratio.

11.3.2 BISPHENOL-AN EFFECT ON THE SEXUAL FUNCTION OF MALE

BPA exposure was regulated by studying the ancient histories of the industrial unit, carrying out spot air samples, and accomplishing particular air monitoring (Li et al., 2010). Recent data revealed that BPA significantly lowers sexual function, such as sexual desire and erectile function. BPA contact in dose-dependent manner results in a reduction of sexual function (Li et al., 2010). The noteworthy, control group who were showed a significant correlation with a few parameters, such as sexual function environmentally but not occupationally. BPA exposure indicates and reduces the male sexual function in the control group (Li et al., 2010). Earlier industrial chemicals and dense metals contacts had no effect on the consequences (Li et al., 2010).

11.3.3 ABNORMALITIES IN MALE GENITAL

Abnormalities in male genital, like less distance from the anus to genitals (shorter anogenital distance (AGD)), have been liked with contact to anti-androgenic EDCs found in individuals (Marsee et al., 2006). The previous study, measured the AGD (0–17 years) occupationally contact with BPA and without BPA contacts (Miao et al., 2011). AGD was regulated for weight and height, and pre-pubertal and post-pubertal boys were grouped distinctly. Contacts with high BPA concentration showing smaller AGD and indicates BPA (anti-androgenic effects in utero) (Miao et al., 2011). Similarly, during embryonic development small adjusted AGD may show anti-androgenic contacts (Swan et al., 2005; Miao et al., 2011), even though more research needs to confirm this in human beings. Additionally, one abnormality in male genitals, cryptorchidism was not linked by the evolutional BPA contact in infants, levels parallel to those that produce unfavorable effects such as developmental diseases (Fenichel et al., 2012). In newborn babies (boys) were found interlinked BPA in the controlled study but not found cryptorchidism and hypospadias, and during the gestation period of the maternal urine samples were together (Chevrier et al., 2012). According to the present situation, there is no connection between BPA contact and "cryptorchidism" (Brucker-Davis et al., 2011). Prenatal BPA unveiling like AGD has been interlinked with anti-androgenic activity, through ER/AR-mediated mechanisms. Therefore, anti-androgenic activity is sometimes affected (Fenichel et al., 2012).

11.3.4 TYPE-2 DIABETES MELLITUS (T2DM)

In many research works, "type-2 diabetes mellitus (T2DM)" has been connected with BPA. Lang et al. (2008) were the first ones to analyze the connection between T2DM and BPA in individual, and data for the studies were taken (NHANES) 2003–2004. Adults between the age gap of 18–74 years old were inquired whether they had a history of suffering from certain diseases. On conducting serum analysis, blood glucose was discovered. It was later on revealed that high concentration of BPA in urine was notably linked to the inclined identification of "T2DM"; alteration in other chemical disclosure did not show any variation in observed outcomes (Lang et al., 2008). Notably, BPA concentrations did not show any relation with the measured glucose (Lang et al., 2008), signifying the possibility that diabetes medication may have a special role in modifying the glucose, therefore resulting in no linkage with diagnosed type-2 diabetes (Ning et al., 2011). BPA level declined in 2005–2006, explaining a fragile connection. Although, in the pooled sample of all the years, the level of BPA in urine showed association with a diabetes mellitus analysis (Melzer et al., 2003). Shankar et al. (2003) took NHANES data from 2003 to 2008, investigating applicants above the age of 20 who were already detected with the disease diabetes. Endpoints of serum glucose were examined, contradictory to self-statement diabetes mellitus in the preceding research. A positive linkage "between BPA and diabetes mellitus was observed among usual weight and overweight/obese patients, and smokers along with non-smokers" (Shankar et al., 2003). In last, Silver et al. (2011) also analyzed the relation between T2DM and level of BPA in urine, using the data from NHANES (2003–2008). They described "T2DM" in patients by self-statement utilize of T2DM medicine. They also calculated hemoglobin A1c (HbA1c), which may be a better alternative than fasting glucose to diagnose type-2 diabetes. They discovered that high urinary BPA showed close linkage to the inclined occurrence of type-2 diabetes in individuals along with inclined blood HbA1c levels. These noteworthy correlations were propelled by one study cycle (2003–2004) (Silver et al., 2011). "The NHANES studies are strong in that they provide a robust sample size, follow a standardized methodology, and can control for many demographic and other factors" (Lang et al., 2008; Silver et al., 2011). Additionally, the extent of these effects is huge and the pretentious consequences are certain. Ning et al. (2011) in their studies evaluated adults from China inclined level of BPA in urine was an unrelated connection by inclined diabetes mellitus, as calculated through glucose level in blood, through the probabilities of having T2DM somewhat inclined in the second and fourth quartile of BPA contact, but non the third. Based on the above studies, the authors claimed that no connection exists between total BPA and diabetes mellitus. This outcome was more convincing as it had a huge sample size. Although, the major drawback emphasized by the author was that in general, the connection between BPA and glucose may have been attenuated as 1,087 people were being treated for diabetes, and blood glucose levels may have declined in these individuals (Ning et al., 2011). There is a possibility that civilization affects the association between BPA contact in adults and type-2 diabetes, as a transversion Korean study observed no association between BPA and diagnosis of diabetes. Although the analysis on diabetes was solely dependent on self-reporting in these studies, which may under signify the actual occurrence of diabetes (Kim et al., 2011). Various studies from many Asian countries stated valuable relation between BPA and IR (Hong et al., 2009; Wang et al., 2012). Hong et al. (2009) analyzed huge adult individuals of Korea and discovered a noteworthy connection between inclined IRs and increased BPA concentration in urine, accompanied by other endpoints. They also calculated contact with various phthalate compounds and discovered a relation between one di-2-Ethylhexyl phthalate, MEHP, and inclined blood sugar level, signifying that BPA cannot be entirely accountable for the health effect (Hong et al., 2009). Others discovered identical results too. In the preceding stated study, Kandaraki et al. (2011) discovered an optimistic association between serums BPA with low insulin sensitivity. Wang et al. (2012) analyzed a large number of individuals and discovered that inclined urinary BPA showed an optimistic correlation with amplified IR.

11.3.5 HEART DISEASES, HIGH BLOOD PRESSURE, AND CHOLESTEROL LEVELS

Contacts with BPA also influence heart diseases and high blood. Lang et al. (2008) investigated that concentration of increased BPA in urine was correlated with an additional recurrent diagnosis of heart disease like coronary heart attack (Melzer et al., 2003) analyzed person between the age gap of 18–74 years old and discovered a major rise in "myocardial infarction, angina, coronary heart disease (CHD), CVD") with rise in urine BPA level in 2003–2004 records, but in the 2005-record, the only noteworthy cardiovascular endpoints were inclined myocardial infarction. Although, mutual record was noteworthy in every group (Melzer et al., 2003). Shankar et al. (2012) discovered that increased BPA in urine was correlated with the inclined occurrence of "peripheral arterial disease" in human. In a different paper, Shankar et al. (2012) found an individual that was detected with high blood pressure, either by calculating blood pressure or if they were testified to be on blood pressure plummeting medicine. Increased BPA in urine was correlated with high blood pressure in Korean individuals. BPA contact is also linked to the declined "heart rate variability (HRV)." HRV is also essential for 'proper working of the heart to match up by the blood requirement in the body. Reduced HRV and inclined blood pressure are equally danger issues in favor of heart diseases. Melzer et al. (2012) discovered that person through critical and median coronary artery disease (CAD) had remarkably increased BPA level in urine in contrast to normal controls. The authors found individuals with the age gap of 40–74 with CAD and monitored them for 10 years and 8 months. They analyzed the level of BPA from a particular one-day sample of urine, collected at research record. They discovered that increased levels of BPA in urine from preceding samples showed a correlation with increased prevalence of CAD throughout the 10 years and 8 months of research, signifying initial adult contact may have a long-term effect (Melzer et al., 2012). Olsen et al. (2012) discovered a fragile association between contact with BPA and CHD. The authors analyzed 70-year-old persons. However, they discovered that increased total serum BPA was correlated with increased LDL and HDL cholesterol levels. This importance gradually diminished when improvements were made during numerous trials. Other features like BMI diastolic and systolic blood pressure concentration of blood glucose levels also responsible for CHD risk showed no correlation with BPA contact. Indicating that NHANES has its drawback, but in general, it is strong (Olsen et al., 2012). Multivariate-adjusted effects were normally huge, and various independent laboratories confirmed results. It is fascinating to know that in the NHANES analysis, which analyzed participant's reported health outcomes, there were noteworthy correlations between the level of BPA in urine and CHD. The reason that merely these consequences were noteworthy recommends certainty of BPA as a factor responsible for heart-related problems. The further work analyzing heart diseases was too firm, by huge sample sizes and solid measurement and logical techniques. They generally calculated physiological outcomes (rather than analyzing doctor diagnosis), which fortifies the conclusions. In general, there is firm proof that contact of an adult with BPA is firmly related to heart diseases and severe heart health, in numerous individuals (Rochester, 2013).

11.3.6 LIVER FUNCTION

Lang et al. (2008) and Melzer et al. (2012) discovered variation in liver function of adults in accordance to increased BPA levels in urine. Lang et al. (2008), which analyzed the NHANES 2003 record, discovered increased BPA levels in urine showed noteworthy association with the increased level of liver enzymes "alkaline phosphatase," "gamma-glutamyl-transferase," and "lactate dehydrogenase." Melzer et al. (2012) showed a fragile connection between BPA levels in urine, as the data taken from NHANES (2005–2006) study showed no noteworthy relation between total BPA levels in urine and liver enzymes. Possibly the reduced BPA levels in urine as given in NHANES (2005–2006) analysis, may have resulted in fragile connections. BPA exposure effects on liver, increase the glucose synthesis and decrease the glycogen and insulin receptor signaling due to which glucose utilization decrease and fatty acid accumulations increase (Figure 11.3).

11.3.7 BPA EXPOSURE AND THEIR EFFECT ON IMMUNE FUNCTIONS

Overall calculation of immune function revealed a negative correlation between contact with BPA in a divisional examination of adults and children (above the age of six years old). Clayton et al. (2011) analyzed the NHANES (2003) record. They discovered that BPA levels in urine showed an important association with antibody titers to cytomegalovirus (CMV). Higher CMV antibody titers signify a weakened immune system and can be used as a tool to detect immune dysfunction in humans at an early stage. In adults (above the age of 18), increased BPA was connected with a higher CMV antibody titer, whereas in children and adolescents (below the age of 18) increased BPA was correlated with declined CMV antibody titer. The authors recommended that the inconsistency might be due to the length of the contact with BPA and that contact might have a severe effect on immune function with time.

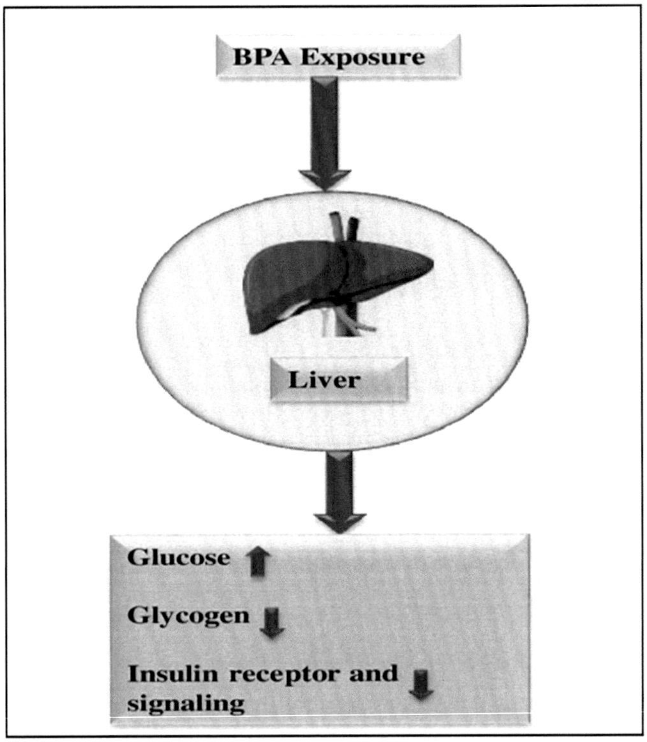

FIGURE 11.3 BPA exposure effects on liver, increase the glucose synthesis and decrease the glycogen and insulin receptor signaling due to which glucose utilization decrease and fatty acid accumulations increase.

11.3.8 EPIGENETIC, GENE EXPRESSION, AND SISTER CHROMATID EXCHANGE (SCE)

However, it is still not known how variations in the gene expression connect straight to severe health problems in individuals, some learning has discovered that BPA brings variation in these limitations. Another time, however, all of them were divisional learnings, few of these problems may be instant than the chronic growth of further sickness. Therefore, spot samples have proven to be a fine procedure of current BPA contact, which may be a better alternative than long-term studies to analyze these effects. Eichenlaub-Ritter and Pacchierotti (2015); Hann et al. (2018) analyzed women from a divisional study consisting of IVF patients from the inhabitants. They discovered that increased unrelated serum BPA was importantly connected with fewer methylations at the TSP50 P137CpG SITE, especially at the TSP50 gene promoter, in entire blood. The TSP50 gene codes for "testes-specific protease 50" whose function is not known. A decline in methylation signifies inclined gene expression at this rate. In breast malignancy cells an identical reduction in methylation of this gene is observed. Therefore, BPA is related to changed methylation, but its biological outcomes are still not known in this case. They discovered that increased total BPA in urine was related to the expression of two estrogen-responsive genes: "ESR2 (ER beta) and ESRRA (estrogen-related receptor [ERR] alpha)." ERR alpha is an orphan receptor, identical in structure to ER alpha, shows no affinity for estradiol, and is perhaps participates in ligand-nondependent transcriptional activity. Variations in ESR2 (ER beta) and ESRRA genes were calculated by RT-PCR. These genes may be over-expressed to control the estrogenic action of BPA. Even though these genes are linked to estrogenic activation/response, the biological functions are still known. At last, Yang et al. (2015) looked to found any connection between inclined sister chromatid exchange (SCE) and BPA contact. SCE frequency can be used as a measure to find chromosomal stability regarding contact to a mutagenic or carcinogen.

11.4 IMPACT OF EDCS EXPOSURE ON FEMALE REPRODUCTIVE HEALTH

Xenoestrogen is the main endocrine-disrupting chemical and has the capability of bringing out estrogenic effects and it is of global concern that xenoestrogen has put female reproduction at risk. Xenoestrogens are capable of exerting their influences at the genomic as well as the non-genomic level and activate various signaling pathways and downstream kinases (Özen and Darcan, 2001). It has been reported that contact to xenoestrogen impaired the whole hypothalamic-pituitary-gonadal system by elevating the generation of gonadotrophin-releasing hormone causing an elevated production of endogenous estrogen and premature puberty (Mantovan, 2002). Estrogen is a female steroid hormone responsible for regulating a broad range of events occurring in the female reproductive system, which includes endometrial growth, regulates the menstrual cycle, and the development of the breast.

Interestingly, ovaries and uteri parts of the female reproductive system are the most susceptible to xenoestrogen causing a surplus of female reproductive disorders, for instance, early puberty, premature ovarian failure, dysregulated fertility, and various cancers of the breast, ovaries, and uterus. Given the host of endocrine deformities related to a huge number of ovarian diseases like polycystic ovarian syndrome, endometriosis, ovarian cancer, etc., endocrine-disrupting chemicals in the environment are significantly considered to have possible relation with such Pathophysiology. It has been reported that there are 70% of women are diagnosed with sporadic breast cancer, and this threat is elevated at an exponential rate with the elevated contact to estrogenic compounds (Fenton, 2006). The other cause of infertility in reproductive-aged women is endometriosis; it is an estrogen-dependent female reproductive system disease, in which endometrium-like tissue grows outside the cavity of the uterus (Giudice and Kao, 2004). Various surveys have been done and showed that women who underwent estrogen replacement therapy are more prone to having endometrial cancer than those who never used estrogen replacement therapy (Sherman et al., 1995). Furthermore, the development of ovarian tumor cell lines expressing the estrogen receptor stimulated by estrogen, making ovarian cancer is having more prevalent among women experiencing different hormone therapy treatments (Cunat et al., 2004). So, it has been suggested in various reports that endocrine-disrupting chemicals can disrupt various functions of the reproductive system via hormone-driven processes causing increased risks for a variety of diseases. Though, endocrine-disrupting chemicals can enter inside the body of an organism throughout its whole life, Although, EDCs might enter into living organisms throughout their life. But the predominantly susceptible time for exposure is the prenatal period, covering the whole journey from the production of gametes, and fertilization through to intrauterine and postpartum development of progeny (Barrett and Sobolewski, 2014). The endocrine disruption and female reproductive health are the main area of concern because exposure during development can cause serious changes that will remain in adult life. BPA is a broadly used plasticizer that has properties like xenoestrogen, and it has been reported that it is present in a higher amount in the circulation of women with the polycystic ovarian syndrome. It has been reported that bisphenols-A has the potential to displace several bound androgens or hinders the catabolism of androgen and ultimately causing excess androgen in polycystic women (Zama et al., 2016). Bisphenols-A maybe moves from mother to child via two routes, i.e., lactational or transplacental. It has also been shown that BPA affects the morphology of the uterus and production of androgen by interfering with the genes related to RNA methylation

(Prokešová et al., 2020). Furthermore, di (2-Ethylhexyl) phthalate (DEHP) is a phthalate ester that is used in medical bags and tubing, has been reported to cause a short duration of pregnancy and elevated incidences of miscarriage in women who are exposed to it (Latini et al., 2003). It has been documented in a study in which mice were treated with genistein which is an isoflavone phytoestrogen and acts as a potent inhibitor of angiogenesis that causes various uterine defects such as precocious menstruation, unbalanced estrous cycle, and short reproductive lifespan (Jefferson et al., 2005).

11.5 ALTERATION IN HORMONE RECEPTOR SIGNALING, REPRODUCTIVE HORMONE PRODUCTION, SECRETION, AND METABOLISM

The hormonal balance inside the reproductive system of the female is maintained by the hypothalamic-pituitary-ovarian (HPO) axis. A peptide hormone GnRH is formed in the hypothalamus, which activate the "anterior pituitary (AP) through particular receptors present on the plasma membrane to synthesize and release gonadotropins, LH, and follicle-stimulating hormone (FSH)" (Padmanabhan et al., 2018). LH and FSH acting through the receptors present on the plasma membrane consecutively promote the development of follicles, ovulation corpus luteum maintenance, and the production of steroid hormones by the ovary (Padmanabhan et al., 2018). In response to the stimulation of LH and FSH, the ovary synthesizes two main steroid hormones such as progesterone and estradiol, and peptide hormones such as inhibin and activin. These hormones of the ovary consecutively regulate the production and secretion of gonadotropins via "positive and negative feedback to the pituitary and hypothalamus" (Padmanabhan et al., 2018). Moreover, progesterone and estradiol regulate the roles of target organs, for example, the uterus and mammary gland, typically through intracellular receptors. The AP releases the prolactin (PRL) which promotes the growth of the "mammary gland" and the production of milk (Seachrist et al., 2018). In the laboratory rodents, PRL also plays a function in controlling the functions of the ovary (BenJonathan et al., 2008). Another hormone that is released from the posterior pituitary gland is oxytocin; it is a hypothalamic protein hormone that stimulates the secretion of milk during parturition and contraction of the uterus (Seachrist et al., 2018). Toxicants might hinder one or many steps that are described above, causing many harmful effects to the developmental and reproduction of organisms. Toxicants might interfere with the HPO axis. For instance, female rats when exposed to the atrazine herbicide, it decreases the secretion of LH or PRL hormone by the pituitary gland (Cooper et al., 2007). There are various shreds of evidence that support that these effects are arbitrated through a "hypothalamic site of action." Pre-ovulatory content of gonadotrophin-releasing hormone is elevated upon exposure to atrazine in the hypothalamic site of action, reliable with reduced release of the gonadotrophin-releasing hormone as the reason of reduced preovulatory secretion of LH, and elevated "dopamine" for PRL secretion (Cooper et al., 2000, 2007; Stoker et al., 1999). There is a piece of evidence that "2,3,7,8-Tetrachlorodibenzo-p-dioxin (TCDD) changes the estradiol positive feedback to the hypothalamus which causes the preovulatory LH and FSH surge by reducing hypothalamic sensitivity to estradiol in rats" (Gao et al., 2001). It has been observed that the mice which are exposed to PFOA have decreased amount of GnRH, LH, and FSH. The numerous neurons and the amount of Kiss1 mRNA in hypothalamic nuclei which activate the liberate of the gonadotrophin-releasing hormone were decreased constantly with a reduced level of gonadotrophin-releasing hormone (Feng et al., 2015). The production of steroid hormones and metabolism might be directly disrupted by chemicals or these chemicals might interact with the receptor of hormones or serum binding proteins. The secretion of estradiol is reduced by "TCDD" without changing the activity of aromatase in cultured female granulosa cells, probably by reducing the enzymes that generate substrate of androgen for aromatization (Morán et al., 2000, 2003; Sanderson, 2006). In mice, the exposure of Perfluorooctanesulfonic acid inhibited the synthesis of progesterone and estradiol by down-regulating the steroidogenic acute regulatory (StAR) protein (Feng et al., 2015). BPA may change the proportion of steroid hormone that is bound to the serum binding protein (Hodgert et al., 2000).

11.6 CONTACT WITH BISPHENOLS A AFFECTS THE STEROID SYNTHESIS OF OVARIES

In all organisms, various types of hormones are secreted by endocrine glands which regulate the development, physiological processes, and homeostasis. According to the resemblance hormones, these activities together with different receptors present on target cells. BPA is EDCs it binds to the receptors of estrogen (α and β) with a binding affinity 10,000 times less than that of "endogenous Estradiol" (Kuiper et al., 1997; Mlynarcíková et al., 2005). BPA binds to the membrane receptors of γ and G-protein and the 17β-ethylandrostane "X receptor," therefore, "activating ion channels and inducing pro-inflammatory responses of cytokines and chemokines" (Chapin et al., 2008; Huang and Leung, 2009). 17β-ethylandrostane is acted as the substrate needed for enzyme "CYP450scc" to complete its cleavage and initiate the conversion from 17β-ethylandrostane to "pregn-5-en-3β-of-20-one." It is then converted into dehydroepiandrosterone (an androgen precursor), is then converted into an androgenic precursor, "Dehydroepiandrosterone," including the intermediary product, "17-hydroxipregnenolone," involving two enzymes in this conversion: "17 α-hydroxylase and 17,20-lyase." Then, dehydroepiandrosterone is converted into "4-androstenedione," which is change into testosterone. Because in the granulosa cells there is aromatase CYP450 is present, both androgens are converted into estrone and estradiol (Hillier et al., 1994). An exposure of 100 μg/ml of bisphenols-A shows decrease in the estradiol, estrone, testosterone, androstenedione, and dehydroepiandrosterone sulfate synthesis in a female rat (Peretz and Flaws, 2013). This high level of bisphenols-A compromised the growth of follicles, but following a dose decrease of 1 μg/ ml, no effect was observed (Takayanagi et al., 2006). Various studies are there which showed that bisphenols-A inhibits the "aromatase CYP450" mechanism in the granulosa cells, therefore decreasing the estradiol production (Mlynarcíková et al., 2005; Huang and Leung, 2009; Watanabe et al., 2012). Merely a small number of studies in humans are connected BPA with the synthesis of the ovarian follicle. Lee et al. (2014) found that young people on quick puberty unprotected to BPA had substantial proliferation in estradiol, pregnenolone, and testosterone concentrations. Instead, Mínguez-Alarcón et al. (2015) institute no statistical importance in the middle of BPA concentrations found in urine, serum E2 levels, and endometrial thickness dignified by ultrasound studies after regulating the outcomes for body mass index (BMI), age, race, and infertility diagnosis. Simply a small number of investigations assessing the damaging consequences of disclosure to di-2-Ethylhexyl phthalate and the harmful effects on steroid synthesis were accomplished, with unsatisfactory information assortment and poor numerical techniques. Hart et al. (2014) on "The Western Australian Pregnancy Cohort Study," defined the adverse effects of phthalate metabolic products on motherly serum SHBG concentrations, whereas the connotation among those similar metabolites with the motherly androgen concentrations was unpredictable. Further *in vitro* and animal investigations defined the effect of phthalates on regular steroid synthesis. Xu et al. (2010) displayed inhibition of aromatase in the granulosa cells of rats headed to reduce E2 concentrations. Svechnikova et al. (2007); and Liu et al. (2014) recounted reduced progesterone concentration in rats uncovered to diethylhexyl phthalate (DEHP) and in female rats at 20 days of age found reduced E2 amounts, however, also declined concentration of sexual hormone in adult rats although the results are quite debatable. It has been reported that disclosure to pesticides may alter the steroid synthesis process in females as well as have their limits. According to a study by Luderer et al. (2013) containing 457 Hawaiian members uncovered to heptachlor epoxide, detected a smaller luteal phase, and reduced serum amounts of estradiol and progesterone metabolites. About age, dose, and experimental model, atrazine impacts in steroid synthesis appear different. According to *in vitro* studies by Fa et al. (2013) concluded that in the case of female rats, enzyme expression in E2 level which are associated with immature Granulosa cells and steroid synthesis are changed due to Atrazine. Various other researchers: Taketa et al. (2011); Quignot et al. (2012); Tinfo et al. (2011); and Buck et al. (2014) displayed those recurrent dosages of Atrazine leads to increment in the enzyme expression in biosynthesis of steroid and amounts of sexual hormone. Therefore, there is a need for research in order to know about the pesticide's effects on the biosynthesis of steroids and fertility. Buck et al. (2014) displayed some data impacts of dissimilar DEC to female fertility. Ecological Toxicants: Su et al. (2012) accompanied investigations in humans presenting that high amount of PCBs and dioxins by-products reduced plasma estradiol concentrations. Experiment carried on dissimilar animals concluded that there is a negative effect of dioxins on the ovarian steroidogenesis. Additional consequences comprise reductions in the production of estradiol and antral follicles in case of female rats which subsequently lead to loss in enzyme production (Karman et al., 2012a, b). Though there are unblemished restrictions in these trials and more research is required in humans.

11.7 BISPHENOL-A (BPA) AND POLYCYSTIC OVARIAN DISORDER

Polycystic ovary disorder (PCOS) is a type of endocrine syndrome which is consists of many clinical disorders like IR linked with hypercholesterolemia, high androgen concentration, anovulatory cycles, weight gain, dyslipidemia, and metabolic variations. Nowadays, above 800 chemical products considered as ETs, which can intensely influence hormone receptors leads to infertility. An assessment of dissimilar phenotypes of patients with PCOS (polycystic ovary syndrome) discovered an extensive geographical, ethnic, and familial variety even between twin relations. Current investigations exhibited several women visible to chemical composites have an inherited vulnerability to emerging PCOS and several associated metabolic diseases. Time of disclosure to these EDs is essential to govern their influence, particularly throughout fetal progress, given their prospective dangerous impacts on gestation hormones and fetal cellular encoding (Palioura and Diamanti-Kandarakis, 2015; Diamanti-Kandarakis et al., 2007). Prenatal androgenization gives rise to epigenetic deviations and future progress of PCOS phenotypes (Xu et al., 2011). In case of animals, DNA methylation is triggered by plastic constituents and contact to phthalates and biphenyl in the F0-generation Zero give rise to trans-generational variation equipped to the third generation (F3) (Nilsson et al., 2012; Manikkamet al., 2012b, 2014). It is observed in a study that, female rats when given high doses of BPA in the newborn phase give rise to adult PCOS phenotypes with augmented plasma estradiol and testosterone concentrations, declined progesterone concentrations, and formation of ovarian cysts (Fernández et al., 2010). Fernández et al. (2009) also defined modifications in the excretion of gonadotrophin-releasing hormone and LH secretion from the pituitary gland. Despite having an enormous number of outcomes in the case of animal studies, associating their inference to individual is quite tough when the cystic part and the dissimilar phenotypes establish in animals diverge from the outcomes in humans of PCOS antral follicles. Disclosure to testosterone at the start of pregnancy in rhesus (RH) monkeys aggravate metabolic illnesses like many women with PCOS. Women with PCOS contact with EDCs cause irregular periods and high androgen (Abbott et al., 2013). Approximately 50–80% of women with PCOS were found to have IR which leads to the reduction in insulin sensitivity, hyperglycemia, and hyperinsulinemia, which subsequently leads to hyperandrogenism and chronic anovulation. More research is needed to know the mechanisms related to IR. BPA has reflected a contributing factor of child fatness associated with reduced adiponectin levels, the beginning of swelling, and more risk of generating diabetes type 2 and cardiovascular diseases (Menale et al., 2017). "Other mechanisms have been connected to BPA-related hyperglycemia and hyperinsulinemia with a direct effect on pancreatic cells; although no alterations on the pancreas islets have been documented" (Alonso-Magdalena et al., 2005, 2006).

11.8 ENDOCRINE-DISRUPTING CHEMICALS AND ENDOMETRIOSIS

Various etiological factors and the number of EDCs disturbing the cells of endometrial layers and cause endometriosis. EDs such as "TCDD"and "dioxin-like (DL) PCBs" can bring the growth of "peritoneal endometriosis in female RH monkeys, and the immensity of effect determined by infectivity level" (Rier et al., 2001). Bredhult et al. (2007) via estrogenization action angiogenic "effect of TCDD" the endometrial tissue production is activated. Though, further researcher was not able to authenticate or illustrate the association between endometriosis and DL PCB (Porpora et al., 2009; Buck Louis et al., 2012). Endometrial cells prevalence may also be affected by phthalates. In a women study with complex pelvic endometriosis is having risen in plasma level of "DEHP (di-(2-Ethylhexyl) phthalate)" and "MEHP (mono-(2-Ethylhexyl) phthalate)" in contrast with non-endometriosis women (Kim et al., 2011). A supplementary control case study shows that the elevated concentration of "mono-n-butyl-phthalates" in urine sample in women with endometriosis in contrast with non-endometriosis women (Huang et al., 2010). Women affected with pelvic endometriosis are having double level of 6 phthalate metabolites (Buck et al., 2013). A women study in USA northeast region demonstrates a contrary connection between the growth of endometriosis and concentration of MEHP (Upson et al., 2013). In a case study of non-fertile women (Itoh et al., 2009) verified these results, even though the authors incorporated only 80 endometriosis-free controls and 57 cases of endometriosis. Endometriosis growth is triggered by which pathway with the help of phthalates still indistinct. An *in vitro* experiment indicates that "DEHP" triggers endometrial cell stroma and also enhances the feasibility of Ishikawa cells (Kim et al., 2010).

11.9 CONTACT WITH ENDOCRINE TOXICANTS (ETS) AFFECTS THE PITUITARY GLAND, FERTILITY, AND MENSTRUAL CYCLES

In human studies, it has been found that there is an association between endocrine-disrupting chemicals (EDCs), i.e., "bisphenols, phthalates, and pesticides" and bad affects pituitary gland. Other hand, Xi et al. (2011); and Brannick et al. (2012) illustrate the rise in replication of pituitary gonadotrophin in female rats if there is exposure to pre-fetal and post-fetal to BPA. On the third day of the IVF cycle, it was studied that there is no association between vulnerability to BPA and FSH level (Souter et al., 2013). Toxicants like DES and dioxins having sometimes bad effects described in animal studies (Yoshida et al., 2011; Ishikawa et al., 2014) as having effects on the "pituitary gland and human gonadotropins." In women who are willing to conceive if exposed to organochlorine pollutants, it increases the three extra days on the gap of periods (Buck et al., 2011). Some studies showed irregularity in periods and estrous cycle in rats if expose to phthalates and BPA, respectively (Fernández et al., 2009; Nah et al., 2011; Vélez et al., 2015),

At the time of conception a study is done by the "Maternal-Infant Research on Environmental Chemicals" which incorporate pregnant women with the first trimester and pregnable to bisphenols and phthalates. A bacteria-killing phenolic complex TCS by elevated concentrations (>72 ng/ml), reduced the fertility, however, there is a negative association was established through phthalates and bisphenols.

11.10 ENDOCRINE DISRUPTORS (EDS) AFFECT THE OVARIES, UTERINE MYOMAS, VAGINA, AND UTERINE STRUCTURE

Recently, numerous studies show deformity in the growth of ovaries associated with EDs chemicals. The BPA effect on the growth of human beings ovaries still doubtful. A study shows ovarian steroidogenesis might be altered and ovarian follicles with multiple oocytes are increased due to the short doses of BPA (Rivera et al., 2011; Veiga-Lopez et al., 2013). Hunt et al. (2012) described the harmful effect of BPA in embryo growth of female monkey ovaries disturbing the early meiosis process, "recombination between homologous chromosomes," and synaptic variations. "Insufficient data is available on the impact of phthalates, pesticides, and other environmental toxicants in human prenatal and postnatal ovarian development." In the female rats, with the low dose of BPA, the number of follicles reduced whereas there is an increment in follicular atresia. Meanwhile, if the high dose was administered to the female rats, then an increment in "follicular cystification," "corpus luteum" evisceration, and decrement of "antral follicle" counts was observed (Rodríguez et al., 2010; Delclos et al., 2014). Chen et al. (2012) observed that if after the birth, human ovaries were exposed to "benzyl butyl phthalate" then chances would be high of growing granulosa cell apoptosis. The decreased follicle counts and increased follicular atresia were caused due to the postnatal ovary anomalies which were caused by the pesticides like malathion chlorpyriphos, endosulfan, and cypermethrin (Koç et al., 2009; Nandi et al., 2011). The lower fertilization rates, as well as under-development of oocytes in humans, may be attributed to the enhanced levels of ovarian follicular liquid and chlorinated bisphenols (Petro et al., 2012). The endometrium and cervix are the two main elements of the uterus which are significantly affected by the hormonal influence since the early stages of their development. Due to the periodical cycles after puberty, expansion, as well as hormonal variation, has been seen in the uterus, whereas after menopause, contraction, and complication of the uterus were observed. While the detailed effect of pesticide and BPA on the uterine structure of human beings is subjected to detailed research but in the case of animals, abnormalities have been observed. Development of stroma and endometrial glands was observed in adipose tissues of type Balb-c-adult female rats and may be attributed to the prolonged exposure to the BPA during the gestational and neonatal stages (Signorile et al., 2010, 2012). It was concluded that prolonged exposure of mono (carboxy-isooctyl) phthalate to human beings leads to variation in the uterine volume (Hart et al., 2014). It was concluded by Troisi et al. (2013); Harris and waring (2012); Herbst et al. (1971); and Dieckmann et al. (1953) that a higher number of uterine abnormalities were observed due to diethyl-stilbestrol (DES) in the daughters of pregnant women that were treated for the prevention of miscarriage during the gestation period. More recently, the recent studies done on the female rats and hamsters showed that there is an increased chance of endometrial adenocarcinoma and uterine anomalies due to the exposure of subjects to DES-induced endometrial hyperplasia/dysplasia (Alwis et al., 2011; Yoshida et al., 2011). Su et al. (2012) studied the effect of other environmental toxicants like dioxin and polychlorinated aromatic by-products of biphenyls on 33 young girls and concluded that prolonged exposure of these environmental toxicants leads to irregularity in structure of uterine, purpose of uterine, and fundus of these girls. Approximately 75–80% female population is affected by fibromyomas or Uterine myomas, which are a form of benign tumors. There is a hormone-dependent growth of nodular tumors and multiple myomas, which are further connected to the receptors such as estradiol and progesterone in the myometrium. The estradiol, which is the product of granulosa cells, activates the alpha and beta cellular receptors, which in turn regulate the production of endometrium and myometrium cells. The proteins are recruited for cellular reproduction in the cell nucleus, which may be due to the binding affinity between estradiol and its receptor. Ovulation, the corpus luteum produces progesterone, binding to progesterone receptors. These receptors support the manifestation of various genes, as a result of apart cellular responses (Figure 11.4).

The epigenetic term consigns to stable alterations in gene expression "that can be heritable during cell divisions" (Jirtle and Skinner, 2007). It includes "DNA methylation, posttranslational histone modifications, and chromatin remodeling, and expression of non-coding RNAs such as microRNAs (miRNAs)." All these epigenetic actions co-coordinately escort usual growth and allow active feedback to environmental stimuli. Contact with toxicants may cause resolute epigenetic alteration, which leads to changes in gene manifestation and the proteins which are necessary for growth and normal function (Bommarito et al., 2017). Prenatal Mice contact with BPA changes "methylation of maternally imprinted genes in oocytes" (Chao et al., 2012) and placentas and embryos (Susiarjo et al., 2013), resultant in irregular oocyte growth, placental growth, and formation of follicles. Women living in a polluted area having BPA gathering in the placenta of pregnant women, which increases the expression of miRNA (miR)-146a (De Felice et al., 2015). In mice rise in protein level and functional role of histone methyltransferase EZH2 mammary gland if prenatal mice expose to BPA (Doherty et al., 2010). Neonatal and prenatal mice contact with BPA prompted in resolute hypomethylation of sites in "several uterine gene promoters" (Li et al., 1997; Tang et al., 2008), and "hypermethylation of uterine homeobox A10 gene" (Hoxa10) respectively (Bromer et al., 2009). Site-specific tenure of customized "histones in oncogenic SIX homeobox 1 (Six1)," and enduringly change the "expression of Six1 in adult mice" via histone deacetylases, and reduced uterine levels of EZH2 if the neonatal comes in contact with DES (Jefferson et al., 2013). "miRNAs targeting the transforming growth factor-beta (TGFβ) pathway were detailed to be dysregulated in preeclamptic placentas from women with rising placental cadmium levels and in a human trophoblast cell model treated with cadmium *in vitro* (Brooks et al., 2016)." *In vitro* cadmium contacts miRNAs mediated an increase in TGF-b pathway signaling by which placental trophoblast migration decreased (Brooks and Fry, 2017). The elevated placental cadmium levels are linked with a rise in the expression of inflammatory signals and the genes, and 17 methylated CPG sites were revealed from an epigenome-wide linked study of placental DNA from two U.S. birth cohorts (Table 11.1) (Everson et al., 2018).

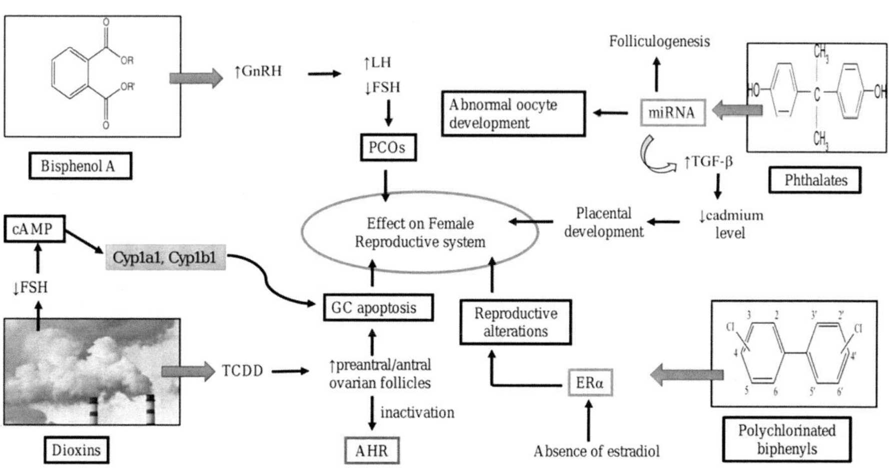

FIGURE 11.4 Effect of different endocrine toxicants on female reproductive system induces epigenetic alterations.

TABLE 11.1 Endocrine Disrupting Chemicals (EDCs) and Their Impact on Female Reproductive Health

SL. No.	Endocrine Disrupting Chemicals (EDCs)	Effect on Female Reproductive System (Parameters)	Adverse Effects	References
1.	Phthalates	↑MEHP and inhibits FSH-stimulated cAMP production; ↓Estradiol and LH; ↓FSH-stimulated cAMP production and progesterone hormone; Activation of PPARs	Multiple reproductive parameters, such as estrous cyclicity, pubertal age, litter size Alterations in serum hormone levels Ovarian morphology	Goldberg and Falcone (1999); Latini et al. (2003); Jefferson et al. (2005)
2.	Bisphenol-A	Ovarian expression of IGF-I; ↑IGF-I receptor, GDF9 and BMP-15.	Cystic endometrial hyperplasia Adenomyosis Leiomyomas (fibroids) Hyperplasia Stromal polyps	Zama et al. (2016); Zhang et al. (2020); Prokešová et al. (2020); Jiao et al. (2020)
3.	Polychlorinated biphenyls	↓FSH-stimulated cAMP production; ↓Estradiol and LH.	Alteration in serum hormones and uterine/ovarian weight Late pregnancy	Fenton (2006); Diamanti-Kandarakis et al. (2007); Palioura and Diamanti-Kandarakis (2015)
4.	Dioxins	↓LH in granulosa cells (GC), inhibit estradiol secretion, inhibits *Cyp11a1* and aromatase *Cyp19a1* gene	Fertility-disrupting complications Organ cancer Hepatotoxicity Gonadal Immune changes	Morán et al. (2000, 2003); Sanderson (2006); Fenton (2006); Feng et al. (2015)

KEYWORDS

- **consequences**
- **dibutyl phthalate**
- **diphenyl-dichloro-trichloroethane**
- **endocrine toxicants**
- **endocrine-disrupting cells**
- **females**
- **males**
- **reproduction**

REFERENCES

Abbott, D., Nicol, L., Levine, J., Xu, N., Goodarzi, M., & Dumesic, D., (2013). Nonhuman primate models of polycystic ovary syndrome. *Mol. Cell. Endocrinol., 373*(1, 2), 21–28.

Akash, M., & Rehman, K., (2021). Exposure to endocrine disrupting chemicals is a risk factor for the pathogenesis and development of metabolic disorders. *Metab., 116*, 154498.

Akintunde, J., Akintola, T., Aliu, F., Fajoye, M., & Adimchi, S., (2020). Naringin regulates erectile dysfunction by abolition of apoptosis and inflammation through NOS/Cgmp/PKG signaling pathway on exposure to bisphenol-A in hypertensive rat model. *Reprod. Toxicol., 95*, 123–136.

Alonso-Magdalena, P., Laribi, O., Ropero, A., Fuentes, E., Ripoll, C., Soria, B., & Nadal, A., (2005). Low doses of bisphenol A and diethylstilbestrol impair Ca^{2+} signals in pancreatic α-cells through a nonclassical membrane estrogen receptor within intact islets of Langerhans. *Environ. Health Perspect., 113*, 969–977.

Alonso-Magdalena, P., Morimoto, S., Ripoll, C., Fuentes, E., & Nadal, A., (2006). The estrogenic effect of bisphenol a disrupts pancreatic β-cell function *in vivo* and induces insulin resistance. *Environ. Health Perspect., 114*, 106–112.

Alwis, I., Maroni, D., Hendry, I., Roy, S., May, J., Leavitt, W., & Hendry, W., (2011). Neonatal diethylstilbestrol exposure disrupts female reproductive tract structure/function via both direct and indirect mechanisms in the hamster. *Reprod. Toxicol., 32*, 472–483.

Amaral, M. J., (2002). The endocrine disruptors: A major medical challenge. *Food. Chem. Toxicol., 40*, 781–788.

Bae, S., Kim, J., & Hong, Y., (2011). The effect of bisphenol a exposure on heart rate variability and blood pressure. *Epidemiol., 22*, S80.

Bauer, E., Bitsch, N., Brunn, H., Sauerwein, H., & Meyer, H., (2002). Development of an immuno-immobilized androgen receptor assay (IRA) and its application for the characterization of the receptor binding affinity of different pesticides. *Chemosphere, 46*, 1107–1115.

Ben-Jonathan, N., LaPensee, C., & LaPensee, E., (2007). What can we learn from rodents about prolactin in humans? *Endocr. Rev., 29*, 1–41.

Bhatia, R., Shiau, R., Petreas, M., Weintraub, J., Farhang, L., & Eskenazi, B., (2005). Organochlorine pesticides and male genital anomalies in the child health and development studies. *Environ. Health Perspect., 113*, 220–224.

Biles, J., McNeal, T., Begley, T., & Hollifield, H., (1998). Determination of bisphenol-A in reusable polycarbonate food-contact plastics and migration to food-simulating liquids. *J. Agr. Food Chem., 46*, 2894.

Bommarito, P., Martin, E., & Fry, R., (2017). Effects of prenatal exposure to endocrine disruptors and toxic metals on the fetal epigenome. *Epigenomics, 9*, 333–350.

Brannick, K., Craig, Z., Himes, A., Peretz, J., Wang, W., Flaws, J., & Raetzman, L., (2012). Prenatal exposure to low doses of bisphenol a increases pituitary proliferation and gonadotroph number in female mice offspring at birth. *Bio. Reprod., 87*.

Bredhult, C., Backlin, B., & Olovsson, M., (2007). Effects of some endocrine disruptors on the proliferation and viability of human endometrial endothelial cells *in vitro*. *Reprod. Toxicol., 23*, 550–559.

Bromer, J., Wu, J., Zhou, Y., & Taylor, H., (2009). Hypermethylation of homeobox A10 by in utero diethylstilbestrol exposure: An epigenetic mechanism for altered developmental programming. *Endocrinol., 150*, 3376–3382.

Brooks, S., Martin, E., Smeester, L., Grace, M., Boggess, K., & Fry, R., (2016). miRNAs as common regulators of the transforming growth factor (TGF)-B pathway in the preeclamptic placenta and cadmium-treated trophoblasts: Links between the environment, the epigenome and preeclampsia. *Food. Chem. Toxicol., 98*, 50–57.

Brucker-Davis, F., Ferrari, P., Boda-Buccino, M., Wagner-Mahler, K., Pacini, P., Gal, J., Azuar, P., & Fenichel, P., (2011). Cord blood thyroid tests in boys born with and without cryptorchidism: Correlations with birth parameters and in uteroxenobiotics exposure. *Thyroid, 21*, 1133–1141.

Buck, L. G., Sundaram, R., Sweeney, A., Schisterman, E., Maisog, J., & Kannan, K., (2014). Urinary bisphenol A, phthalates, and couple fecundity: The longitudinal investigation of fertility and the environment (LIFE) study. *Fertil. Steril., 101*, 1359–1366.

Chao, H., Zhang, X., Chen, B., Pan, B., Zhang, L., Li, L., Sun, X., Shi, Q., & Shen, W., (2011). Bisphenol A exposure modifies methylation of imprinted genes in mouse oocytes via the estrogen receptor signaling pathway. *Histochem. Cell Biol., 137*, 249–259.

Chapin, R., Adams, J., Boekelheide, K., Hayward, S., McIntyre, B., Portier, K., Lees, P., et al., (2004). NTP-CERHR expert panel report on the reproductive and developmental toxicity of ethylene glycol. *Reprod. Toxicol., 18*, 457–532.

Chen, H., Chiang, P., Wang, Y., Kao, M., Shieh, T., Tsai, C., & Tsai, E., (2012). Benzyl butyl phthalate induces necrosis by Ahr mediation of CYP1B1 expression in human granulosa cells. *Reprod. Toxicol., 33*, 67–75.

Chevrier, C., Petit, C., Philippat, C., Mortamais, M., Slama, R., Rouget, F., Calafat, A., et al., (2012). Maternal urinary phthalates and phenols and male genital anomalies. *Epidemiol., 23*, 353–356.

Cooper, R., Laws, S., Das, P., Narotsky, M., Goldman, J., Lee, T. E., & Stoker, T., (2007). Atrazine and reproductive function: Mode and mechanism of action studies. *Birth defects Res. B. Dev. Reprod. Toxicol., 80*, 98–112.

Costa, E., Spritzer, P., Hohl, A., & Bachega, T., (2014). Effects of endocrine disruptors in the development of the female reproductive tract. *Arq. Bras. Endocrinol. Metabol., 58*, 153–161.

Cunat, S., Hoffmann, P., & Pujol, P., (2004). Estrogens and epithelial ovarian cancer. *Gynecol. Oncol., 94*, 25–32.

De Coster, S., & Van, L. N., (2012). Endocrine-disrupting chemicals: Associated disorders and mechanisms of action. *J. Environ. Public Health.,* 1–52.

De Felice, B., Manfellotto, F., Palumbo, A., Troisi, J., Zullo, F., Di Carlo, C., Sardo, A., et al., (2015). Genome-wide microRNA expression profiling in placentas from pregnant women exposed to BPA. *BMC Medical Genom., 8.*

Delclos, K., Camacho, L., Lewis, S., Vanlandingham, M., Latendresse, J., Olson, G., Davis, K., et al., (2014). Toxicity evaluation of bisphenol a administered by gavage to Sprague Dawley rats from gestation day 6 through postnatal day 90. *Toxicol. Sci., 139,* 174–197.

Diamanti-Kandarakis, E., Piperi, C., Patsouris, E., Korkolopoulou, P., Panidis, D., Pawelczyk, L., Papavassiliou, A., & Duleba, A., (2007). Immunohistochemical localization of advanced glycation end-products (AGEs) and their receptor (RAGE) in polycystic and normal ovaries. *Histochem. Cell Bio., 127,* 581–589.

Dieckmann, W., Davis, M., Rynkiewicz, L., & Pottinger, R., (1999). Does the administration of diethylstilbestrol during pregnancy have therapeutic value? *Am. J. Obstet. Gynecol., 181,* 1572–1573.

Doherty, L., Bromer, J., Zhou, Y., Aldad, T., & Taylor, H., (2010). In utero exposure to diethylstilbestrol (DES) or bisphenol-A (BPA) increases EZH2 expression in the mammary gland: An epigenetic mechanism linking endocrine disruptors to breast cancer. *Horm. Cancer., 1,* 146–155.

Edwards, T., Moore, B., & Guillette, L., (2006). Reproductive dysgenesis in wildlife: A comparative view. *Int. J. Androl., 29,* 109–121.

Eichenlaub-Ritter, U., Vogt, E., Cukurcam, S., Sun, F., Pacchierotti, F., & Parry, J., (2008). Exposure of mouse oocytes to bisphenol a causes meiotic arrest but not aneuploidy. *Mutat. Res. Genet. Toxicol. Environ. Mutagen., 651,* 82–92.

Everson, T., Punshon, T., Jackson, B., Hao, K., Lambertini, L., Chen, J., Karagas, M., & Marsit, C., (2018). Cadmium-associated differential methylation throughout the placental genome: Epigenome-wide association study of two U.S. birth cohorts. *Environ. Health Perspect., 126,* 017010.

Fa, S., Pogrmic-Majkic, K., Samardzija, D., Glisic, B., Kaisarevic, S., Kovacevic, R., & Andric, N., (2013). Involvement of ERK1/2 signaling pathway in atrazine action on FSH-stimulated LHR and CYP19A1 expression in rat granulosa cells. *Toxicol Applied Pharmacol., 270,* 1–8.

Feng, X., Wang, X., Cao, X., Xia, Y., Zhou, R., & Chen, L., (2015). Chronic exposure of female mice to an environmental level of perfluorooctane sulfonate suppresses estrogen synthesis through reduced histone H3K14 acetylation of the star promoter leading to deficits in follicular development and ovulation. *Toxicol. Sci., 148,* 368–379.

Fenichel, P., Dechaux, H., Harthe, C., Gal, J., Ferrari, P., Pacini, P., Wagner-Mahler, K., Pugeat, M., & Brucker-Davis, F., (2012). Unconjugated bisphenol A cord blood levels in boys with descended or undescended testes. *Human. Reprod., 27,* 983–990.

Fenton, S., (2006). Endocrine-disrupting compounds and mammary gland development: Early exposure and later life consequences. *Endocrinol., 147,* s18–s24.

Fernández, M., Bianchi, M., Lux-Lantos, V., & Libertun, C., (2009). Neonatal exposure to bisphenol A alters reproductive parameters and gonadotropin-releasing hormone signaling in female rats. *Environ. Health Perspect., 117,* 757–762.

Fernández, M., Bourguignon, N., Lux-Lantos, V., & Libertun, C., (2010). Neonatal exposure to bisphenol A and reproductive and endocrine alterations resembling the polycystic ovarian syndrome in adult rats. *Environ. Health Perspect., 118,* 1217–1222.

Frye, C., Bo, E., Calamandrei, G., Calzà, L., Dessì-Fulgheri, F., Fernández, M., Fusani, L., et al., (2011). Endocrine disruptors: A review of some sources, effects, and mechanisms of actions on behavior and neuroendocrine systems. *J. Neuroendocrinol., 24*, 144–159.

Gao, X., Mizuyachi, K., Terranova, P., & Rozman, K., (2001). 2,3,7,8-tetrachlorodibenzo-p-dioxin decreases responsiveness of the hypothalamus to estradiol as a feedback inducer of preovulatory gonadotropin secretion in the immature gonadotropin-primed rat. *Toxicol. Applied Pharmacol., 170*, 181–190.

Giudice, L., & Kao, L., (2004). Endometriosis. *Lancet, 364*, 1789–1799.

Goldberg, J., & Falcone, T., (1999). Effect of diethylstilbestrol on reproductive function. *Fertil. Steril., 72*, 1–7.

Guillette, L., & Moore, B., (2006). Environmental contaminants, fertility, and multioocytic follicles: A lesson from wildlife? *Semin. Reprod. Med., 24*, 134–141.

Hanna, C., Bloom, M., Robinson, W., Kim, D., Parsons, P., Vom, S. F., Taylor, J., et al., (2012). DNA methylation changes in whole blood are associated with exposure to the environmental contaminants, mercury, lead, cadmium and bisphenol A, in women undergoing ovarian stimulation for IVF. *Human Reprod., 27*, 1401–1410.

Harris, R., & Waring, R., (2012). Diethylstilboestrol—A long-term legacy. *Maturitas., 72*, 108–112.

Hart, R., Doherty, D., Frederiksen, H., Keelan, J., Hickey, M., Sloboda, D., Pennell, C., et al., (2014). The influence of antenatal exposure to phthalates on subsequent female reproductive development in adolescence: A pilot study. *Reprod, 147*, 379–390.

Hauser, R., Chen, Z., Pothier, L., Ryan, L., & Altshul, L., (2003). The relationship between human semen parameters and environmental exposure to polychlorinated biphenyls and P,P'-DDE. *Environ. Health Perspect., 111*, 1505–1511.

Herbst, A., Ulfelder, H., & Poskanzer, D., (1976). Summary of the changes in the human female genital tract as a consequence of maternal diethylstilbestrol therapy. *J. Toxicol. Environ. Health., 1*, 13–20.

Herbst, A., Ulfelder, H., & Poskanzer, D., (1999). Adenocarcinoma of the vagina. Association of maternal stilbestrol therapy with tumor appearance in young women. *Am. J. Obstet. Gynecol., 181*, 1574–1575.

Hillier, S., Whitelaw, P., & Smyth, C., (1994). Follicular oestrogen synthesis: The 'two-cell, two-gonadotrophin' model revisited. *Mol. Cell. Endocrinol., 100*, 51–54.

Hong, Y., Park, E., Park, M., Ko, J., Oh, S., Kim, H., Lee, K., Leem, J., & Ha, E., (2009). Community level exposure to chemicals and oxidative stress in adult population. *Toxicol. Lett., 184*, 139–144.

Huang, P., Tsai, E., Li, W., Liao, P., Chung, M., Wang, Y., & Wang, S., (2010). Association between phthalate exposure and glutathione S-transferase M1 polymorphism in adenomyosis, leiomyoma and endometriosis. *Human Reprod., 25*, 986–994.

Ishikawa, M., Murai, E., Hashiguchi, Y., Iguchi, T., & Sato, T., (2014). Effects of diethylstilbestrol on luteinizing hormone-producing cells in the mouse anterior pituitary. *Exp. Biol. Med., 239*, 311–319.

Itoh, H., Iwasaki, M., Hanaoka, T., Sasaki, H., Tanaka, T., & Tsugane, S., (2009). Urinary phthalate monoesters and endometriosis in infertile Japanese women. *Sci. Total Environ., 408*, 37–42.

Jefferson, W., Chevalier, D., Phelps, J., Cantor, A., Padilla-Banks, E., Newbold, R., Archer, T., et al., (2013). Persistently altered epigenetic marks in the mouse uterus after neonatal estrogen exposure. *Mol. Endocrinol., 27*, 1666–1677.

Jefferson, W., Padilla-Banks, E., & Newbold, R., (2005). Adverse effects on female development and reproduction in CD-1 mice following neonatal exposure to the phytoestrogen genistein at environmentally relevant doses. *Biol. Reprod., 73*, 798–806.

Jiao, X., Ding, Z., Meng, F., Zhang, X., Wang, Y., Chen, F., Duan, Z., et al., (2019). The toxic effects of fluorene-9-bisphenol on porcine oocyte *in vitro* maturation. *Environ. Toxicol., 35*, 152–158.

Jirtle, R., & Skinner, M., (2007). Environmental epigenomics and disease susceptibility. *Nat. Rev. Genet., 8*, 253–262.

Joensen, U., Bossi, R., Leffers, H., Jensen, A., Skakkebæk, N., & Jørgensen, N., (2009). Do perfluoroalkyl compounds impair human semen quality? *Environ. Health Perspect., 117*, 923–927.

Jury, H., Zacharewski, T., & Hammond, G., (2000). Interactions between human plasma sex hormone-binding globulin and xenobiotic ligands. *J. Steroid Biochem. Mol. Biol., 75*, 167–176.

Kandaraki, E., Chatzigeorgiou, A., Livadas, S., Palioura, E., Economou, F., Koutsilieris, M., Palimeri, S., et al., (2011). Endocrine disruptors and polycystic ovary syndrome (PCOS): Elevated serum levels of bisphenol a in women with PCOS. *J. Clin. Endocrinol. Metab., 96*, 480–484.

Kania-Korwel, I., El-Komy, M., Veng-Pedersen, P., & Lehmler, H., (2009). Clearance of polychlorinated biphenyl atropisomers is enantioselective in female C57bl/6 mice. *Environ. Sci. Technol., 44*, 2828–2835.

Karman, B., Basavarajappa, M., Hannon, P., & Flaws, J., (2012). Dioxin exposure reduces the steroidogenic capacity of mouse antral follicles mainly at the level of HSD17B1 without altering atresia. *Toxicol. Appl. Pharmacol., 264*, 1–12.

Kilian, E., Delport, R., Bornman, M., & Jager, C., (2007). Simultaneous exposure to low concentrations of dichlorodiphenyltrichloroethane, deltamethrin, nonylphenol and phytoestrogens has negative effects on the reproductive parameters in male Sprague-Dawley rats. *Androl., 39*, 128–135.

Kim, S., Chun, S., Jang, J., Chae, H., Kim, C., & Kang, B., (2011). Increased plasma levels of phthalate esters in women with advanced-stage endometriosis: A prospective case-control study. *Fertil. Steril., 95*, 357–359.

Koc, N., Kayhan, F., Sesal, C., & Muslu, M., (2009). Dose-dependent effects of endosulfan and malathion on adult Wistar albino rat ovaries. *Pak. J. Biol. Sci., 12*, 498–503.

Kuiper, G., Carlsson, B., Grandien, K., Enmark, E., Häggblad, J., Nilsson, S., & Gustafsson, J., (1997). Comparison of the ligand binding specificity and transcript tissue distribution of estrogen receptors A and B. *Endocrinol., 138*, 863–870.

Lang, I., (2008). Association of urinary bisphenol a concentration with medical disorders and laboratory abnormalities in adults. *JAMA, 300*, 1303.

Latini, G., De Felice, C., Presta, G., Del, V. A., Paris, I., Ruggieri, F., & Mazzeo, P., (2003). In utero exposure to di-(2-ethylhexyl)phthalate and duration of human pregnancy. *Environ. Health Perspect., 111*, 1783–1785.

Lee, S., Kang, S., Choi, M., Lee, J., Park, M., Kim, S., Lee, W., Hong, J., & Chung, B., (2014). Changes in steroid metabolism among girls with precocious puberty may not be associated with urinary levels of bisphenol A. *Reprod. Toxicol., 44*, 1–6.

Li, D., Zhou, Z., Miao, M., He, Y., Qing, D., Wu, T., Wang, J., et al.,(2010). Relationship between urine bisphenol-A level and declining male sexual function. *J. Androl., 31*, 500–506.

Li, S., Washburn, K., Moore, R., Uno, T., Teng, C., Newbold, R., Negishi, M., & Mclachlan, J., (1997). Developmental exposure to diethylstilbestrol elicits demethylation of estrogen-responsive lactoferrin gene in mouse uterus. *Cancer Res., 57*, 4356–4359.

Liu, T., Li, N., Zhu, J., Yu, G., Guo, K., Zhou, L., Zheng, D., et al., (2014). Effects of di-(2-ethylhexyl) phthalate on the hypothalamus-pituitary-ovarian axis in adult female rats. *Reprod. Toxicol., 46*, 141–147.

Luderer, U., Eskenazi, B., Hauser, R., Korach, K., McHale, C., Moran, F., Rieswijk, L., et al., (2019). Proposed key characteristics of female reproductive toxicants as an approach for organizing and evaluating mechanistic data in hazard assessment. *Environ. Health Perspect., 127*, 075001.

Luderer, U., Kesner, J., Fuller, J., Krieg, E., Meadows, J., Tramma, S., Yang, H., & Baker, D., (2013). Effects of gestational and lactational exposure to heptachlor epoxide on age at puberty and reproductive function in men and women. *Environ. Res., 121*, 84–94.

Mahalingaiah, S., Meeker, J., Pearson, K., Calafat, A., Ye, X., Petrozza, J., & Hauser, R., (2008). Temporal variability and predictors of urinary bisphenol a concentrations in men and women. *Environ. Health Perspect., 116*, 173–178.

Manikkam, M., Guerrero-Bosagna, C., Tracey, R., Haque, M., & Skinner, M., (2012). Transgenerational actions of environmental compounds on reproductive disease and identification of epigenetic biomarkers of ancestral exposures. *Plos One, 7*, 31901.

Manikkam, M., Haque, M., Guerrero-Bosagna, C., Nilsson, E., & Skinner, M., (2014). Pesticide methoxychlor promotes the epigenetic transgenerational inheritance of adult-onset disease through the female germline. *PLoS One, 9*, 102091.

Mantovani, A., (2002). Hazard identification and risk assessment of endocrine disrupting chemicals with regard to developmental effects. *Toxicol., 181, 182*, 367–370.

Marcoccia, D., Pellegrini, M., Fiocchetti, M., Lorenzetti, S., & Marino, M., (2017). Food components and contaminants as (Anti)androgenic molecules. *Genes Nutr., 12*.

Marsee, K., Woodruff, T., Axelrad, D., Calafat, A., & Swan, S., (2006). Estimated daily phthalate exposures in a population of mothers of male infants exhibiting reduced anogenital distance. *Environ. Health Perspect., 114*, 805–809.

Meeker, J., Ehrlich, S., Toth, T., Wright, D., Calafat, A., Trisini, A., Ye, X., & Hauser, R., (2010). Semen quality and sperm DNA damage in relation to urinary bisphenol A among men from an infertility clinic. *Reprod. Toxicol., 30*, 532–539.

Melzer, D., Gates, P., Osborn, N., Henley, W., Cipelli, R., Young, A., Money, C., et al., (2012). Correction: Urinary bisphenol A concentration and angiography-defined coronary artery stenosis. *Plos One, 7*.

Melzer, D., Rice, N., Lewis, C., Henley, W., & Galloway, T., (2010). Association of urinary bisphenol a concentration with heart disease: Evidence from NHANES 2003/06. *PLoS One, 5*, 8673.

Menale, C., Grandone, A., Nicolucci, C., Cirillo, G., Crispi, S., Di Sessa, A., Marzuillo, P., et al., (2016). Bisphenol A is associated with insulin resistance and modulates adiponectin and resistin gene expression in obese children. *Pediatr. Obes., 12*, 380–387.

Mendiola, J., Jørgensen, N., Andersson, A., Calafat, A., Ye, X., Redmon, J., Drobnis, E., et al., (2010). Are environmental levels of bisphenol a associated with reproductive function in fertile men? *Environ. Health Perspect., 118*, 1286–1291.

Miao, M., Yuan, W., He, Y., Zhou, Z., Wang, J., & Li, D., (2011). Parental exposure to bisphenol a during pregnancy and the anogenital distance of male offspring. *Epidemiol., 22*, S242.

Mínguez-Alarcón, L., Gaskins, A., Chiu, Y., Souter, I., Williams, P., Calafat, A., Hauser, R., & Chavarro, J., (2016). Dietary folate intake and modification of the association of urinary bisphenol a concentrations with *in vitro* fertilization outcomes among women from a fertility clinic. *Reprod. Toxicol., 65*, 104–112.

Mlynarčíková, A., Kolena, J., Ficková, M., & Scsuková, S., (2005). Alterations in steroid hormone production by porcine ovarian granulosa cells caused by bisphenol A and bisphenol A dimethacrylate. *Mol. Cell. Endocrinol., 244*, 57–62.

Morán, F., Conley, A., Corbin, C., Enan, E., VandeVoort, C., Overstreet, J., & Lasley, B., (2000). 2,3,7,8-tetrachlorodibenzo-p-dioxin decreases estradiol production without altering the enzyme activity of cytochrome P450 Aromatase Of human luteinized granulosa cells *in vitro*. *Biol. Reprod., 62*, 1102–1108.

Morán, F., VandeVoort, C., Overstreet, J., Lasley, B., & Conley, A., (2003). Molecular target of endocrine disruption in human luteinizing granulosa cells by 2,3,7,8-tetrachlorodibenzo-p-dioxin: Inhibition of estradiol secretion due to decreased 17α-hydroxylase/17,20-lyase cytochrome P450 expression. *Endocrinol., 144*, 467–473.

Nagel, S., Vom, S. F., Thayer, K., Dhar, M., Boechler, M., & Welshons, W., (1997). Relative binding affinity-serum modified access (RBA-SMA) assay predicts the relative *in vivo* bioactivity of the xenoestrogens bisphenol A and octylphenol. *Environ. Health Perspect., 105*, 70–76.

Nah, W., Park, M., & Gye, M., (2011). Effects of early prepubertal exposure to bisphenol a on the onset of puberty, ovarian weights, and estrous cycle in female mice. *Clin. Exp. Reprod. Med., 38*, 75.

Nandi, S., Gupta, P., Roy, S., Selvaraju, S., & Ravindra, J., (2011). Chlorpyrifos and endosulfan affect buffalo oocyte maturation, fertilization, and embryo development *in vitro* directly and through cumulus cells. *Environ. Toxicol., 26*, 57–67.

Nilsson, E., Larsen, G., Manikkam, M., Guerrero-Bosagna, C., Savenkova, M., & Skinner, M., (2012). Environmentally induced epigenetic transgenerational inheritance of ovarian disease. *Plos One, 7*, 36129.

Ning, G., Bi, Y., Wang, T., Xu, M., Xu, Y., Huang, Y., Li, M., et al., (2011). Relationship of urinary bisphenol a concentration to risk for prevalent type 2 diabetes in Chinese adults. *Ann. Intern. Med., 155*, 368.

Nordkap, L., Joensen, U., Blomberg, J. M., & Jørgensen, N., (2012). Regional differences and temporal trends in male reproductive health disorders: Semen quality may be a sensitive marker of environmental exposures. *Mol. Cell. Endocrinol., 355*, 221–230.

Norström, K., Czub, G., McLachlan, M., Hu, D., Thorne, P., & Hornbuckle, K., (2010). External exposure and bioaccumulation of PCBS in humans living in a contaminated urban environment. *Environ. Int., 36*, 855–861.

Olsén, L., Lind, L., & Lind, P., (2012). Associations between circulating levels of bisphenol a and phthalate metabolites and coronary risk in the elderly. *Ecotoxicol. Environ. Saf., 80*, 179–183.

Özen, S., & Darcan, Ş., (2011). Effects of environmental endocrine disruptors on pubertal development. *J. Clin. Res. Pediatr. Endocrinol., 3*, 1–6.

Padmanabhan, V., Puttabyatappa, M., & Cardoso, R., (2018). *Encyclopedia of Reprod.* (pp. 121–129). UK: Elsevier: Oxford.

Palioura, E., & Diamanti-Kandarakis, E., (2015). Polycystic ovary syndrome (PCOS) and endocrine disrupting chemicals (EDCS). *Rev. Endocr. Metab. Disord., 16*, 365–371.

Panzica, G., Aste, N., Viglietti-Panzica, C., & Ottinger, M., (1995). Structural sex differences in the brain: Influence of gonadal steroids and behavioral correlates. *J. Endocrinol. Investig., 18*, 232–252.
Patterson, Jr. D., Wong, L., Turner, W., Caudill, S., DiPietro, E., McClure, P., Cash, T., et al., (2009). Levels in the U.S. population of those persistent organic pollutants (2003−2004) included in the Stockholm convention or in other long-range transboundary air pollution agreements. *Environ. Sci. Technol., 43*, 1211–1218.
Peretz, J., & Flaws, J., (2013). Bisphenol A down-regulates rate-limiting Cyp11a1 to acutely inhibit steroidogenesis in cultured mouse antral follicles. *Toxicol. App. Pharmacol., 271*, 249–256.
Petro, E., Leroy, J., Covaci, A., Fransen, E., De Neubourg, D., Dirtu, A., De Pauw, I., & Bols, P., (2012). Endocrine-disrupting chemicals in human follicular fluid impair *in vitro* oocyte developmental competence. *Human Reprod., 27*, 1025–1033.
Piazza, M., & Urbanetz, A., (2019). Environmental toxins and the impact of other endocrine disrupting chemicals in women's reproductive health. *JBRA Assist. Reprod., 23*, 154–164.
Porpora, M., Medda, E., Abballe, A., Bolli, S., De Angelis, I., Di Domenico, A., Ferro, A., et al., (2009). Endometriosis and organochlorinated environmental pollutants: A case-control study on Italian women of reproductive age. *Environ. Health Perspect., 117*, 1070–1075.
Prokešová, Š., Ghaibour, K., Liška, F., Klein, P., Fenclová, T., Štiavnická, M., Hošek, P., et al., (2020). Acute low-dose bisphenol s exposure affects mouse oocyte quality. *Reprod. Toxicol., 93*, 19–27.
Quignot, N., Arnaud, M., Robidel, F., Lecomte, A., Tournier, M., Cren-Olivé, C., Barouki, R., & Lemazurier, E., (2012). Characterization of endocrine-disrupting chemicals based on hormonal balance disruption in male and female adult rats. *Reprod. Toxicol., 33*, 339–352.
Rees, C. E., Todd, M., Dowd, J., & Aiello, A., (2011). The impact of bisphenol a and triclosan on immune parameters in the U.S. population, NHANES 2003–2006. *Environ. Health Perspect., 119*, 390–396.
Rier, S., (2001). Serum levels of TCDD and dioxin-like chemicals in rhesus monkeys chronically exposed to dioxin: Correlation of increased serum PCB levels with endometriosis. *Toxicol. Sci., 59*, 147–159.
Rignell-Hydbom, A., (2004). Exposure to CB-153 and P,P'-DDE and male reproductive function. *Hum. Reprod., 19*, 2066–2075.
Rivera, O., Varayoud, J., Rodríguez, H., Muñoz-de-Toro, M., & Luque, E., (2011). Neonatal exposure to bisphenol A or diethylstilbestrol alters the ovarian follicular dynamics in the lamb. *Reprod. Toxicol., 32*, 304–312.
Rochester, J., (2013). Bisphenol A and human health: A review of the literature. *Reprod. Toxicol., 42*, 132–155.
Rodríguez, H., Santambrosio, N., Santamaría, C., Muñoz-de-Toro, M., & Luque, E., (2010). Neonatal exposure to bisphenol a reduces the pool of primordial follicles in the rat ovary. *Reprod. Toxicol., 30*, 550–557.
Sanderson, J., (2006). The steroid hormone biosynthesis pathway as a target for endocrine-disrupting chemicals. *Toxicol. Sci., 94*, 3–21.
Seachrist, D., Donaubauer, E., & Keri, R., (2018). *Encyclopedia of Reprod* (pp. 798–807). UK: Elsevier: Oxford.
Shankar, A., & Teppala, S., (2011). Relationship between urinary bisphenol A levels and diabetes mellitus. *J. Clin. Endocrinol. Metab., 96*, 3822–3826.

Sherman, M., Bur, M., & Kurman, R., (1995). P53 in endometrial cancer and its putative precursors: Evidence for diverse pathways of tumorigenesis. *Hum. Pathol., 26*, 1268–1274.

Signorile, P., (2012). Endocrine disruptors in utero cause ovarian damage linked to endometriosis. *Front. Biosci., 4*, 1724.

Signorile, P., Spugnini, E., Mita, L., Mellone, P., D'Avino, A., Bianco, M., Diano, Net al., (2010). Pre-natal exposure of mice to bisphenol A elicits an endometriosis-like phenotype in female offspring. *Gen. Comp. Endocrinol., 168*, 318–325.

Silva, M., Barr, D., Reidy, J., Malek, N., Hodge, C., Caudill, S., Brock, J., et al.,(2004). Urinary levels of seven phthalate metabolites in the U.S. population from the national health and nutrition examination survey (NHANES) 1999–2000. *Environ. Health Perspect., 112*, 331–338.

Sobolewski, M., & Barrett, E., (2014). Polycystic ovary syndrome: Do endocrine-disrupting chemicals play a role? *Sem. Reprod. Med., 32*, 166–176.

Souter, I., Smith, K., Dimitriadis, I., Ehrlich, S., Williams, P., Calafat, A., & Hauser, R., (2013). The association of bisphenol-A urinary concentrations with antral follicle counts and other measures of ovarian reserve in women undergoing infertility treatments. *Reprod. Toxicol., 42*, 224–231.

Steinmetz, R., Brown, N., Allen, D., Bigsby, R., & Ben-Jonathan, N., (1997). The environmental estrogen bisphenol a stimulates prolactin release *in vitro* and *in vivo*. *Endocrinol., 138*, 1780–1786.

Stoker, T., (1999). Maternal exposure to atrazine during lactation suppresses suckling-induced prolactin release and results in prostatitis in the adult offspring. *Toxicol. Sci., 52*, 68–79.

Su, P., Huang, P., Lin, C., Ying, T., Chen, J., & Wang, S., (2012). The effect of in utero exposure to dioxins and polychlorinated biphenyls on reproductive development in eight-year-old children. *Environ. Int., 39*, 181–187.

Susiarjo, M., Sasson, I., Mesaros, C., & Bartolomei, M., (2013). Bisphenol A exposure disrupts genomic imprinting in the mouse. *PLoS Genet., 9*, e1003401.

Svechnikova, I., Svechnikov, K., Söder, O., (2007). The influence of di-(2-ethylhexyl) phthalate on steroidogenesis by the ovarian granulosa cells of immature female rats. *J. Endocrinol., 194*, 603–609.

Swan, S., (2006). Prenatal phthalate exposure and anogenital distance in male infants. *Environ. Health Perspect., 11*.

Takayanagi, S., Tokunaga, T., Liu, X., Okada, H., Matsushima, A., & Shimohigashi, Y., (2006). Endocrine disruptor bisphenol A strongly binds to human estrogen-related receptor Γ (Errγ) with high constitutive activity. *Toxicol. Lett., 167*, 95–105.

Taketa, Y., Yoshida, M., Inoue, K., Takahashi, M., Sakamoto, Y., Watanabe, G., Taya, Ket al., (2011). Differential stimulation pathways of progesterone secretion from newly formed corpora lutea in rats treated with ethylene glycol monomethyl ether, sulpiride, Or atrazine. *Toxicol. Sci., 121*, 267–278.

Tang, W., Newbold, R., Mardilovich, K., Jefferson, W., Cheng, R., Medvedovic, M., & Ho, S., (2008). Persistent hypomethylation in the promoter of nucleosomal binding protein 1 (Nsbp1) correlates with overexpression of Nsbp1 in mouse uteri neonatally exposed to diethylstilbestrol or genistein. *Endocrinol., 149*, 5922–5931.

Tinfo, N., Hotchkiss, M., Buckalew, A., Zorrilla, L., Cooper, R., & Laws, S., (2011). Understanding the effects of atrazine on steroidogenesis in rat granulosa and H295R adrenal cortical carcinoma cells. *Reprod. Toxicol., 31*, 184–193.

Troisi, R., Hyer, M., Hatch, E., Titus-Ernstoff, L., Palmer, J., Strohsnitter, W., Herbst, A., et al., (2013). Medical conditions among adult offspring prenatally exposed to diethylstilbestrol. *Epidemiol., 24*, 430–438.

Turusov, V., Rakitsky, V., & Tomatis, L., (2002). Dichlorodiphenyltrichloroethane (DDT): Ubiquity, persistence, and risks. *Environ. Health Perspect., 110*, 125–128.

Upson, K., Sathyanarayana, S., De Roos, A., Thompson, M., Scholes, D., Dills, R., & Holt, V., (2013). Phthalates and risk of endometriosis. *Environ. Res., 126*, 91–97.

Veiga-Lopez, A., Luense, L., Christenson, L., & Padmanabhan, V., (2013). Developmental programming: Gestational bisphenol-a treatment alters the trajectory of fetal ovarian gene expression. *Endocrinol., 154*, 1873–1884.

Vélez, M., Arbuckle, T., & Fraser, W., (2015). Female exposure to phenols and phthalates and time to pregnancy: The maternal-infant research on environmental chemicals (MIREC) study. *Fertil. Steril., 103*, 1011–1020.e2.

Vom, S. F., Cooke, P., Buchanan, D., Palanza, P., Thayer, K., Nagel, S., Parmigiani, S., & Welshons, W., (1998). A physiologically based approach to the study of bisphenol A and other estrogenic chemicals on the size of reproductive organs, daily sperm production, and behavior. *Toxicol. Ind. Health., 14*, 239–260.

Wang, T., Li, M., Chen, B., Xu, M., Xu, Y., Huang, Y., Lu, J., et al., (2012). Urinary bisphenol A (BPA) concentration associates with obesity and insulin resistance. *J. Clin. Endocrinol. Metab., 97*, E223-E227.

Watanabe, M., Ohno, S., & Nakajin, S., (2012). Effects of bisphenol A on the expression of cytochrome P450 aromatase (CYP19) in human fetal osteoblastic and granulosa cell-like cell lines. *Toxicol. Lett., 210*, 95–99.

Xi, W., Lee, C., Yeung, W., Giesy, J., Wong, M., Zhang, X., Hecker, M., & Wong, C., (2011). Effect of perinatal and postnatal bisphenol A exposure to the regulatory circuits at the hypothalamus–pituitary-gonadal axis of CD-1 mice. *Reprod. Toxicol., 31*, 409–417.

Xu, C., Chen, J., Qiu, Z., Zhao, Q., Luo, J., Yang, L., Zeng, H., et al., (2010). Ototoxicity and PPAR-mediated aromatase downregulation in female Sprague–Dawley rats following combined oral exposure to benzo[A]pyrene and di-(2-ethylhexyl) phthalate. *Toxicol. Lett., 199*, 323–332.

Yang, M., Kim, S., Chang, S., Lee, I., & Kawamoto, T., (2006). Urinary concentrations of bisphenol A in relation to biomarkers of sensitivity and effect and endocrine-related health effects. *Environ. Mol. Mutagen., 47*, 571–578.

Yang, O., Kim, H., Weon, J., & Seo, Y., (2015). Endocrine-disrupting chemicals: Review of toxicological mechanisms using molecular pathway analysis. *J. Cancer Prev., 20*, 12–24.

Yoshida, M., Takahashi, M., Inoue, K., Hayashi, S., Maekawa, A., & Nishikawa, A., (2011). Delayed adverse effects of neonatal exposure to diethylstilbestrol and their dose dependency in female rats. *Toxicol. Pathol., 39*, 823–834.

Zama, A., Bhurke, A., & Uzumcu, M., (2015). Effects of endocrine-disrupting chemicals on female reproductive health. *Open Biotechnol. J., 9*.

CHAPTER 12

Therapeutic Effects of Medicinal Plants on Endocrine Disorders in Humans

MUSADIQ HUSSAIN BHAT[1], RIJUPALIKA ROY[1], TAWSEEF AHMAD MIR[2], YOUNIS AHMAD HAJAM[3], and DINESH KUMAR AGRAWALA[4]

[1]*Central National Herbarium, Botanical Survey of India, Howrah, Kolkata, West Bengal–711103, India, E-mail: musadiqali131@gmail.com (M. H. Bhat)*

[2]*Center of Research for Ethnobotany, GMS College, Jiwaji University, Gwalior, Madhya Pradesh–474009, India*

[3]*Department of Life Sciences and Allied Health Sciences, Sant Baba Bhag Singh University, Padhiana, Jalandhar, Punjab, India*

[4]*Botanical Survey of India, CGO Complex, 3rd MSO Building, 5th and 6th Floor DF Block, Sector-I, Salt Lake, Kolkata, West Bengal–700064, India*

ABSTRACT

Non-communicable diseases (NCDs) are a serious concern, as their prevalence may escalate in both developing and developed nations mandating an urgent need of curbing them. Disorders like diabetes, thyroid, obesity, and allied complications may put tremendous pressure on global healthcare systems. Endocrine disorders, primarily related to thyroid, pituitary, and adrenal glands, account for more than 8% of all diseases worldwide. Traditional medicine systems have always played a vital role in the world healthcare system. A thorough literature review shows that a vast number of medicinal plants have been exploited for their potential against various endocrine

Environmental Endocrine Toxicants: Biology, Effects, and Management. Younis Ahmad Hajam, Rouf Ahmad Bhat, Khalid Rehman Hakeem, & Rajesh Kumar (Eds.)
© 2023 Apple Academic Press, Inc. Co-published with CRC Press (Taylor & Francis)

disorders. In this chapter, we will emphasis on some renowned medicinal plants reportedly used to treat various endocrine disorders.

12.1 INTRODUCTION

Plant resources have been utilized for the well-being of mankind ever since the dawn of civilization for the treatment of various disorders, the earliest evidence being traced out from Mesopotamian civilizations around 2600 BC (Gurib-Fakim, 2006). According to the World Health Organization (WHO), about 80% of the world population rely upon the conventional medicinal remedies for their basic healthcare. The practice of traditional medical practices is mostly found in developing countries. The use of medicinal plants to various disorders is predominantly found in most of Asian countries like India, Pakistan, China, Nepal, etc. (Ahmad, 2007). From the very beginning, people living in rural and forest areas have relied on the utilization of native plant species for various medicinal purposes (Shinwari and Khan, 2000). The dependence of these people on the herbal remedies may be attributed to the underdeveloped healthcare sector and other socio-economic issues. More than 20,000 plant species are being used worldwide for various medicinal purposes (Hamayun, 2007). Furthermore, about 50% of the modern medicines are acquired from natural sources (Gurib-Fakim, 2006).

Non-communicable diseases (NCDs) are a serious concern, as their prevalence may escalate in both developing and developed nations. Disorders like diabetes, thyroid, obesity, and allied complications may put tremendous pressure on global healthcare systems (Hasani-Ranjbar and Larijani, 2014). Moreover, in developed nations, these problems receive little attention leading to an increased mortality rate (Lopez et al., 2006; Anyanwu et al., 2010). This is despite the fact that number of NCDs in such nations is on the rise. Due to the lack of funding sources, infectious, and diet-related diseases are given much priority in the developed countries, and limited focus and resources are devoted to the alarmingly increasing endocrine and other NCDs (Robinson and Hort, 2012). Endocrine disorders, primarily related to thyroid, pituitary, and adrenal glands account for more than 8% of all the diseases worldwide (Lopez et al., 2006). Obesity, diabetes, and metabolic disorders are becoming more common across the world, indicating a connection to major health issues such as ischemic heart disease, stroke, and cancer (WHO, 1997; Sicree et al., 2006). Diabetes, obesity, reduced fasting glucose, osteoporosis, erectile dysfunction, thyroiditis, hypovitaminosis, dyslipidemia, osteopenia, and impaired glucose tolerance were among the endocrine disorders with prevalence in at about 5% throughout adults in the United States. Males and females have the highest rates of erectile dysfunction and osteopenia/osteoporosis, respectively. Pituitary adenoma and diabetes among children are the least common health problems affecting less than 1% of the population of the United States. Women are more likely to have hyperparathyroidism and thyroid disorders. Diabetes mellitus is found to be most common among ethnic minorities, as predicted. Thyroid gland (TG) enlargement affects about 20% of the population in several countries, mostly the regions once covered by glaciers (DeLange and Ermans, 1991). Iodine deficiency, goitrogens, and other factors cause another endocrine disorder – goiter. Despite significant advancement in public health sector, iodine deficiency continues to be a major burden on the health system of many nations (Ma et al., 1993). As per the WHO report of 2012, diabetes affects one out of every 10 adults worldwide, and about 12% of the world population is obese. PCOS is a highly prevalent endocrine ailment characterized by symptoms like polycystic ovaries, hyperandrogenism, etc., leading to menstrual irregularity, infertility, and hirsutism. According to the WHO estimate, 116 million women have been affected worldwide by PCOS in 2012 (Treatment and Kabel, 2016). Another prevalent disorder in humans is the imbalance in thyroid hormone production due to improper functioning of the TG or the pituitary gland. Thyroid disorder include hyperthyroidism, hypothyroidism, and thyroiditis.

Because of a less skilled workforce and lack of necessary health facilities for endocrinological amenities in developing countries, especially investigation tools, the endocrine disorders are mostly neglected and diagnoses delayed (Vos et al., 2010). In light of this, research on endocrinology practice is required to focus on endocrine disorders and the treatment available for patients with these disorders. In addition, the burden and scope of disorders in the community and hospitals must be characterized to improve the healthcare system of developing countries (Famuyiwa, 1990).

12.2 PLANTS AND THE TREATMENT OF ENDOCRINE DISORDERS

The key causes of these conditions and complications are behavioral changes, poor nutrition, and a sedentary lifestyle. Medications and surgical procedures aren't always the right alternatives. Currently, some herbal remedies are adopted by diabetic patients, by both who are candidates for insulin therapy and also who have been unsuccessfully controlled. When modern medicine fails in the effective treatment of chronic diseases and disorders like obesity, goiter, PCOS, thyroid, etc., many people turn to alternative remedies such as herbal therapy (Ziai, 2003). The earliest evidence of therapeutic plants in humans can be found in the Middle East, dating back to the Stone Age. However, in recent years, alternative medicine has regained favor. According to the survey in the United States, over 34% respondents taken one or another form of any alternative medication such as herbal remedy. As safe modern medications are lacking, WHO has urged that potential medicinal plants be evaluated for their effectiveness against various disorders like diabetes and other endocrine disorders (Sadighi et al., 2004). Herbal medicine is the most accepted alternative and complementary therapy in several nations (Klepser and Klepser, 1999). Many herbs have been recognized as hazardous, however there are also plants that are potentially harmless, and clinical trial data suggest that some herbal remedies are effective for certain illnesses (Hasani-Ranjbar et al., 2008). Some medicinal plants are utilized to treat a numerous endocrine disorders for example reduced blood sugar levels. These include *Citrullus colocynthis*, *Teucrium polium*, *Punica granatum*, *Silybum marianum*, *Aloe vera*, and *Trigonella foenum-graecum*.

A large number of medicinal plants possess anti-thyroid activity in both thyroid diseases – hypothyroidism and hyperthyroidism. Various medicinal plants possess the potential for treating hyperthyroidism, like *Lycopus virginicus*, *Lycopus europaeus*, *Lithospermum ruderale*, *Lycopus lucidus*, *Melissa officinalis*, *Rosmarinus officinalis*, and *Salvia officinalis*.

A thorough literature review shows that a vast number of medicinal plants have been exploited for their potential against various endocrine disorders. This chapter highlights some renowned medicinal plants used to treat various endocrine disorders. The plant parts found to be effective for the treatment of such disorders are precisely summarized in Table 12.1.

12.3 CONCLUSION

Medicinal herbs are efficient, economical, and safe with minimum side effects (Rahmawati et al., 2006). Moreover, they are easily available (Feng et al., 2005), compatible, and highly acceptable (Ghimire et al., 2005). Numerous medicinal plants have been reported to have the potential to treat various endocrine disorders that make them beneficial in the management of diabetes, goiter, PCOS, thyroid, etc. In the present chapter, traditionally used medicinal plants with such potential have been summarized. However, medicinal plants which are used for the treatment of endocrine disorders need to be studied in detail for their pharmacological potential as only some of these plant species have been clinically tested. Also, there are limited

TABLE 12.1 Medicinal Plants and Their Effective Parts Used for the Management of Various Endocrine Disorders

Medicinal Plant	Effective Part(s)	Ailments	Mode of Administration	References
Abelmoschus esculentus (L.) Moench Malvaceae	Fruit	Diabetes	Maceration	Sabitha et al. (2011); Hachi et al. (2016)
Acacia segenegalensis (Houtt.) Roberty Leguminosae	Bark	Diabetes	Oral drink	Danjuma et al. (2018)
Adansonia digitata L. Malvaceae	Pod	Goiter	Oral as infusion	Danjuma et al. (2018)
Ajuga iva (L.) Schreb. Lamiaceae	Stem, leaves, flower	Thyroid disorders	Powder	Taibi et al. (2020)
	Stem, leaves	Diabetes	Powder, decoction	Barkaoui et al. (2017)
Allium ampeloprasum L. Amaryllidaceae	Bulb	Diabetes	Raw	Skalli et al. (2019)
Allium cepa L. Amaryllidaceae	Bulb	Thyroid disorders	Juice, decoction	Taibi et al. (2020)
		Diabetes	Raw, juice	Hachi et al. (2016)
Allium sativum L. Amaryllidaceae	Bulb	Thyroid disorders	Juice, decoction, raw	Taibi et al. (2020)
	Cloves	Diabetes	Raw	Mrabti et al. (2019)
Aloe barbadensis Mill. Xanthorrhoeaceae	Gel, juice	Thyroid disorders	Compress	Taibi et al. (2020)
Aloe vera L. Aloaceae	Stem	PCOS	Extract	Radha and Laxmipriya (2016)
Ammi visnaga (L.) Lam.Apiaceae	Inflorescence	Diabetes	Decoction	Alami et al. (2015)
	Fruits	Diabetes	Aqueous extract	Jouad et al. (2002)
Ammodaucus leucotrichus Coss. Apiaceae	Seed	Diabetes	Infusion, decoction	Laadim et al. (2017)
Apteranthes europea (Guss.) Murb. Apocynaceae	Stem	Diabetes	Decoction, infusion, raw	Barkaoui et al. (2017)
Anacyclus pyrethrum (L.) Lag. Asteraceae	Flowers	Thyroid disorders	Infusion	Taibi et al. (2020)

TABLE 12.1 *(Continued)*

Medicinal Plant	Effective Part(s)	Ailments	Mode of Administration	References
Anastatica hierochuntica L. Brassicaceae	Stem, leaves	Diabetes	Powder	Hachi et al. (2016)
Angelica Gigas Nakai Apiaceae	Root	PCOS	Powder	Ding et al. (2015)
Annona muricata L. Annonaceae	Fruit, seed	Thyroid disorders	Raw, syrup	Taibi et al. (2020)
Annona senegalensis Pers. Annonaceae	Leaves	Goiter	Infusion	Danjuma et al. (2018)
Apium graveolens L. Apiaceae	Seed	Diabetes	Infusion	Laadim et al. (2017)
Aquilaria malaccensis Lam. Thymelaeaceae	Stem, leaves, flower	Thyroid disorders	Oil, decoction	Taibi et al. (2020)
Argania spinosa (L.) Skeels Sapotaceae	Seeds	Diabetes	Raw, powder	Alami et al. (2015)
Aristolochia baetica L. Aristolochiaceae	Root	Diabetes	Powder	Katiri et al. (2017)
Aristolochia fontanesii Boiss. and Reut. Aristolochiaceae	Seed	Diabetes	Decoction	Alami et al. (2015)
Artemisia absinthium L. Compositae	Aerial part	Diabetes	Infusion	Laadim et al. (2017)
Artemisia herba-alba Asso. Compositae	Aerial part	Thyroid disorders	Infusion	Taibi et al. (2020)
	Stem, leaves, root	Diabetes	Decoction, infusion	Laadim et al. (2017)
Asparagus albus L. Asparagaceae	Young sprout	Diabetes	Raw	Hachi et al. (2016)
Atriplex halimus L. Amaranthaceae	Aerial part	Thyroid disorders	Decoction, infusion, powder	Taibi et al. (2020)
Avena sativa L. Poaceae	Seed	Thyroid disorders	Powder	Taibi et al. (2020)
Azadirachta indica A. Juss. Meliaceae	Bark	Diabetes	Oral drink	Danjuma et al. (2018)

TABLE 12.1 *(Continued)*

Medicinal Plant	Effective Part(s)	Ailments	Mode of Administration	References
Bacopa monnieri (L.) Wettst. Plantaginaceae	Leaves	Thyroid disorders	Infusion	Taibi et al. (2020)
Bambusa vulgaris Schrad. Poaceae	Stem	PCOS	Extract	Soumya et al. (2016)
Bauhinia purpurea L. Caesalpiniaceae	Bark	Thyroid disorders	Extract	Kumar and Chandrashekar (2011)
Bauhinia variegate L. Fabaceae	Bark	Thyroid disorders	Decoction	Verma and Jameel (2014)
Berberis vulgaris L. Berberidaceae	Peel, bark	Thyroid disorders	Powder	Taibi et al. (2020)
Beta vulgaris L. Amaranthaceae	Leaves	Thyroid disorders	Infusion, compress	Taibi et al. (2020)
Brassica oleracea L. Brassicaceae	Aerial part, fruit	Diabetes	Raw, maceration	Laadim et al. (2017)
Brassica rapa L. Brassicaceae	Root, leaves	Diabetes	Decoction	Hachi et al. (2016)
Bunium incrassatum (Boiss.) Amo Apiaceae	Tuber	Thyroid disorders	Powder	Taibi et al. (2020)
Buxus balearica Lam. Buxaceae	Leaves	Diabetes	Decoction	Hachi et al. (2016)
Calotropis procera (Aiton) Dryand. Apocynaceae	Leaves	Diabetes	Powder	Ghourri et al. (2013)
Capparis spinosa L. Capparaceae	Aerial part, fruit, root	Diabetes	Powder, decoction, infusion	Hachi et al. (2016); Tahraoui et al. (2007)
Capsicum annuum L. Solanaceae	Fruit	Thyroid disorders	Raw, decoction	Taibi et al. (2020)
Caralluma europaea (Guss.) N. E. Br. Apocynaceae	Aerial part	Diabetes	Maceration	Laadim et al. (2017)
Carum carvi L. Apiaceae	Seed	Thyroid disorders	Infusion	Taibi et al. (2020)
	Seed	Diabetes	Decoction	Tahraoui et al. (2007)

TABLE 12.1 *(Continued)*

Medicinal Plant	Effective Part(s)	Ailments	Mode of Administration	References
Catharanthus roseus (L.) G. Don Apocynaceae	Leaves	Diabetes	Juice	Ahmed et al. (2010)
Centaurium erythraea Rafn. Gentianaceae	Flower, aerial parts	Diabetes	Infusion, decoction	Hachi et al. (2016)
Chamaemelum nobile (L.) All. Asteraceae	Flower	PCOS	Extract	Zangeneh et al. (2010)
	Leaves	Diabetes	Decoction	Alami et al. (2015)
Chamaerops humilis L. Arecaceae	Root	Diabetes	Raw, cooked	Hachi et al. (2016)
Cistus albidus L. Cistaceae	Leaves	Diabetes	Decoction	Hachi et al. (2016)
Cistus creticus L. Cistaceae	Leaves	Diabetes	Decoction, powder	Barkaoui et al. (2017)
Cicer arietinum L. Fabaceae	Seed	Thyroid disorders	Powder	Taibi et al. (2020)
Cinnamomum cassia (L.) J. Presl Lauraceae	Bark	Diabetes	Decoction	Hachi et al. (2016)
	Peel, bark	Thyroid disorders	Powder	Taibi et al. (2020)
Cinnamomum zeylanicum Blume Lauraceae	Stem	PCOS	Extract	Wang et al. (2007)
Citrullus colocynthus (L.) Schrad. Cucurbitaceae	Seed, fruit	Diabetes	Decoction, cataplasm, and powder	Barkaoui et al. (2017)
Citrullus vulgaris Schrad. Cucurbitaceae	Fruit peel	Thyroid disorders	Raw	Sharma et al. (2015); Verma and Jameel (2014)
Citrus limon (L.) Osbeck Rutaceae	Fruit	Thyroid disorders	Juice, infusion, decoction	Taibi et al. (2020)
Cladanthus scariosus (Ball) Oberpr. and Vogt Compositae	Flower	Diabetes	Decoction	Katiri et al. (2017)
Clinopodium alpinum (L.) Kuntze Lamiaceae	Leaves	Diabetes	Decoction	Skalli et al. (2019)

TABLE 12.1 *(Continued)*

Medicinal Plant	Effective Part(s)	Ailments	Mode of Administration	References
Clinopodium nepeta subsp. *glandulosum* (Req.) Govaerts Lamiaceae	Aerial parts	Diabetes	Infusion, decoction	Alami et al. (2015)
Cocos nucifera L. Arecaceae	Flower	PCOS	Extract	Soumya et al. (2014)
Commiphora mukul (Hook. ex Stocks) Engl. Burseraceae	Gum	Thyroid disorders	Extract	Bharthi et al. (2017)
Convolvulus pluricaulis Choisy Convolvulaceae	Flowers, whole plant	Thyroid disorders	Extract	Bhowmik et al. (2012)
Convolvulus sabatius subsp. *mauritanicus* (Boiss.) Murb. Convolvulaceae	Aerial parts	Thyroid disorders	Decoction, infusion	Taibi et al. (2020)
Coriandrum sativum L Apiaceae	Seed	Diabetes	Infusion	Skalli et al. (2019)
Crocus sativus L. Iridaceae	Flower, pistil	Thyroid disorders	Raw, infusion	Taibi et al. (2020)
Cucumis melo L. Cucurbitaceae	Fruit peel	Thyroid disorders	Raw	Sharma et al. (2015); Verma and Jameel (2014)
Cucumis sativus L. Cucurbitaceae	Fruit	Diabetes	Raw	Barkaoui et al. (2017)
Curcuma longa L. Zingiberaceae	Rhizome	Thyroid disorders	Powder, decoction	Taibi et al. (2020)
Cydonia oblonga Mill. Rosaceae	Fruit	Diabetes	Raw	Tahraoui et al. (2007)
Cynara cardunculus L. Compositae	Aerial parts	Diabetes	Decoction	Alami et al. (2015)
Daucus carota L. Apiaceae	Root, seed	Thyroid disorders	Decoction, infusion	Taibi et al. (2020)
	Root	Diabetes	Juice, puree	Hachi et al. (2016)
Diospyros mespiliformis Hochst. ex A. DC. Ebenaceae	Bark	Goiter	Infusion	Danjuma et al. (2018)

TABLE 12.1 *(Continued)*

Medicinal Plant	Effective Part(s)	Ailments	Mode of Administration	References
Dittrichia viscosa (L.) Greuter Compositae	Leaves	Diabetes	Decoction	Katiri et al. (2017)
Dysphania ambrosioides (L.) Amaranthaceae	Leaves	Diabetes	Infusion	Laadim et al. (2017)
Eichhornia crassipes (Mart.) Solms Pontederiaceae	Whole plant	Thyroid disorders	Extract	Sharma et al. (2015)
Echinops spinosissimus Turra. Compositae	Flower	Diabetes	Decoction	Katiri et al. (2017)
Ephedra altissima Desf. Ephedraceae	Stem, leaves, whole plant	Diabetes	Decoction	Mrabti et al. (2019)
Ephedra fragilis Desf. Ephedraceae	Green stem	Diabetes	Decoction	Katiri et al. (2017)
Eruca sativa Mill Brassicaceae	Leaves	Thyroid disorders	Raw, compress	Taibi et al. (2020)
Eryngium ilicifolium Lam. Apiaceae	Stem, leaves	Diabetes	Decoction	Barkaoui et al. (2017)
Eucalyptus globulus Labill. Myrtaceae	Leaves, fruit	Diabetes	Decoction	Alami et al. (2015)
Euphorbia officinarum L. Euphorbiaceae	Stem, leaves	Diabetes	Powder	Barkaoui et al. (2017)
	Fruit	Thyroid disorders	Raw	Taibi et al. (2020)
	Fruit, leaves	Diabetes	Decoction	Tahraoui et al. (2007)
Foeniculum vulgare Mill. Apiaceae	Seed	Diabetes	Decoction	Barkaoui et al. (2017)
Fraxinus angustifolia Vahl Oleaceae	Leaves	Diabetes	Infusion	Ziyyat et al. (1997)
Fragaria vesca L. Rosaceae	Fruit	Diabetes	Raw	Jouad et al. (2001)

TABLE 12.1 *(Continued)*

Medicinal Plant	Effective Part(s)	Ailments	Mode of Administration	References
Globularia alypum L. Plantaginaceae	Flower, leaves, stem	Diabetes	Infusion, decoction	Bousta et al. (2014)
Globularia repens Lam. Plantaginaceae	Leaves	Diabetes	Decoction	Laadim et al. (2017)
Glycyrrhiza glabra L. Fabaceae	Rhizome, root	Thyroid disorders	Powder	Taibi et al. (2020)
Glycine max (L.) Merr. Fabaceae	Seed	PCOS	Extract	Rajan and Balaji (2017)
Haloxylon scoparium Pomel Amaranthaceae	Aerial parts	Thyroid disorders	Powder	Taibi et al. (2020)
*Hammada scoparia (*Pomel) Iljin Amaranthaceae	Seed	Diabetes	Decoction	Barkaoui et al. (2017)
Heracleum persicum Desf Apiaceae	Fruit	PCOS	Extracts	Alizadeh et al. (2015)
Herniaria glabra L. Caryophyllaceae	Aerial part	Diabetes	Decoction	Jouad et al. (2001)
Hordeum vulgare L. Poaceae	Seed	Thyroid disorders	Raw, juice, powder	Taibi et al. (2020)
Hyphaene thebaica (L.) Mart. Arecaceae	Fruit	Diabetes	Powder	Ghourri et al. (2013)
Juglans regia L. Juglandaceae	Leaves, bark	Diabetes	Infusion, decoction	Ziyyat et al. (1997)
Juniperus phoenica L. Cupressaceae	Leaves, fruit	Thyroid disorders	Decoction, infusion	Taibi et al. (2020)
	Leaves, aerial parts	Diabetes	Powder, decoction, maceration	Hachi et al. (2016)
Juniperus thurifera L. Cupressaceae	Leaves	Diabetes	Decoction	Hachi et al. (2016)
Lactuca sativa L. Compositae	Leaves	Diabetes	Raw	Eddouks et al. (2002)

TABLE 12.1 *(Continued)*

Medicinal Plant	Effective Part(s)	Ailments	Mode of Administration	References
Launaea arborescens (Batt.) Murb. Compositae	Whole plant	Diabetes	Powder, infusion, decoction	Giovannini et al. (2016)
Lavandula dentata L. Lamiaceae	Stem, leaves	Diabetes	Decoction, powder, raw, infusion	Barkaoui et al. (2017)
Lavandula maroccana Murb. Lamiaceae	Stem, leaves	Diabetes	Decoction	Barkaoui et al. (2017)
Lavandula stoechas L. Lamiaceae	Leaves	Diabetes	Decoction	Laadim et al. (2017)
Lawsonia inermis L. Lythraceae	Leaves	Thyroid disorders	Powder, compress	Taibi et al. (2020)
Leonurus cardiaca L. Lamiaceae	Leaves, flowers	Thyroid disorders	Extract	Bove (2012)
Lepidium sativum L. Brassicaceae	Seed	Diabetes	Decoction, maceration, infusion	Skalli et al. (2019)
	Seed	Diabetes	Powder	Chauhan et al. (2012)
Lepidium sativum L. Brassicaceae	Seed	Thyroid disorders	Raw, powder	Taibi et al. (2020)
Linum usitatissimum L. Linaceae	Seed	Thyroid disorders	Raw, powder, compress	Taibi et al. (2020)
	Seed	Diabetes	Decoction, powder	Hachi et al. (2016)
	Seed	Thyroid disorders	Extract	Sharma et al. (2015); Verma and Jameel (2014)
Lippia citriodora (Palau) Kunth Verbenaceae	Leaves	Thyroid disorders	Infusion, decoction	Taibi et al. (2020)
Lithodora fruticosa (L.) Griseb. Boraginaceae	Aerial parts	Thyroid disorders	Infusion, decoction	Taibi et al. (2020)
Lithospermum officinale L. Boraginaceae	Root	Thyroid disorders	Extract	Bove (2012)

TABLE 12.1 *(Continued)*

Medicinal Plant	Effective Part(s)	Ailments	Mode of Administration	References
Lithospermum ruderale Douglas ex Lehm. Boraginaceae	Root	Thyroid disorders	Extract	Verma (2012)
Lupinus luteus L. Fabaceae	Seed	Thyroid disorders	Powder	Taibi et al. (2020)
Lycopus europaeus L. Lamiaceae	Leaves, stem, flower	Thyroid disorders	Infusion, decoction	Taibi et al. (2020)
Lycopus virginicus L. Lamiaceae	Leaves, flowers	Thyroid disorders	Extract	Skvaa (2011)
Malus communis (L.) Poir. Rosaceae	Fruit	Diabetes	Juice	Katiri et al. (2017)
Mangifera indica L. Anacardiaceae	Fruit peel	Thyroid disorders	Extract	Parmer and Kar (2009)
Marrubium vulgare L. Lamiaceae	Leaves, aerial part	Diabetes	Decoction, infusion	Laadim et al. (2017)
	Aerial parts	PCOS	Extract	Mokhtari et al. (2014)
Matricaria chamomilla Linn. Compositae	Leaves, flower	Diabetes	Decoction, infusion	Mrabti et al. (2019)
Medicago sativa Linn. Leguminosae	Aerial parts, seed	Diabetes	Infusion, maceration	Hachi et al. (2016)
Melissa officinalis L. Lamiaceae	Leaves, stem, flowers	Thyroid disorders	Extract	Hameed et al. (2014)
	Aerial parts	Thyroid disorders	Infusion, decoction	Taibi et al. (2020)
Mentha pulegium L. Lamiaceae	Leaves, aerial parts	Diabetes	Infusion, decoction	Skalli et al. (2019)
Mentha spicata L. Lamiaceae	Leaves, stem	Diabetes	Infusion, decoction	Hachi et al. (2016)
	–	PCOS	Tea	Grant (2010)
Moringa oleifera Lam. Moringaceae	Root bark, seeds	Thyroid disorders	Extract	Tabassum et al. (2013)
	Leaves	Diabetes	Infusion	Danjuma et al. (2018)

TABLE 12.1 *(Continued)*

Medicinal Plant	Effective Part(s)	Ailments	Mode of Administration	References
Morus alba L. Moraceae	Leaves	Diabetes	Beverages, powder	Naowaboot et al. (2009); Park et al. (2009)
	Leaves	Thyroid disorders	Extract	Sharma et al. (2015); Verma and Jameel (2014)
Morus nigra L. Moraceae	Fruit	Thyroid disorders	Raw	Taibi et al. (2020)
Murraya koenigii (L.) Spreng. Rutaceae	Leaves	Diabetes	Aqueous extracts, dry powder	Kesari et al. (2007); Yadav et al. (2002)
Myristica fragrans Houtt. Myristicaceae	Seed	Diabetes	Powder	Barkaoui et al. (2017)
Myrtus communis L. Myrtaceae	Leaves	Thyroid disorders	Infusion	Taibi et al. (2020)
	Leaves, fruit	Diabetes	Decoction, infusion	Alami et al. (2015)
Nerium oleander L. Apocynaceae	Leaves	Diabetes	Fumigation, decoction	Laadim et al. (2017)
Nigella sativa L. Ranunculaceae	Seed	Thyroid disorders	Raw, powder	Taibi et al. (2020)
Ocimum basilicum L. Lamiaceae	Stem	Diabetes	Infusion	Hachi et al. (2016)
	Leaves	Thyroid disorders	Infusion, decoction	Taibi et al. (2020)
Oenothera biennis L. Onagraceae	Seed	PCOS	Oil	Ried (2015); Arentz et al. (2014)
Olea europaea L. Oleaceae	Leaves	Thyroid disorders	Infusion	Taibi et al. (2020)
Origanum elongatum (Bonnet) Emb. and Maire Lamiaceae	Leaves	Diabetes	Infusion	Skalli et al. (2019)
Origanum majorana L. Lamiaceae	Leaves	PCOS	Tea	Haj-Husein et al. (2016)
		Thyroid disorders	Infusion	Taibi et al. (2020)
Origanum vulgare L. Lamiaceae	Leaves	Thyroid disorders	Infusion	Taibi et al. (2020)
		Diabetes	Infusion	Laadim et al. (2017)

TABLE 12.1 *(Continued)*

Medicinal Plant	Effective Part(s)	Ailments	Mode of Administration	References
Paronychia argentea Lam Caryophyllaceae	Leafy stem	Diabetes	Infusion	Katiri et al. (2017)
Pastinaca sativa L. Apiaceae	Root	Diabetes	Raw	Mrabti et al. (2019)
Peganum harmala L. Nitrariaceae	Seeds	Diabetes	Powder, infusion	Alami et al. (2015)
		Thyroid disorders	Raw, powder	Taibi et al. (2020)
Periploca laevigata (Labill.) Markgr. Apocynaceae	Fruit	Diabetes	Decoction	Barkaoui et al. (2017)
Petroselinum crispm (Mill.) Fuss Apiaceae	Seeds	Diabetes	Infusion	Mrabti et al. (2019)
	Stem, leaves, fruit	Thyroid disorders	Decoction	Taibi et al. (2020)
Phaseolus vulgaris L. Leguminosae	Fruit	Diabetes	Decoction, powder, juice	Hachi et al. (2016)
Phoenix dactylifera L. Arecaceae	Pollen	PCOS	Extract	Jashni et al. (2016)
	Fruit	Diabetes	Raw, decoction	Laadim et al. (2017)
	Fruit, nut	Thyroid disorders	Raw, powder	Taibi et al. (2020)
Pimpinella anisum L. Apiaceae	Seed	Diabetes	Decoction, powder	Mrabti et al. (2019)
Pistacia atlantica Desf. Anacardiaceae	Fruit	Diabetes	Decoction	Barkaoui et al. (2017)
Pistacia lentiscus L. Anacardiaceae	Leaves, gum	Diabetes	Infusion, decoction	Hachi et al. (2016)
	Gum		Crude gum	Rehman et al. (2015)
Prunus amygdalus Batsch Rosaceae	Root, leaves, seed	Thyroid disorders	Raw, powder, decoction	Taibi et al. (2020)
Prunus persica (L.) Batsch Rosaceae	Leaves, seed, fruit	Thyroid disorders	Raw, powder	Taibi et al. (2020)

TABLE 12.1 *(Continued)*

Medicinal Plant	Effective Part(s)	Ailments	Mode of Administration	References
Ptilotrichum spinosum (L.) Boiss. Brassicaceae	Stem, leaves	Diabetes	Decoction	Hachi et al. (2016)
Ptychotis verticillata Duby Apiaceae	Aerial part	Diabetes	Infusion	Hachi et al. (2016)
Punica granatum L. Punicaceae	Leaves, fruit, seed	Thyroid disorders	Raw, juice, powder	Taibi et al. (2020)
	Fruit	PCOS	Juice extract	Hossein et al. (2015)
	Pericarp	Diabetes	Decoction, infusion, powder	Hachi et al. (2016)
Quercus ilex L. Fagaceae	Fruit peel	Thyroid disorders	Infusion, maceration	Taibi et al. (2020)
Raphanus sativus L. Brassicaceae	Roots	Diabetes	Juice	Shukla et al. (2011)
			Raw	Barkaoui et al. (2017)
			Juice	Shukla et al. (2011)
		Thyroid disorders	Raw	Taibi et al. (2020)
Rosmarinus officinalis L. Lamiaceae	Stems, leaves, flowers	Thyroid disorders	Extract	Verma (2012)
	Leaves, stem, flower	Thyroid disorders	Infusion	Taibi et al. (2020)
	Leaves	Diabetes	Powder, decoction, infusion	Barkaoui et al. (2017)
Salvia officinalis L. Lamiaceae	Leaves	Diabetes	Decoction, infusion	Hachi et al. (2016)
Salvia verbenaca L. Lamiaceae	Aerial parts	Thyroid disorders	Infusion, decoction	Taibi et al. (2020)
Saussurea costus (Falc.) Lipsch. Asteraceae	Rhizome	Thyroid disorders	Infusion, decoction, powder	Taibi et al. (2020)
Saussurea lappa (Decne.) Sch. Bip. Compositae	Root	Thyroid disorders	Extract	Sharma et al. (2015); Verma and Jameel (2014)

TABLE 12.1 *(Continued)*

Medicinal Plant	Effective Part(s)	Ailments	Mode of Administration	References
Scolymus hispanicus L. Compositae	Stem, leaves	Diabetes	Raw, decoction	Katiri et al. (2017)
Scorzonera undulate Vahl Compositae	Flower	Diabetes	Raw	Barkaoui et al. (2017)
Seronoa repens (W. Bartram) Small Arecaceae	Whole plant	PCOS	Extract	Ried (2015); Arentz et al. (2014); Kamel (2013)
Sesamum indicum L. Pedaliaceae	Seeds	Diabetes	Powder, decoction, infusion	Tahraoui et al. (2007)
Sorghum bicolor (L.) Moench Poaceae	Seeds	Diabetes	Decoction, infusion	Hachi et al. (2016)
Sonchus arvensis L. Compositae	Leaves	Diabetes	Infusion	Alami et al. (2015)
Stachys lavandulifolia Vahl. Lamiaceae	Leaves	PCOS	Tea	Jalilian et al. (2013)
Syzygium cumini (L.) Skeels Myrtaceae	Seeds	Diabetes	Powder	Willatgamuwa et al. (1998)
Tamarindus indica L. Leguminosae	Stem	Goiter	Decoction	Danjuma et al. (2018)
Taraxacum officinale (L.) Weber ex. FHWigg Asteraceae	Root	PCOS	Extract	Ried (2015); Arentz et al. (2014); Kamel (2013)
Tetraclinis articulata (Vahl.) Mast. Cupressaceae	Leaves, aerial parts	Diabetes	Maceration, infusion	Alami et al. (2015)
Teucrium polium L. Lamiaceae	Leaves	Diabetes	Decoction, powder	Katiri et al. (2017)
	Leaves	Thyroid disorders	Extract	Taibi et al. (2020)
Thymelaea hirsute (L.) Endl. Thymelaeaceae	Stem	Diabetes	Powder	Ziyyat et al. (1997)

TABLE 12.1 *(Continued)*

Medicinal Plant	Effective Part(s)	Ailments	Mode of Administration	References
Thymelaea tartonraira (L.) All. Thymelaeaceae	Leaves	Diabetes	Decoction	Tahraoui et al. (2007)
Thymus algeriensis Boiss. and Reut. Lamiaceae	Stem, leaves	Diabetes	Infusion, decoction	Hachi et al. (2016)
Thymus broussonetii Boiss. Lamiaceae	Stem, leaves	Diabetes	Maceration, infusion	Skalli et al. (2019)
Thymus munbyanus Boiss. and Reut Lamiaceae	Leaves, stem	Diabetes	Decoction, infusion	Hachi et al. (2016)
Thymus vulgaris L. Lamiaceae	Aerial parts	Thyroid disorders	Infusion	Taibi et al. (2020)
Trigonella foenumgraecum L. Fabaceae	Seed	Thyroid disorders	Raw, infusion, compress	Taibi et al. (2020)
	Seed capsule	PCOS	Extract	Bashtian et al. (2013)
	Stem, leaves	Diabetes	Decoction, infusion	Alami et al. (2015)
Vicia faba L. Fabaceae	Seed	Thyroid disorders	Powder	Taibi et al. (2020)
Viscum album L. Santalaceae	Seed	Diabetes	Infusion	Alami et al. (2015)
Vitis vinifera L. Vitaceae	Leaves, fruit	Thyroid disorders	Raw, powder	Taibi et al. (2020)
Vitex agnus-castus L. Verbenaceae	Whole plant	PCOS	Extract	Jelodar and Askari (2012)
Withania somnifera (L.) Dunal Solanaceae	Leaves	Thyroid disorders	Infusion, decoction	Taibi et al. (2020)
Zea mays L. Poaceae	Stigmas	Diabetes	Powder	Mrabti et al. (2019)
Zingiber officinale Roscoe Zingiberaceae	Rhizome	Thyroid disorders	Powder, infusion	Taibi et al. (2020)

reports on the toxicological effects of these plants. The literature showed that people using these medicinal plants have scanty information regarding the possible toxicity of these plants. Regardless of their therapeutic properties, these medicinal plants may have a toxicity risk because of contamination, misidentification, consumption of the plant part of the wrong species, improper dosage, etc. In brief, this review provides basic data of numerous plant species having the potential to treat various endocrine disorders. However, many of the plant species mentioned in this chapter demand more clinical and pharmacological studies for validation of any effective remedy to treat these disorders.

KEYWORDS

- **endocrine disorders**
- **herbal medicine**
- ***Lithospermum ruderale***
- ***Lycopus lucidus***
- **medicinal plants**
- ***Melissa officinalis***
- **polcystic ovary syndrome**

REFERENCES

Ahmad, S. S., (2007). Medicinal wild plants from Lahore-Islamabad motorway (M-2). *Pak. J. Bot., 39*, 355–375.

Ahmed, M. F., Kazim, S. M., Ghori, S. S., Mehjabeen, S. S., Ahmed, S. R., Ali, S. M., & Ibrahim, M., (2010). Antidiabetic activity of Vinca rosea extracts in alloxan-induced diabetic rats. *Int. J. Endocrinol.,* 1–6.

Alami, Z., Aynaou, H., Alami, B., Hdidou, Y., & Latrech, H., (2015). Herbal medicines use among diabetic patients in oriental morocco. *J. Pharmacogn. Phytother., 7*, 9–17.

Alizadeh, F., Azarnia, M., Mirabolghasemi, G., & Karampoor, P., (2015). Effect of fruit *Heracleum persicum* extract on changes in serum levels of sex hormones in rats with polcystic ovary syndrome (PCOS). *Armaghane Danesh., 20*, 31–42.

Anyanwu, A. C., Odeniyi, I. A., Fasanmade, O. A., Adewunmi, A. J., Adegoke, O., Mojeed, A. C., Olofin, K. E., & Ohwovoriole, A. E., (2010). Endocrine-related diseases in the emergency

unit of a tertiary health care center in Lagos: A study of the admission and mortality patterns. *Niger. Med. J., 54*, 254–257.
Arentz, S., Abbott, J. A., Smith, C. A., & Bensoussan, A., (2014). Herbal medicine for the management of polycystic ovary syndrome (PCOS) and associated oligo/amenorrhoea and hyperandrogenism; a review of the laboratory evidence for effects with corroborative clinical findings. *BMC Complement Altern. Med., 14*, 511.
Barkaoui, M., Katiri, A., Boubaker, H., & Msanda, F., (2017). Ethnobotanical survey of medicinal plants used in the traditional treatment of diabetes in Chtouka Ait Baha and Tiznit (Western anti-atlas), Morocco. *J. Ethnopharmacol., 198*, 338–350.
Bashtian, M. H., Emami, S. A., Mousavifar, N., Esmaily, H. A., Mahmoudi, M., & Poor, A. H. M., (2013). Evaluation of fenugreek (*Trigonella Foenum-graceum* L.), effects seeds extract on insulin resistance in women with polycystic ovarian syndrome. *Iranian J. Pharm. Res., 12*, 475.
Bharthi, K. N., Shubhashree, M. N., & Bhat, S., (2017). Herbal approach to management of thyroid disease. *J. Ayurvedic Herb. Med., 3,* 51–55.
Bhowmik, D. K., Kumar, P. S., Paswan, S., Srivatava, S., Yadav, A., & Dutta, A., (2012). Traditional Indian herbs *Convolvulus pluricaulis* and its medicinal importance. *J. Pharmacog. Phytochem., 1*, 51–44.
Bousta, D., Boukhira, S., Aafi, A., Ghanmi, M., & El-Mansouri, L., (2014). Ethnopharmacological study of anti-diabetic medicinal plants used in the middle-atlas region of Morocco (Sefrou region). *Int. J. Pharm. Res. Health. Sci., 2*, 75–79.
Bove, M., (2012). *Botanical insights into Autoimmune Thyroid Disease.* http://cdn.natural-dispensary.com/downloads/White_Paper-Botanical_Insights_into_Autoimmune_Thyroid_Disease.pdf (accessed on 26 October 2022).
Chauhan, K., Sharma, S., Agarwal, N., Chauhan, S., & Chauhan, B., (2012). A study on potential hypoglycemic and hypolipidemic effects of *Lepidium sativum* (garden cress) in Alloxan induced diabetic rats. *Am. J. Pharm. Tech. Res., 2*, 522–535.
Danjuma, M. N., Abdullahi, I., & Shittu, U., (2018). Ethnobotanical assessment of plants used for the treatment of endocrine disorder in Daura Emirate, Katsina, Nigeria. *J. Res. Forest, Wildl. Environ., 10*, 28–41.
DeLange, F. E., (1991). Iodine deficiency. In: Braverman, L. E., & Utiger, R. D., (eds.), *Werner and Ingbar's the Thyroid* (6th edn., pp. 368–389). Philadelphia, Pa: JB Lippincott Co.
Eddouks, M., Maghrani, M., Lemhadri, A., Ouahidi, M. L., & Jouad, H., (2002). Ethnopharmacological survey of medicinal plants used for the treatment of diabetes mellitus, hypertension and cardiac diseases in the south-east region of Morocco (Tafilalet). *J. Ethnopharmacol., 82*, 97–103.
Famuyiwa, O. O., (1990). Problems and challenges in practice of endocrinology in developing countries—An overview. *Niger. Med. Pract., 20*, 3–6.
Feng, X. U., Zhang, S., Shao, R., & Zhen, Y., (2005). Anticancer activity of sodium caffeate and its mechanism. *Acta. Pharmacol. Sin., 26*, 1248–1252.
Ghimire, S. K., McKey, D., & Aumeeruddy-Thomas, Y., (2005). Conservation of Himalayan medicinal plants: Harvesting patterns and ecology of two threatened species nardostachys grandiflora and *Neopicrorhiza scrophulariiflora. Biol. Conserv., 124*, 463–475.
Ghourri, M., Zidane, L., & Douira, A., (2013). Use of medicinal plants in the treatment of diabetes in the Moroccan Sahara (Tan-Tan). *J. Anim. Plant. Sci., 17*, 2388–24411.
Giovannini, P., Howes, M. J. R., & Edwards, S. E., (2016). Medicinal plants used in the traditional management of diabetes and its sequelae in Central America: A review. *J. Ethnopharmacol., 184*, 58–71.

Grant, P., (2010). Spearmint herbal tea has significant anti-androgen effects in polycystic ovarian syndrome. A randomized controlled trial. *Phytother. Res., 24*, 186–188.

Gurib-Fakim, A., (2006). Medicinal plants: Traditions of yesterday and drugs of tomorrow. *Mol. Aspects. Med., 27*, 1–93.

Hachi, M., Ouafae, B., Hachi, T., Imane, B., Atmane, R., & Zidane, L., (2016). Contribution to the ethnobotanical study of antidiabetic medicinal plants of the central middle atlas region (Morocco). *Lazaroa., 37*, 1.

Haj-Husein, I., Tukan, S., & Alkazaleh, F., (2016). The effect of marjoram (*Origanum majorana*) tea on the hormonal profile of women with polycystic ovary syndrome: A randomized controlled pilot study. *J. Human Nutrition Dietetics, 29*, 105–111.

Hamayun, M., (2007). Traditional uses of some medicinal plants of swat valley, *Pakistan. Ind J Trad Know, 06*, 636–641.

Hameed, R. F., Atya, K. F., & Mohameed, M. A., (2014). Role of *Fucus vesiculosus* L extract in the regulation of thyroid hormones status in adult male rabbits Kerbala. *J. Pharma. Sci., 7*, 84–67.

Hasani-Ranjbar, S., & Larijani, B., (2014). Medicinal plants as potential new target drugs in endocrine disorders- review article. *Iranian J. Publ. Health., 43*, 24–34.

Hasani-Ranjbar, S., Larijani, B., & Abdollahi, M., (2008). A systematic review of Iranian medicinal plants useful in diabetes mellitus. *Arch. Med. Sci., 43*, 285–292.

Hossein, K. J., Leila, K. J., Koukhdan, E. T., Nazanin, S. J., Farzad, P., & Elham, R., (2015). The effect of pomegranate juice extract on hormonal changes of female Wistar rats caused by polycystic ovarian syndrome, *Biomed. Pharmacol. J., 8*, 971–977.

Jalilian, N., Modarresi, M., Rezaie, M., Ghaderi, L., & Bozorgmanesh, M., (2013). Phytotherapeutic management of polycystic ovary syndrome: Role of aerial parts of wood betony (*Stachys lavandulifolia*). *Phytother. Res., 27*, 1708–1713.

Jashni, H. K., Jahromi, H. K., & Bagheri, Z., (2016). The effect of palm pollen extract on polycystic ovary syndrome (POS) in rats. *Int. J Med. Res. Health Sci., 5*, 317–321.

Jelodar, G., & Askari, K., (2012). Effect of vitex agnus-castus fruits hydroalcoholic extract on sex hormones in rat with induced polycystic ovary syndrome (PCOS). *Physiol. Pharmacol., 16,* 62–69.

Jie, D., Ying, X., Xiaoping, M., Jinna, A., Xiudong, Y., Zhiqiang, L., & Na, L., (2015). Estrogenic effect of the extract of renshen (radix ginseng) on reproductive tissues in immature mice. *J. Tradit. Chin. Med., 35*, 460–467.

Jouad, H., Haloui, M., Rhiouani, H., El-Hilaly, J., & Eddouks, M., (2001). Ethnobotanical survey of medicinal plants used for the treatment of diabetes, cardiac and renal diseases in the North Centre region of Morocco (fez–Boulemane). *J. Ethnopharmacol., 77*, 175–182.

Kamel, H. H., (2013). Role of phyto-oestrogens in ovulation induction in women with polycystic ovarian syndrome. *Eur. J. Obstet. Gynecol. Reprod. Biol., 168*, 60–63.

Katiri, A., Barkaoui, M., Msanda, F., & Boubaker, H., (2017). Ethnobotanical survey of medicinal plants used for the treatment of diabetes in the Tizi n'Test region (Taroudant province, Morocco). *J. Pharmacogn. Nat. Prod., 3*, 1–10.

Kesari, A. N., Kesari, S., Singh, S. K., Gupta, R. K., & Watal, G., (2007). Studies on the glycemic and lipidemic effect of *Murraya koenigii* in experimental animals. *J. Ethnopharmacol., 112*, 305–311.

Klepser, T. B., & Klepser, M. E., (1999). Unsafe and potentially safe herbal therapies. *Am. J. Health. Syst. Pharm., 56,* 125–138.

Kumar, T., Chandrashekar, K. S., (2011). *Bauhinia purpurea Linn.* A review of its ethnobotany, phytochemical and pharmacological profile. *Res. J. Med. Plant., 5*, 420–431.

Laadim, M., Ouahidi, M., Zidane, L., El-Hessni, A., Ouichou, A., & Mesfioui, A., (2017). Ethnopharmacological survey of plants used for the treatment of diabetes in the town of Sidi Slimane (Morocco). *J. Pharmacogn. Phytother., 9*, 101–110.

Lopez, A. D., Mathers, C., & Ezzati, M., (2006). *Global Burden of Disease and Risk Factors*. The World Bank and Oxford University Press, Washington, DC, USA.

Ma, T., Guo, J., & Wang, F., (1993). The epidomology of iodine deficiency diseases in China. *Am. J. Clin. Nutr., 57*, 264–266.

Mokhtari, M., Ebrahimpoor, M. R., & Harfsheno, S., (2014). The effects of alcoholic extract of *Marrubium vulgare* on hormonal parameters in female rat model of polycystic ovarian syndrome. *Med. Sci. J. Islamic Azad Univ.-Tehran Med. Branch., 24*, 74–80.

Mrabti, H. N., Jaradat, N., Kachmar, M. R., Ed-Dra, A., Ouahbi, A., Cherrah, Y., & Faouzi, M. E. A., (2019). Integrative herbal treatments of diabetes in Beni Mellal region of Morocco. *J. Integr. Med., 17*, 93–99.

Naowaboot, J., Pannangpetch, P., Kukongviriyapan, V., & Kongyingyoes, B., (2009). Antihyperglycemic, antioxidant and antiglycation activities of mulberry leaf extract in streptozotocin induced chronic diabetic rats. *Plant. Foods Hum. Nutr., 64*, 116–121.

Park, J. M., Bong, H. Y., Jeong, H. I., Kim, Y. K., Kim, J. Y., & Kwon, O., (2009). Postprandial hypoglycemic effect of mulberry leaf in Goto-Kakizaki rats and counterpart control Wistar rats. *Nutr. Res. Pract., 3*, 272–278.

Parmer, H., & Kar, A., (2009). Protective role of *Mangifera indica*, *Cucumis melo* and *Citrullus vulgaris* peel extracts in chemically induced hypothyroidism. *Chemico. Biol. Inter., 177*, 254–258.

Radha, M., & Laxmipriya, N., (2016). Efficacy of non polar extract (NPE) of aloe barbadensis mill. in polycystic ovarian syndrome (PCOS) rodent model-an "*in vivo*" study. *Int. J. Pharm. Sci. Res., 7*, 4933.

Rahmawati, E., Dewoto, H. R., & Wuyung, P. E., (2006). Anticancer activity study of ethanol extract of mahkota dewa fruit pulp (phaleria macrocarpa (Scheff.) boerl.) in c3h mouse mammary tumor induced by transplantation. *Med. J. Indonesia., 15*, 217–222.

Rajan, R. K., & Balaji, B., (2017). Soy isoflavones exert beneficial effects on letrozole induced rat polycystic ovary syndrome (PCOS) model through the anti-androgenic mechanism. *Pharm. Biol., 55*, 242–251.

Rehman, M. S. U., Kamran, S. H., Ahmad, M., & Akhtar, U., (2015). Anti-diabetic activity of crude *Pistacia lentiscus* in alloxan-induced diabetes in rats. *Bangl. J. Pharmacol., 10*, 543–547.

Ried, K., (2015). Chinese herbal medicine for female infertility: An updated meta-analysis. *Complement Ther. Med., 23*, 116–128.

Robinson, H. M., & Hort, K., (2012). Non-communicable diseases and health system reforms in low-and-middle-income countries. *Pacific. Health. Dialog., 18*, 179–190.

Sabitha, V., Ramachandran, S., Naveen, K., & Panneerselvam, K., (2011). Antidiabetic and antihyperlipidemic potential of *Abelmoschus esculentus* (L.) Moench. in streptozotocin-induced diabetic rats. *J. Pharm. Bioallied. Sci., 3*, 397.

Sadighi, J., Maftoon, F., & Moshrefi, M., (2004). Complementary and alternative medicine (CAM): Knowledge, attitude and practice in Tehran, Iran. *Payesh., 4*, 289–279.

Sharma, T., Gupta, M., & Nathani, S., (2015). Review of ayurvedic drugs acting on hypothyroidism. *Intl. Ayurvedic Med. J., 3*, 2579–2572.

Shinwari, M. I., & Khan, M. A., (2000). Folk use of medicinal herbs of Margalla hills national park, Islamabad. *J. Ethnopharmacol., 69*, 45–56.

Sicree, R., Shaw, J., & Zimmet, P., (2006). Diabetes and impaired glucose tolerance. Delice Gan. Diabetes Atlas (3rd edn.). *Internat. Diabetes Fed., 43*, 24–34.

Skalli, S., Hassikou, R., & Arahou, M., (2019). An ethnobotanical survey of medicinal plants used for diabetes treatment in Rabat, Morocco. *Heliyon., 5*, e01421.

Skvaa, K., (2011). Therapeutic uses of *Withania somnifera* (ashwagandha) with a note on withanolides and its pharmacological actions. *Asian. J. Pharm. Clin. Res., 4*, 1–4.

Soumya, V., Muzib, Y. I., & Venkatesh, P., (2016). A novel method of extraction of bamboo seed oil (Bambusa bamboos Druce) and its promising effect on metabolic symptoms of experimentally induced polycystic ovarian disease. *Indian J. Pharmacol., 48*, 162.

Tabassum, W., Arunakullu, M. P., & Sinha, (2013). Effect of leaf extracts of *Moringa oleifera* on regulation of hypothyroidism and lipid profile. *The Biascan., 8*, 665–669.

Tahraoui, A., El-Hilaly, J., Israili, Z., & Lyoussi, B., (2007). Ethnopharmacological survey of plants used in the traditional treatment of hypertension and diabetes in South–Eastern Morocco (*Errachidia province*). *J. Ethnopharmacol., 110*, 105–117.

Taïbi, K., Abderrahim, L. A., Helal, F., & Hadji, K., (2021). Ethnopharmacological study of herbal remedies used for the management of thyroid disorders in Algeria. *Saudi Pharmaceut. J., 29*, 43–52. https://doi.org/10.1016/j.jsps.2020.12.004.

Treatment, N., & Kabel, A. M., (2016). Polycystic ovarian syndrome: Insights into pathogenesis, diagnosis. *J. Pharmacol. Reports., 1*, 1–5.

Verma, K. J., (2012). Studies on traditional treatment of thyroid by the tribals of Chitrakoot district, Uttar Pradesh. *Int. J. Sci. Res., 3*, 2319–7064.

Verma, P., & Jameel, K., (2014). Studies on traditional treatment of thyroid by the tribals of Chitrakoot district, Uttar Pradesh. *Intl. J. Sci. Res., 3*, 1373–1370.

Vos, T., Flaxman, A. D., Naghavi, M., Lozano, R., Michaud, C., Ezzati, M., Shibuya, K., et al., (2010). Years lived with disability (YLDs) for 1160 sequelae of 289 diseases and injuries 1990–2010: A systematic analysis for the global burden of disease study 2010. *The Lancet, 380*, 2163–2196.

Wang, J. G., Anderson, R. A., Graham, G. M., Chu, M. C., Sauer, M. V., Guarnaccia, M. M., & Lobo, R. A., (2007). The effect of cinnamon extract on insulin resistance parameters in polycystic ovary syndrome: A pilot study. *Fertil. Steril., 88*, 240–243.

WHO, (2012). *The World Health Statistics; 2012 Report*. Geneva World Health Organization.

Willatgamuwa, S. A., Platel, K., Saraswathi, G., & Srinivasan, K., (1998). Antidiabetic influence of dietary cumin seeds (*Cuminum cyminum*) in streptozotocin induced diabetic rats. *Nutr. Res., 18*, 131–142.

World Health Organization, (1997). *Obesity: Preventing and Managing the Global Epidemic*. Report of a WHO Consultation on Obesity Geneva.

Yadav, S., Vats, V., Dhunnoo, Y., & Grover, J. K., (2002). Hypoglycemic and antihyperglycemic activity of *Murraya koenigii* leaves in diabetic rats. *J. Ethnopharmacol., 82*, 111–116.

Zangeneh, F. Z., Minaee, B., Amirzargar, A., Ahangarpour, A., & Mousavizadeh, K., (2010). Effects of chamomile extract on biochemical and clinical parameters in a rat model of polycystic ovary syndrome, *J. Reprod. Infert., 11*, 169.

Ziai, S. A., (2003). Traditional herbal medicine. *J. Med. Plants., 2*, 51–43.

Ziyyat, A., Legssyer, A., Mekhfi, H., Dassouli, A., Serhrouchni, M., & Benjelloun, W., (1997). Phytotherapy of hypertension and diabetes in oriental Morocco. *J. Ethnopharmacol., 58*, 45–54.

CHAPTER 13

Preventive Measures to Control the Pathogenic Action of Endocrine Toxicants

RITIKA ANTHAL and RAMAN JASROTIA

Department of Zoology, University of Jammu, Jammu–180006, Jammu and Kashmir, India, E-mail: ramanjasrotia18@gmail.com (R. Jasrotia)

ABSTRACT

The increased contamination of our environment by endocrine-disrupting chemicals (EDCs) has recently emerged as a global problem. EDCs are chemicals that alter the normal endocrine system of organisms by having agonistic, antagonistic, or both effects. They have been shown to act by disturbing hormone synthesis, metabolism, transport, excretion, and their receptor binding. EDCs are highly heterogeneous, including some natural substances also, but most of them are manmade compounds used as industrial solvents, plastic compounds, plasticizers, pesticides, pharmaceutical agents, and heavy metals such as cadmium and lead, released into the environment without prior knowledge of their implications for the environment and wildlife. In humans, EDCs have been found to cause various endocrine disorders affecting male and female reproductive systems along with developmental disorders, cancer of reproductive tissues, thyroid dysfunction, adrenal dysfunction, metabolic disorders like obesity and diabetes, disorders of bone, cardio-vascular disorders, etc. Considering the vast number of adverse effects that these chemicals have on the environment and human health, it was only after 1990 that international actions were directed toward this problem. The world Wildlife Fund Wingspread conference in Wisconsin,

Environmental Endocrine Toxicants: Biology, Effects, and Management. Younis Ahmad Hajam, Rouf Ahmad Bhat, Khalid Rehman Hakeem, & Rajesh Kumar (Eds.)
© 2023 Apple Academic Press, Inc. Co-published with CRC Press (Taylor & Francis)

USA in 1991, the Weybridge meeting, in 1996, the Rotterdam convention of UNEP, the Stockholm Convention, etc., were some of the international efforts to address this global problem. Over the years, research related to EDC has emerged as a new multidisciplinary research area encompassing a large number of disciplines. In 2013, WHO and UNEP launched their joint report on EDCs highlighting their implication in deteriorating human health and recommended improved testing as well as reduced exposures to endocrine toxicants (ETs). The present work highlights the necessity of a strategy for prevention of exposure to EDC and development of remediation technologies. This requires participation by all the stake holders from the individual level to the administration. More efforts should be put into identifying and studying these chemicals and Green Chemistry should be encouraged as it aims at identifying and developing substitutes not having endocrine disrupting activity. Apart from that, recommendations at the authority and individual levels have been suggested.

13.1 INTRODUCTION

It was about some decades ago that wildlife studies suggested a correlation between environmental chemicals and many endocrine disorders by a possible interference with the hormone system of these organisms (Carson, 1962). Research was subsequently directed towards determining the toxicity of most commonly used chemicals and a wide range of chemicals including polychlorinated biphenyls (PCBs), pharmaceutical products, dioxin, and similar compounds, organo-chlorinated pesticides and plasticizers that are part of our everyday life in the form of toys, cosmetic products, food plastic bags and bottles, metal food cans, detergents, pesticides, insecticides, and flame retardants were found to have endocrine disrupting activity (Gore, 2001; McKinlay et al., 2008). This phenomenon was called endocrine disruption and such chemicals were termed as endocrine disrupting chemicals (EDCs) that may include "any exogenous substance or material that alters the function(s) of the endocrine system and consequently causes adverse health effects in an intact organism, its progeny or its (sub)population" (Damstra et al., 2002).

EDCs mostly include chemicals made by man for various purposes, but some natural substances are also known to cause similar effects. When released into the environment, they have been shown to alter the normal endocrine processes of living organisms which ensures the coordinated action of hormones, ultimately regulating the physiological and behavioral activities of these organisms. EDCs disrupt the normal endocrine activity

(EA) of organisms by a variety of mechanisms affecting hormone production, synthesis, transport, metabolism, and excretion and by binding with their receptors (Acerini and Hughes, 2006). In humans, earlier experimental and epidemiological studies related to EDC mainly focused on their estrogenic effects, acting as xenoestrogens, antiestrogens, xenoandrogens, and anti-androgens, causing many reproductive disorders in the effected organisms (Colborn and Clement, 1992; Davis et al., 1993; Uslu et al., 2013; Marty and O'Connor, 2014; Rachon, 2015; Pinto et al., 2016; Giulivo et al., 2016; Morgan et al., 2017; Lymperi and Giwercman, 2018). Later studies reported their effect on other organs like thyroid, adrenal, and a rise in the incidence of diabetes, obesity, and many other metabolic disorders (Kilic et al., 2005; Calsolaro et al., 2007; Agas et al., 2013; Sharpe and Drake, 2013; Carpenter and Sly, 2016; Darbre, 2017; Li et al., 2018).

However, with the rise in scientific research and media awareness in the 19th century, the focus of many national and international health organizations, including the World Wildlife Fund, World Health Organization (WHO), and the Greenpeace, was shifted towards this alarming situation. With the help of many non-governmental organizations and media, the process of creating awareness among the general public has started. This chapter highlights the main threats posed by EDCs to the environment and human health, proposals for future research, and the preventive measures at the administrative and individual levels.

13.2 A BRIEF HISTORY OF THE PROCESS OF ENDOCRINE DISRUPTION

The phenomenon of endocrine disruption is very old, although it gained publicity only recently. In 1920, pig farmers reported a lack of fertility in pig herds after feeding on moldy grains. Later this was shown to result from myco-estrogens present in the mold (McNutt et al., 1928; Bennett and Klich, 2003). In the 1940s, similar reports started to come out from Western Australia where lack of fertility was reported in sheep after they fed on clover fields. Later, it was shown that the phyco-estrogens present in the plant material were responsible for this (Bennets et al., 1946; Woods, 2003). In these cases, infertility was caused by the natural EDCs present in the environment.

On the other hand, the impact of human-made chemicals acting as EDCs was recognized much later. A breakthrough in EDC awareness was provided when Rachel Carlson published a book titled 'Silent Spring' in 1962. This book cautioned the scientific community and the general public about the consequences of excessive agricultural application of chemical pesticides and herbicides and their long-term implications on wildlife populations (Carson, A follow-up book in 1996 'Our stolen future' by Colborn et al. (1962) highlighted the inadequacy of the actions taken against the problem. These two books made a major contribution in spreading awareness regarding the role of manmade chemicals on the environment (Colborn et al., 1996). However, over the years, the process continued unabated and caused serious consequences for the wildlife species. The aquatic ecosystems were no exceptions, where the addition of chemicals through pesticides, sewage, groundwater, or accidental spills led to huge wildlife loss and genetic abnormalities (Jobling et al., 1998; Guillette and Gunderson, 2001; Horiguchi, 2006). This was followed by reporting of increased incidence of many endocrine diseases in humans also. One typical example is the chemical diethylstilbestrol (DES), one of the synthetic hormones that were used by pregnant women in the 1930s. Later studies showed that it caused vaginal cancers in daughters born to such women (Herbst et al., 1971; Harris and Waring, 2012), suggesting the vulnerability of humans to EDCs. Later, many male reproductive disorders like decline in sperm quality, higher incidence of infertility, hypospadias, cryptorchidism, testicular cancer, and rise in various metabolic disorders like diabetes, obesity, cardiovascular disorders, disorders of bone, thyroid, and adrenal, shifted the attention of the world community towards the effect of EDC on human health.

13.3 CHEMICALS LISTED AS ENDOCRINE DISRUPTORS (EDS)

Industrialization and urbanization over the last few decades have led to an increase in the synthesis and usage of many different types of chemicals, some of which have been polluting the environment. A few of them have also been shown to have endocrine disrupting activity, including the common groups of chemicals contained in industrial chemicals, pharmaceuticals, metals, pesticides, fungicides, and some naturally occurring chemicals like mycoestrogens and phytoestrogens. A list of chemicals acting as xenoestrogens, anti-estrogens, xenoandrogens, and anti-androgens is provided (Table 13.1).

13.4 EDC TESTING

The advancement in technology over the years has led to the development of new tests for determining the toxicity of various chemicals. A detailed list is provided in Table 13.2.

TABLE 13.1 Some Chemicals Acting as Xenoestrogens, Xenoandrogens, Antiestrogens, and Anti-Androgens (De Coster and van Larebeke, 2012)

Category	Substances	References
Naturally occurring substances	20S-Protopanaxadiol, alpha-zearalenol, apigenin, ferutinine, formononetin, genistein, kaempferol, methoxylated brominated diphenyl ethers, zearalenone	Kuiper et al. (1998); Ikeda et al. (2002); Guevel and Pakdel (2001); Beck et al. (2005); Viglietti-Panzica et al. (2007); Yu et al. (2007); Hu et al. (2011)
Industrial products and pollutants	4-nonylphenol, 4-n-propylphenol, benzyl butyl phthalate, bisphenol-A, dibutylphthalate, dichlorostyrene, dodecylphenol, nonylphenols, polybrominated biphenyl ethers (PBDEs), polychlorinated biphenyls (PCBs), 3,9-dihydroxy Benz(a) anthracene, 3-methylcholanthrene, benzotriazole,	Schneider et al. (1976); Pasqualini et al. (1990); Chaloupka et al. (1992); White et al. (1994); Crews et al. (1995); Jobling et al. (1995); Fischer et al. (1998); Bonefeld-Jorgensen et al. (2001); Meerts et al. (2001); Yamasaki et al. (2003); Pillon et al. (2005); Abdelrahim et al. (2006); Hamers et al. (2006); Harris et al. (2007); Mnif et al. (2007); Oh et al. (2007, 2009); Vom Saal et al. (2007, 2012); Svobodova et al. (2009); Preuss et al. (2010); Liu et al. (2011); Winneke et al. (2002); Buckman et al. (2011); Takeuchi et al. (2011); Recio-Vega et al. (2011); Sanchez et al. (2011); Lai et al. (2012)
Personal care items	4,4-Dihydroxy-benzophenone, 4-hydroxybenzoic acid ethyl ester, 4-hydroxybenzophenone, benzophenone-1, benzophenone-2, benzophenone-3, benzyl salicylate, eusolex 2292, eusolex 6300, galaxolide, homosalate, isopentyl-4-methoxycinnamate, octyl salicylate, para-amino-benzoic acid, 3-benzylidene camphor	Schlumpf et al. (2001); Gomez et al. (2005); Suzuki et al. (2005); Kunz and Fent (2006); van Der Burg et al. (2008); Zucchi et al. (2011)
Pesticides	alachlor, aldrin, atrazine, azinphos-methyl, benomyl, bromopropylate, carbaryl, chlordane, chlordecone, cycloprothrin, DDE, DDT, deltamethrin, dichlorvos, dicofol, dieldrin, endosulfan, endrin, fention, lindane, linuron, methiocarb, metolachlor, pentachlorophenol, permethrin, pirimicarb, procymidone, tetramethrin, toxaphene, trans-nonachlor, vinclozolin, triflumizole	Andersen et al. (2002, 2006); Kitamura et al. (2003); Grunfeld and Bonefeld-Jorgensen (2004); Kojima et al. (2004); Lemaire et al. (2004, 2006); Bonefeld-Jorgensen et al. (2005); Kim et al. (2005); Fan et al. (2007); Shen et al. (2010); Kjærstad et al. (2010); Orton et al. (2011)

TABLE 13.2 Various Tests Used for Testing EDC Behavior of Chemicals

Activity	Test	Brief Description	References
Estrogenicity assay	*In vivo* uterotrophic assay using rodents	Growth of uterine tissue under the control of estrogen and recording and evaluating the response	Shelby et al. (1996); OECD. Test No. 440 (2007)
	In vivo vitellogenin assay using fish	Vitellogenin is a hallmark of female protein and its occurrence or high concentration in male fish is considered estrogenic effect of endocrine disruption	Heppell et al. (1995); Denslow et al. (1999); Matozzo et al. (2008)
	In vitro MCF-7 focus assay	MCF-7 cells are known to propagate in response to estrogen and chemicals having similar estrogenic activity	Arcaro and Gierthy (2001); Ssempebwa et al. (2004)
	In vitro estrogen screen assay in yeast	Screening of Lac Z gene activity involving β-galactosidase in yeast *Saccharomyces cerevisiae*	Bistan et al. (2012)
	In vitro luciferase reporter gene assay	Monitoring of transcription of certain genes in cells and generation of light by the action of enzyme luciferase	Naylor (1999)
	In vitro aryl hydrocarbon (dioxin) receptor (AHR) test	Binding of AHR to dioxin like substance that activates cytochrome enzyme and decreases intracellular concentration of estrogen	Denison et al. (2011); EU (2012)
Androgenic assay	Hershberger assay using rats	Increase or decrease in absolute weights in castrated rats on exposure to chemicals	EPA (2011)
Thyroid disrupting chemical	Gene expression using *Xenopus laevis*	Evaluation of endocrine disruption of thyroid axis by analysis of gene expression	Turque et al. (2005); Mughal et al. (2018)
	Signaling through phosphatidylinositol 3 kinase (PI3K)	Exogenous chemicals are used to test disruption of thyroid stimulation of PI3K pathway	Storey et al. (2006)
Obesogens	Adipogenesis model involving T3-L1 cell line	Culture of T3-L1 cell line	Pereira-Fernandes et al. (2014)
	Peroxisome proliferator-activated receptor (PPAR γ)	PPAR γ nuclear receptors bind chemicals and initiate adipogenesis	Casals-Casas and Desvergne (2011); Pereira-Fernandes et al. (2013)
Adreno-cortical disrupting chemicals	H295R cell line	Culture of H295R cell line	Harvey (2016); Ahmed et al. (2018)

13.5 MODE OF ACTION

EDCs have become a part of our environment due to various anthropogenic activities. They are found in air, water, and soil and reach human and other wild animals through various routes (Figure 13.1). These chemicals enter the body and act through various modes, Some EDCs bind to hormone receptors and enhance, reduce or altogether stop their action, showing agonist, antagonist, or sometimes both activities They may also alter the number of various cell receptors, like nuclear or non-steroidal receptors, or may also act through other pathways including ER-mediated non-genomic pathway, Ion channels, co-activator recruitment, epigenesis (Gore et al., 2015) and changes the concentration of circulating hormones. These effects produce alteration in the physiology and phenotype of individual, which may lead to increased disease susceptibility. Long-term exposure to EDC has been found related to increased incidence of four principal non-communicable diseases (NCD) during the last few decades, i.e., cancer, CVD, chronic respiratory diseases, and diabetes mellitus. Although genetics and environment are considered as the main factors regulating these diseases, the rise in their incidence is thought to be contributed by environmental and social factors, including EDC (Barouki et al., 2012) more than any gene variant. This interaction is more vulnerable during the critical life stages like childhood and puberty, during which the exposure to EDCs can modify the developmental processes and lead to adverse consequences (Collman, 2011).

13.6 EDCS AND HUMAN HEALTH

Research over the years has shown many different ways our endocrine system is affected by the EDCs present in the environment causing many disorders.

13.6.1 EDCS AND TYPE 2 DIABETES AND OBESITY

At present, diabetes and obesity are the leading global health issues. However, it was only recently that the role of EDCs in the causation of these and other metabolic disorders was investigated. A study by Baillie-Hamilton in 2002 proposed the hypothesis of Obesogenic activity of EDCs, which emphasizes the impact of prenatal or early post-natal EDC exposure

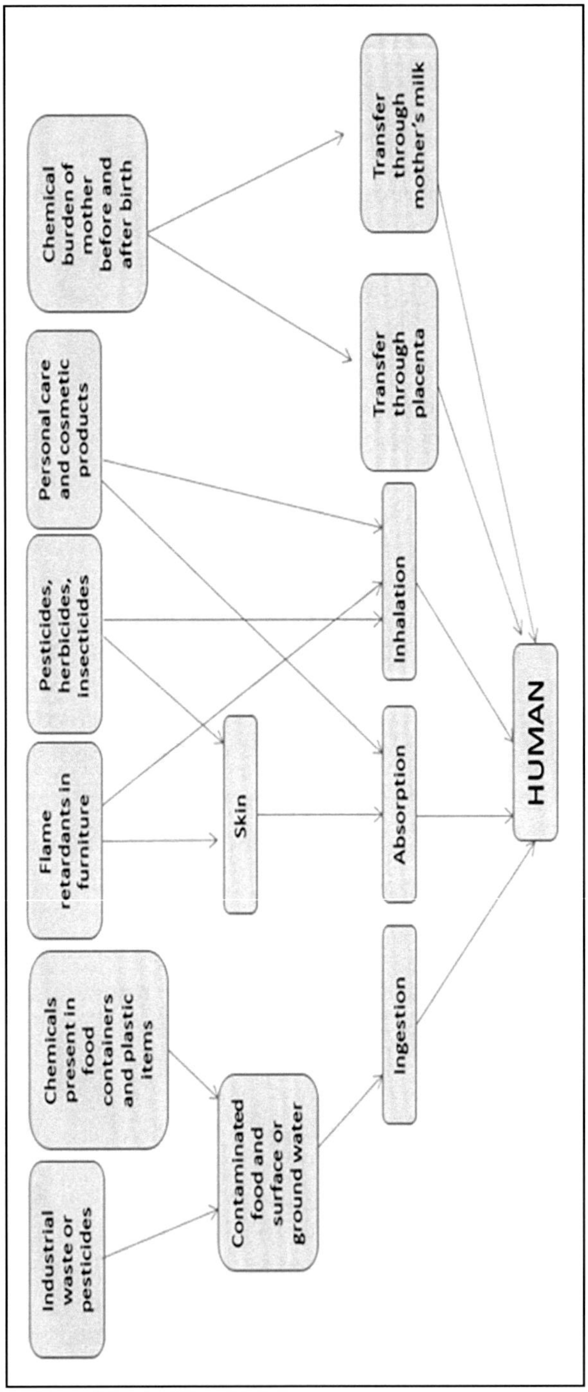

FIGURE 13.1 Routes of EDC exposure in humans EDCs get in the way of the action of hormones and disturb homeostasis and may alter physiology during any time of an individual's life, right from fetal development to adulthood.

in increasing the risk of these individuals to become obese. Later, many experimental studies (Wei et al., 2011; Miyawaki et al., 2007; Patisaul and Bateman, 2008; Rubin and Soto, 2009; Howdeshell et al., 1999) and population studies (Carwile and Michels, 2011; Shankar et al., 2012; Trasande et al., 2012; Wang et al., 2012; Bhandari et al., 2013; Harley et al., 2013; Arsenescu et al., 2008; Kamstra et al., 2014; Regnier and Sargis, 2014) also confirmed this hypothesis. Similarly, one more hypothesis, i.e., the diabetogenic hypothesis was proposed after these chemicals were shown to induce insulin resistance (IR) in some cellular studies and also in animal models. This was supported by many experimental studies (Novelli et al., 2005; Alonso-Magdalena et al., 2006, 2010; Kurita et al., 2009; Lin et al., 2011; Wei et al., 2011; Batista et al., 2012; Ma et al., 2013; Naville et al., 2013). The most common chemicals showing endocrine disrupting activity include BPA, DEHP, DBP, DES, DDT, TCDD, PCB, MEHP, POPs, and TBT. The female reproductive system is also much affected by the EDCs present in the environment and entering the body through various routes. This is manifested in altered structure and functioning causing severe reproductive issues. Various experimental studies (Ashiru and Odusanya, 2009; Martino-Andrade and Chahoud, 2010; Balabani et al., 2011; Bellelis et al., 2011; Craig et al., 2011; Schug et al., 2011; Kadhel et al., 2012; Meeker, 2012; Uzumcu et al., 2012; Weinhold, 2012; Kay et al., 2013; Peretz et al., 2014; Caserta et al., 2014) and epidemiological studies (Caserta et al., 2014; Peretz et al., 2014) have confirmed the adverse effect of gestational and early natal exposure of EDCs (Figure 13.2).

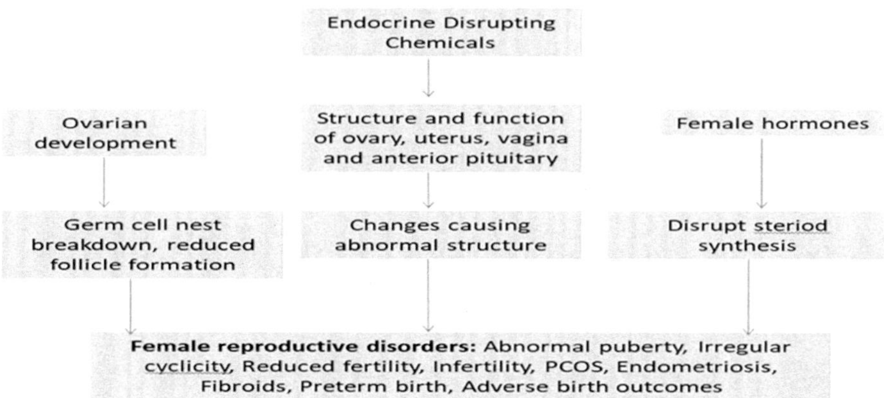

FIGURE 13.2 Various female reproductive disorders caused by EDCs.

13.6.2 EDCS AND MALE REPRODUCTIVE SYSTEM

In humans, the male reproductive system requires a functional endocrine system for its growth and proper functioning. Human population and *in-vitro* animal studies over the years have shown a rise in the number of cases related to a male reproductive system like hypospadias, cryptorchidism, and testicular cancer. Various epidemiological studies (Longnecker et al., 2002; Pierik et al., 2007; McGlynn et al., 2009; Giordano et al., 2010; Cook et al., 2011; Trabert et al., 2012; Giannandrea et al., 2013) and experimental studies (Mably et al., 1992; Kelce et al., 1995; de Jager et al., 1999; Fisher et al., 1999; Gray et al., 1999; Wolf et al., 1999; McIntire et al., 2001; Ohsako et al., 2001; Pocock et al., 2002; Adeeko et al., 2003; Simanainen et al., 2004; Durrer et al., 2007; Kishta et al., 2007; Kristensen et al., 2011) over the years have listed pesticides, PCBs, phthalates, DDT, Dioxin, flame retardants, PBDEs, DDT, DDE, and dibenzofuran as the main chemicals leading to abnormality in its structure and reduction in function as summarized in Figure 13.3.

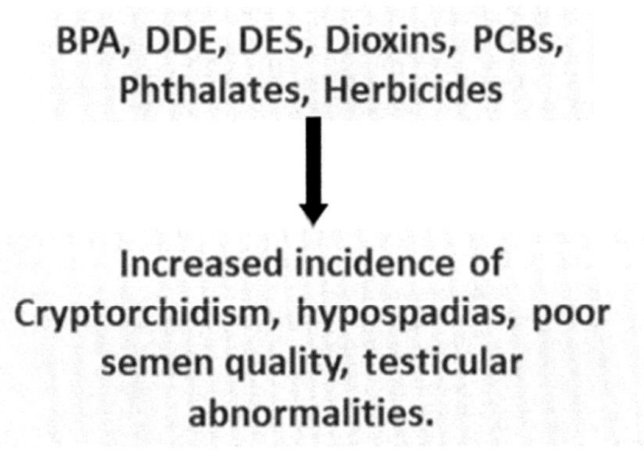

FIGURE 13.3 Various EDCs affecting male reproductive system.

13.6.3 EDCS AND HORMONE SENSITIVE CANCERS

EDC exposure has been linked to increased occurrence of many hormones' sensitive cancers. Table 13.3 summarizes the categories of EDCs known shown to have carcinogenic effect on various body tissues. Usually, the exposure to critical life stages has been shown to be more vulnerable, like in the case of breast cancer, three stages of life are vulnerable for EDC exposure, i.e., pre-natal, puberty, and pregnancy. Exposure to EDCs at these stages can lead to an increased risk of developing disease.

TABLE 13.3 Chemicals Causing Hormone Sensitive Cancers in Humans

Tissue	Chemicals	References	
		Experimental	**Epidemiological**
Breast	Industrial chemicals, pollutants, herbicides, pharmaceuticals, dioxin	Høyer et al. (1998); Ibarluzea et al. (2004); McElroy et al. (2006); Nie et al. (2007); Viel et al. (2008); Lopez-Carrillo et al. (2010); Recio-Vega et al. (2011); Villeneuve et al. (2020);	Mills and Yang (2006); Pesatori et al. (2009); Lopez-Carrillo et al. (2010)
Prostate	BPA, PCBs, alkylphenols, pesticides, insecticides, herbicides, some heavy metals, agent orange chemical	Hardell et al. (2006)	Chen and Wang (1990); Lewis et al. (1999)
Thyroid	PCBs, PBDEs, some phthalates, perchlorate	Hallgren et al. (2001); Ellis-Hutchings et al. (2006); Kuriyama et al. (2007)	Greer et al. (2002); Herbstman et al. (2010); Steinmaus et al. (2013)
Non-Hodgkin lymphoma	DDT, furans, dioxin, dioxin-like PCBs	Viel et al. (2011)	Viel et al. (2008)

13.6.4 EDCS AND NEURO-ENDOCRINE SYSTEM

Gestational and pre-natal exposure to some EDCs has been shown to alter the structure and function of the brain, with non-functional neuroendocrine system (Table 13.4).

13.7 STRATEGY FOR FUTURE PREVENTION

13.7.1 RESEARCH GAPS

Despite international efforts, there is a lack of complete understanding regarding the EDCs present in the environment, their mode of action and their long-term implications for mother Earth and its inhabitants. This is mainly because of the inadequacy of our current research and planning strategies.

TABLE 13.4 Chemicals Affecting Brain and Neuro-Endocrine System

Tissue	Chemicals	Effect	References	
			Experimental	Epidemiological
Brain	BPA, PCB, PCDD, PCDF, PBDEs, chlorpyrifos, organophosphates, phthalates, pesticides	Altered structure and function of brain, and dysfunctional neuro-endocrine system	Desaulniers et al. (2005); Colciago et al. (2006, 2009); Yu et al. (2010)	Engel et al. (2010, 2011); Swan et al. (2010); Bouchard et al. (2011); Kim et al. (2011); Rauh et al. (2011); Fitzgerald et al. (2012); Boucher et al. (2012); Solomon et al. (2013); Braun et al. (2014); Chen et al. (2014); Shelton et al. (2014)

The foremost requirement at this time is to determine the extent of EDC exposure to human and wildlife populations. Also, most of the evaluation for risk assessment of EDC is designed for a certain threshold dose, and this does not account for EDCs present in the environment in low doses and not considered toxic. Moreover, some EDCs do not conform to the safety testing programs and produce an inverted U-shaped curve instead of the classical linear dose-response relationship. EDC assessment is further complicated by the interaction of these compounds with naturally occurring hormones or other chemicals showing additive effects at low, non-toxic concentrations. No assessment exists to determine and quantify such mixture effects that may be synergistic, antagonistic, or additive. Moreover, the research is focused more on the reproductive effects of EDCs, and most of the testing guidelines for EDC are prepared accordingly, not considering the effects of EDCs exposure on other hormones and tissues. These complexities regarding the nature of EDCs and loopholes in the testing protocol makes it difficult to assess their toxicity potential.

13.8 FUTURE RESEARCH PROSPECTS

At present, an urgent requirement is to fully understand the interaction of EDCs with various physiological processes of living organism; however, we are lagging behind in this because of the limited number of trained scientists and clinical. Hence the priority for future research should be to develop scientists, clinical professionals, health care providers, and epidemiologists in the field of EDC research. Here, the role of the funding agencies cannot be ruled out. The funding should be directed towards collaborative research at a global platform, instead of restricted area, and this will ultimately yield substantial savings in treatment and mitigation (Gore et al., 2012). Future research should be aimed towards some goals, which are discussed in subsections.

13.8.1 IDENTIFICATION OF POTENTIAL EDC CHEMICALS

At present, various human developmental activities have led to the addition of thousands of chemicals into the environment with a significant number of new additions every year. Hence, it is imperative to first identify chemicals having EDC activity. The research should be expanded to determine the full array of EDCs by developing new diagnostic and bio-assay techniques. The focus should also be given to the environmental effects of mixtures of EDCs at low non-toxic concentrations. This can be made possible by applying modern technologies like genomics, proteomics, and bioinformatics.

13.8.2 TESTING BEFORE USING

It is a common practice that once a substance is shown to have environment polluting activity, its usage is banned and the search for its alternative starts. However, sometimes, the replacement is made without prior testing and evaluation, and the chemical is released into the environment for example, the chemical bisphenol S was used as a replacement for the BPA, but this was later found to have equivalent endocrine disrupting activity to BPA in experimental. This necessitates proper testing and evaluation of chemicals prior to their inclusion in the open market (Gore et al., 2015).

13.8.3 RESEARCH EXPANDED TO NEW TISSUES AND PROCESSES

Studies focusing on the mode of EDC action should be expanded to novel target tissue and signaling pathways, and include all the known nuclear and membrane hormone receptors also. Enzymes involved in important physiological processes like lipid, fat, protein, and hormone should be investigated for EDC exposure based on human and animal models.

13.8.4 TEST ANIMAL

Most of the *in vitro* research related to EDCs is being carried on rodents, fishes, yeast, etc., which limits the information regarding their interaction with higher organisms, especially humans. Hence more animals should be used for testing especially non-human primates so that a better understanding of their effects of EDCs on hormones and development could be obtained. For this, the use of transgenic animals is also a safe option and should be encouraged. Moreover, human and wildlife reference tissue banks should be established.

13.8.5 NEED FOR MORE AND GLOBAL EPIDEMIOLOGICAL STUDIES

In order to understand the effect of EDC toxicity on various human processes, research should be expanded to include crucial life periods such as prenatal, early post-natal and even adolescence. Short- and long-term effects and trans-generational effects of EDCs should also be considered. Moreover, studies should aim to consider differences based on sex and gender, genetic diversity, and population differences by expanding to include the non-explored countries in Africa, Asia, North, and South America. This will include racial, ethnic, socio-economic, and geographic variables and will generate epidemiology data at international level that can be used for subsequent studies.

13.9 PREVENTIVE MEASURES

Considering the complexity of EDC behavior and the loopholes in the testing and treatment, one possible solution to this problem is to reduce the usage of harmful chemicals to avoid contamination of our environment. However, this requires the integration of all the stakeholders, including the public, media, and the government authorities.

13.9.1 NON-GOVERNMENTAL ORGANIZATIONS AND MEDIA

Since the last decade, non-governmental organizations and media have been playing a very active role in calling global attention towards the adverse effects of EDCs and emphasizing the adoption of precautionary measures.

This has led to increased public awareness and has also exerted much pressure on the policymakers to pass several significant regulatory policies such as BPA, triclosan (TCS) ban in many states of USA, ban of several phthalates (DEHP, DBP, BBP) in Europe (Acerini and Hughes, 2006; Gore et al., 2015).

13.9.2 CLINICAL PROFESSIONALS

Clinical professionals can play a much greater role in educating their patients about minimizing exposure to EDCs. They have the advantage of reaching out to people from different backgrounds, including social, cultural, and ethnic diversity, and can also help in the identification of at-risk individuals.

13.9.3 SCIENTISTS AND GREEN CHEMISTS

The role of Environmental scientists is very important, and they should work together with other green chemists to communicate the effects of EDCs and how to design the next generation of chemicals free of EDC activity. TIPED (tiered protocol for endocrine disruption) is one such team that is working on developing methodology for detecting EDC activity in chemicals and the process is already underway.

13.9.4 INTERNATIONAL ORGANIZATIONS

EDCs are a global problem. Many national and international organizations are playing a significant role in spreading awareness to the general public. The Endocrine Society, the American Public Health Association, the American Chemical Society, International Chemicals Management, and The WHO–United Nations Environment Program have stressed about the need to minimize human exposure to EDCs. The Environment Protection Agency in the USA has started the screening programs for chemicals for endocrine disruption activity under the Tox-Cast program. The establishment of CREDO (Cluster of Research into Endocrine Disruption in Europe) is a coordinated program having the task of collaborating different research institutions having interest in EDCs. Other International initiatives include the Rotterdam Convention of the United Nations Environment Program and the Stockholm Convention (2001) and the Prague declaration on Endocrine disruption in 2005. The Stockholm Convention listed a group of 12 POPs, identified as dirty dozen, namely aldrin, chlordane (CHL), dichlorodiphenyltrichloroethane (DDT), dieldrin, endrin, heptachlor, hexachlorobenzene, mirex, toxaphene, PCBs, polychlorinated dibenzo-p-dioxins (PCDDs), and polychlorinated dibenzofurans (PCDFs). Another significant statement called the San Antonio statement called attention towards the use of harmful brominated and chlorinated flame retardants and recommended their replacement by comparatively safer substitutes. This statement was signed by 150 scientists from more than 20 countries at the 30th International Symposium on Halogenated Persistent Organic Pollutants, 2010 in San Antonio.

13.10 RESEARCH PRIORITIES ACCORDING TO THE PRAGUE DECLARATION ON ENDOCRINE DISRUPTION

The first step should be to determine the endocrine disrupting property of chemicals already present in the environment by developing new methods and techniques. Use of modern technologies like genomics, proteomics, and bioinformatics to be encouraged (Schulte-Oehlmann, 2005).

- This should be followed by research oriented towards finding out their mechanism of action and has to be expanded to include their effect on novel cells and tissues and signaling of major health disorders.
- Various aspects related to endocrine disruption like mixture effect, low dose exposure, exposure time and stage of life, transgenerational effect, etc., should be considered for future research to understand their relation with various human abnormalities. For this, more epidemiological studies and animal studies should be undertaken on a large scale.
- Focus should also be to ascertain the effects of endocrine disruption on environment and wildlife by *in vitro* and *in vivo* studies.
- To integrate the effects at the individual level to its impact on much larger systems like population and ecosystem.

13.11 PREVENTIVE MEASURES AT THE INDIVIDUAL LEVEL

- The emergence and spread of a huge number of harmful chemicals in the environment could not have been possible without demand by the people. Hence, there remains an individual responsibility also to change our habits for the greater good.

- Proper washing of fruits and vegetables before consumption so as to remove the traces of chemical pesticides and herbicides.
- Using plastic cups to serve tea and coffee should be prohibited as high temperature may cause leaching of harmful substances from the plastic material. Similarly, hot milk should not be stored in plastic bottles; instead, glass or steel bottles can be used.
- Not using single use plastic items like disposable bags, cutlery, cups, packing material, and its replacement with bio-degradable material from natural sources.
- Identification of the polluted sources of water and avoid consumption or direct dermal contact with chemical contaminated food and water.
- Vulnerable groups of people like pregnant women and children should especially avoid contact with these chemicals.
- Professional workers working in workplaces with extensive use of harmful chemicals in paints, pesticides, fungicides, etc., are at higher risk and should take adequate precautionary measures to minimize exposure to EDCs.

13.11.1 PREVENTIVE MEASURES AT THE LEVEL OF POLICY MAKERS

There is also a need for the policy makers to step in to develop and implement preventive measures that are beyond individual capacity.

- Strict restrictions prohibiting the contamination of surface water bodies like lakes, rivers, and sea with chemicals should be implemented. This will also reduce ground water contamination to some level. Moreover, contaminated water bodies should be banned for public and the consumption of water and other organisms like fish should be prohibited. Better water treatment technologies should be developed to detect and remove EDCs from drinking contaminated water.
- Strict regulations should be made prohibiting overproduction and use of plastic material. The use of single use plastic should be minimized by substituting it with natural alternatives.
- In order to prevent the contamination of air, burning of harmful chemicals like solvents, plastics, etc., should be prohibited, especially at paint, textile, and plastic manufacturing plants.
- Banning of chemicals having EDC activity and developing remediation technologies for them.

- Chemically clean agriculture should be encouraged that does not cause any damage to the environment.

13.12 CONCLUSION

The problem of the increased incidence of human diseases caused by EDC compounds has highlighted the environmental implications of these compounds. This necessitates rapid action. The present situation can be improved by proper testing and monitoring of commonly used chemicals and banning the use of chemicals having endocrine-disrupting activity. Moreover, efforts should be made to regulate and avoid introducing new EDCs. Better research and communication among scientists and healthcare professionals, along with precaution at the individual level and administrative levels, is the need of the hour.

KEYWORDS

- **dichlorodiphenyltrichloroethane**
- **endocrine**
- **green chemistry**
- **hormone**
- **international actions**
- **non-communicable diseases**
- **polychlorinated biphenyls**
- **reproductive disorders**

REFERENCES

Abdelrahim, M., Ariazi, E., Kim, K., Khan, S., Barhoumi, R., Burghardt, R., & Safe, S., (2006). 3-Methylcholanthrene and other aryl hydrocarbon receptor agonists directly activate estrogen receptor α. *Cancer. Res., 66*, 2459–2467.

Acerini, C. L., & Hughes, I. A., (2006). Endocrine disrupting chemicals: A new and emerging public health problem? *Archives of Disease in Childhood, 91*, 633–641.

Adeeko, A., Li, D., Forsyth, D. S., et al., (2003). Effects of in utero tributyltin chloride exposure in the rat on pregnancy outcome. *Toxicol. Sci., 74*, 407–415.

Agas, D., Sabbieti, M. G., & Marchetti, L., (2013). Endocrine disruptors and bone metabolism. *Arch. Toxicol., 87*, 735–751.

Ahmed, K. E. M., Frøysa, H. G., Karlsen, O. A., Blaser, N., Zimmer, K. E., Berntsen, H. F., et al., (2019). Effects of defined mixtures of POPs and endocrine disruptors on the steroid metabolome of the human H295R adrenocortical cell line. *Chemosphere, 218*, 328–339.

Alonso-Magdalena, P., Morimoto, S., Ripoll, C., Fuentes, E., & Nadal, A., (2006). The estrogenic effect of bisphenol A disrupts pancreatic cell function *in vivo* and induces insulin resistance. *Environ. Health. Perspect., 114*, 106–112.

Alonso-Magdalena, P., Quesada, I., & Nadal, A., (2011). Endocrine disruptors in the etiology of type 2 diabetes mellitus. *Nature Reviews Endocrinol., 7*(6), 346–353.

Alonso-Magdalena, P., Vieira, E., Soriano, S., Menes, L., Burks, D., Quesada, I., & Nadal, A., (2010). Bisphenol A exposure during pregnancy disrupts glucose homeostasis in mothers and adult male offspring. *Environ. Health Perspect., 118*, 1243–1250.

Andersen, H. R., Bonefeld-Jørgensen, E. C., Nielsen, F., Jarfeldt, K., Jayatissa, M. N., & Vinggaard, A. M., (2006). Estrogenic effects *in vitro* and *in vivo* of the fungicide fenarimol. *Toxicol. Lett., 163*, 142–152.

Andersen, H. R., Vinggaard, A. M., Rasmussen, T. H., Gjermandsen, I. M., & Bonefeld-Jørgensen, E. C., (2002). Effects of currently used pesticides in assays for estrogenicity, androgenicity, and aromatase activity *in vitro*. *Toxicol. Appl. Pharmacol., 179*, 1–12.

Arcaro, K. F., & Gierthy, J. F., (2001). Assessing modulation of estrogenic activity of environmental and pharmaceutical compounds using MCF-7 focus assay. *Steroid Receptor Methods.*, 341–351.

Arsenescu, V., Arsenescu, R. I., King, V., Swanson, H., & Cassis, L. A., (2008). Polychlorinated biphenyl-77 induces adipocyte differentiation and proinflammatory adipokines and promotes obesity and atherosclerosis. *Environ. Health Perspect., 116*, 761–768.

Ashiru, O. A., & Odusanya, O. O., (2009). Fertility and occupational hazards: Review of the literature. *Afr. J. Reprod. Health., 13*.

Baillie-Hamilton, P. F., (2002). Chemical toxins: A hypothesis to explain the global obesity epidemic. *J. Altern. Complement. Med., 8*, 185–192.

Balabani, D., Rupnik, M., & Klemeni, A. K., (2011). Negative impact of endocrine-disrupting compounds on human reproductive health. *Reprod. Fertil. Dev., 23*, 403–416.

Barouki, R., Gluckman, P. D., Grandjean, P., Hanson, M., & Heindel, J. J., (2012). Developmental origins of non-communicable disease: Implications for research and public health. *Environ Health., 11*, 42.

Batista, T. M., Alonso-Magdalena, P., Vieira, E., Amaral, M. E. C., Cederroth, C. R., Nef, S., Quesada, I., et al., (2012). Short term treatment with bisphenol-A leads to metabolic abnormalities in adult male mice. *PLoS One, 7*, e33814.

Beck, V., Rohr, U., & Jungbauer, A., (2005). Phytoestrogens derived from red clover: An alternative to estrogen replacement therapy. *J. Steroid Biochem. Mol. Biol., 94*, 499–518.

Bellelis, P., Podgaec, S., & Abrão, M. S., (2011). Environmental factors and endometriosis. *Rev. Assoc. Med. Bras., 57*, 448–452.

Bennett, J. W., & Klich, M., (2003). *Mycotoxins*: *Clin. Microbiol. Rev., 16*, 497–516.

Bennetts, H. W., Underwood, E. J., & Shier, F. L., (1946). A specific breeding problem of sheep on subterranean clover pastures in Western Australia. *Aust. Vet. J., 22*, 2–12.

Bhandari, R., Xiao, J., & Shankar, A., (2013). Urinary bisphenol A and obesity in US children. *Am. J. Epidemiol., 177*, 1263–1270.

Bistan, M., Podgorelec, M., Marinšek, L. R., & Tišler, T., (2012). Yeast estrogen screen assay as a tool for detecting estrogenic activity in water bodies. *Food. Technol. Biotechnol., 50*, 427–433.

Bo, Y. U., Qing-Feng, C. H. E. N., Zhao-Ping, L. I. U., He-Fei, X. U., Zhang, X. P., Xiang, Q., Zhang, W. Z., et al., (2010). Estrogen receptor and expressions in hypothalamus-pituitary-ovary axis in rats exposed lactationally to soy isoflavones and bisphenol A. *Biomed. Environ. Sci., 23*, 357–362.

Bonefeld-Jørgensen, E. C., Andersen, H. R., Rasmussen, T. H., & Vinggaard, A. M., (2001). Effect of highly bioaccumulated polychlorinated biphenyl congeners on estrogen and androgen receptor activity. *Toxicol., 158*, 141–153.

Bonefeld-Jorgensen, E. C., Grünfeld, H. T., & Gjermandsen, I. M., (2005). Effect of pesticides on estrogen receptor transactivation *in vitro*: A comparison of stable transfected MVLN and transient transfected MCF-7 cells. *Mol. Cell. Endocrinol., 244*, 20–30.

Bouchard, M. F., Chevrier, J., Harley, K. G., Kogut, K., Vedar, M., Calderon, N., Trujillo, C., et al., (2011). Prenatal exposure to organophosphate pesticides and IQ in 7-year old children. *Environ. Health Perspect., 119*, 1189–1195.

Boucher, O., Burden, M. J., Muckle, G., Saint-Amour, D., Ayotte, P., Dewailly, É., Nelson, C. A., et al., (2012). Response inhibition and error monitoring during a visual go/no-go task in Inuit children exposed to lead, polychlorinated biphenyls, and methylmercury. *Environ. Health Perspect., 120*, 608–615.

Braun, J. M., Kalkbrenner, A. E., Just, A. C., Yolton, K., Calafat, A. M., Sjödin, A., Hauser, R., et al., (2014). Gestational exposure to endocrine-disrupting chemicals and reciprocal social, repetitive, and stereotypic behaviors in 4- and 5-year-old children: The HOME study. *Environ. Health Perspect., 122*, 513–520.

Buckman, A. H., Veldhoen, N., Ellis, G., Ford, J. K., Helbing, C. C., & Ross, P. S., (2011). PCB-associated changes in mRNA expression in killer whales (Orcinus orca) from the NE Pacific Ocean. *Environ. Sci. Technol., 45*, 10194–10202.

Calsolaro, V., Pasqualetti, G., Niccolai, F., Caraccio, N., & Monzani, F., (2017). Thyroid disrupting chemicals. *Int. J. Mol. Sci., 18*, 2583.

Carpenter, D. O., & Sly, P. D., (2016). Environmental chemicals as endocrine disruptors. *Rev. Environ. Health., 31*, 399.

Carson, R., (2002). *Silent Spring*. Houghton Mifflin Harcourt.

Carwile, J. L., & Michels, K. B., (2011). Urinary bisphenol A and obesity: NHANES 2003–2006. *Environ. Res., 111*, 825–830.

Casals-Casas, C., & Desvergne, B., (2011). Endocrine disruptors: From endocrine to metabolic disruption. *Annu. Rev. Physiol., 73*, 135–162.

Caserta, D., Di Segni, N., Mallozzi, M., Giovanale, V., Mantovani, A., Marci, R., & Moscarini, M., (2014). Bisphenol A and the female reproductive tract: An overview of recent laboratory evidence and epidemiological studies. *Reprod. Biol. Endocrinol., 12*, 37.

Chaloupka, K., Krishnan, V., & Safe, S., (1992). Polynuclear aromatic hydrocarbon carcinogens as antiestrogens in MCF-7 human breast cancer cells: Role of the Ah receptor. *Carcinogenesis, 13*, 2233–2239.

Chen, A., Yolton, K., Rauch, S. A., Webster, G. M., Hornung, R., Sjödin, A., Dietrich, K. N., & Lanphear, B. P., (2014). Prenatal polybrominated diphenyl ether exposures and neurodevelopment in U.S. children through 5 years of age: The HOME study. *Environ. Health Perspect., 122*, 856–862.

Chen, C. J., & Wang, C. J., (1990). Ecological correlation between arsenic level in well water and age-adjusted mortality from malignant neoplasms. *Cancer Res., 50*, 5470–5474.

Colborn, T., & Clement, C., (1992). Wingspread consensus statement. *Chemically Induced Alterations in Sexual, Functional Development: The Wildlife/Human Connection*, 1–8.
Colborn, T., Dumanoski, D., & Myers, J. P., (1996). *Our Stolen Future: Are We Threatening Our Fertility, Intelligence and Survival? A Scientific Detective Story*, *1*, 306
Colciago, A., Casati, L., Mornati, O., Vergoni, A. V., Santagostino, A., Celotti, F., & Negri-Cesi, P., (2009). Chronic treatment with polychlorinated biphenyls (PCB) during pregnancy and lactation in the rat. Part 2: Effects on reproductive parameters, on sex behavior, on memory retention and on hypothalamic expression of aromatase and 5-reductases in the offspring. *Toxicol. Appl. Pharmacol., 239*, 46–54.
Colciago, A., Negri-Cesi, P., Pravettoni, A., Mornati, O., Casati, L., & Celotti, F., (2006). Prenatal Aroclor 1254 exposure and brain sexual differentiation: Effect on the expression of testosterone metabolizing enzymes and androgen receptors in the hypothalamus of male and female rats. *Reprod. Toxicol., 22*, 738–745.
Collman, G. W., (2011). Developmental basis of disease: Environmental impacts. *J. Dev. Orig. Health Dis., 2*, 49–55.
Cook, M. B., Trabert, B., & McGlynn, K. A., (2011). Organochlorine compounds and testicular dysgenesis syndrome: Human data. *Int. J. Androl., 34*, 68–84.
Craig, Z. R., Wang, W., & Flaws, J. A., (2011). Endocrine-disrupting chemicals in ovarian function: Effects on steroidogenesis, metabolism and nuclear receptor signaling. *Reprod., 142*, 633–646.
Crews, D., Bergeron, J. M., & McLachlan, J. A., (1995). The role of estrogen in turtle sex determination and the effect of PCBs. *Environ. Health Perspect., 103*, 73–77.
Damstra, T., Barlow, S., Bergman, A., Kavlock, R., & Van, D. K. G., (2002). *Global Assessment of the State-of-the-Science of Endocrine Disruptors* (pp. 11–32). Geneva: WHO.
Darbre, & Philippa, D., (2017). Endocrine disruptors and obesity. *Curr. Obes. Rep., 6*, 18–27.
Davis, D. L., Bradlow, H. L., Wolff, M., Woodruff, T., Hoel, D. G., & Anton-Culver, H., (1993). Medical hypothesis: Xenoestrogens as preventable causes of breast cancer. *Environ. Health Perspect., 101*, 372–377.
De Coster, S., & Van, L. N., (2012). Endocrine-disrupting chemicals: Associated disorders and mechanisms of action. *Environ. Public Health.*
De Jager, C., Bornman, M. S., & Oosthuizen, J. M., (1999). The effect of p-nonylphenol on the fertility potential of male rats after gestational, lactational and direct exposure. *Andrologia., 31*, 107–113.
Denison, M. S., Soshilov, A. A., He, G., DeGroot, D. E., & Zhao, B., (2011). Exactly the same but different: Promiscuity and diversity in the molecular mechanisms of action of the aryl hydrocarbon (dioxin) receptor. *Toxicol. Sci, 124*, 1–22.
Denslow, N. D., Chow, M. C., Kroll, K. J., & Green, L., (1999). Vitellogenin as a biomarker of exposure for estrogen or estrogen mimics. *Ecotoxicol., 8*, 385–398.
Desaulniers, D., Xiao, G. H., Leingartner, K., Chu, I., Musicki, B., & Tsang, B. K., (2005). Comparisons of brain, uterus, and liver mRNA expression for cytochrome p450s, DNA methyltransferase-1, and catechol-O-methyltransferase in prepubertal female Sprague-Dawley rats exposed to a mixture of aryl hydrocarbon receptor agonists. *Toxicol. Sci., 86*, 175–184.
Du, G., Shen, O., Sun, H., Fei, J., Lu, C., Song, L., & Wang, X., (2010). Assessing hormone receptor activities of pyrethroid insecticides and their metabolites in reporter gene assays. *Toxicol. Sci., 116*, 58–66.
Durrer, S., Ehnes, C., Fuetsch, M., Maerkel, K., Schlumpf, M., & Lichtensteiger, W., (2007). Estrogen sensitivity of target genes and expression of nuclear receptor co-regulators in

rat prostate after pre- and postnatal exposure to the ultraviolet filter 4-methyl benzylidene camphor. *Environ. Health Perspect., 115*, 42–50.

Ellis-Hutchings, R. G., Cherr, G. N., Hanna, L. A., & Keen, C. L., (2006). Polybrominated diphenyl ether (PBDE)-induced alterations in vitamin A and thyroid hormone concentrations in the rat during lactation and early postnatal development. *Toxicol. Appl. Pharmacol., 215*, 135–145.

Engel, S. M., Miodovnik, A., Canfield, R. L., Zhu, C., Silva, M. J., Calafat, A. M., & Wolff, M. S., (2010). Prenatal phthalate exposure is associated with childhood behavior and executive functioning. *Environ. Health Perspect., 118*, 565–571.

Engel, S. M., Wetmur, J., Chen, J., Zhu, C., Barr, D. B., Canfield, R. L., & Wolff, M. S., (2011). Prenatal exposure to organophosphates, paraoxonase 1, and cognitive development in childhood. *Environ. Health Perspect., 119*, 1182–1188.

EPA, (2011). *Endocrine Disruptor Screening Program: Hershberger Assay OCSPP Guideline 890.1400 Standard Evaluation Procedure (SEP)*. Agency UEP, Washington, DC.

Fan, W., Yanase, T., Morinaga, H., Gondo, S., Okabe, T., Nomura, M., & Nawata, H., (2007). Atrazine-induced aromatase expression is SF-1 dependent: Implications for endocrine disruption in wildlife and reproductive cancers in humans. *Environ. Health Perspect., 115*, 720–727.

Fernández, M. F., Santa-Marina, L., Olea-Serrano, M. F., Rivas, A. M., Aurrekoetxea, J. J., Expósito, J., Lorenzo, M., et al., (2004). Breast cancer risk and the combined effect of environmental estrogens. *Cancer Cause Control, 15*, 591–600.

Fischer, L. J., Seegal, R. F., Ganey, P. E., Pessah, I. N., & Kodavanti, P. R. S., (1998). Symposium overview: Toxicity of non-coplanar PCBs. *Toxicol. Sci., 41*, 49–61.

Fisher, J. S., Turner, K. J., Brown, D., & Sharpe, R. M., (1999). Effect of neonatal exposure to estrogenic compounds on development of the ex-current ducts of the rat testis through puberty to adulthood. *Environ. Health Perspect., 107*, 397–405.

Fitzgerald, E. F., Shrestha, S., Gomez, M. I., McCaffrey, R. J., Zimmerman, E. A., Kannan, K., & Hwang, S. A., (2012). Polybrominated diphenyl ethers (PBDEs), polychlorinated biphenyls (PCBs) and neuropsychological status among older adults in New York. *Neurotoxicol., 33*, 8–15.

Giannandrea, F., Paoli, D., Figà-Talamanca, I., Lombardo, F., Lenzi, A., & Gandini, L., (2013). Effect of endogenous and exogenous hormones on testicular cancer: The epidemiological evidence. *Int. J. Dev. Biol., 57*, 255–263.

Giordano, F., Abballe, A., De Felip, E., Di Domenico, A., Ferro, F., Grammatico, P., Ingelido, A. M., et al., (2010). Maternal exposures to endocrine disrupting chemicals and hypospadias in offspring. *Birth Defects Res A Clin. Mol. Teratol., 88*, 241–250.

Giulivo, M., De Alda, M. L., Capri, E., & Barceló, D., (2016). Human exposure to endocrine disrupting compounds: Their role in reproductive systems, metabolic syndrome and breast cancer. A review. *Environ. Res., 151*, 251–264.

Gomez, E., Pillon, A., Fenet, H., Rosain, D., Duchesne, M. J., Nicolas, J. C., Balaguer, P., & Casellas, C., (2005). Estrogenic activity of cosmetic components in reporter cell lines: Parabens, UV screens, and musks. *J. Toxicol. Environ. Health Sci., 68*, 239–251.

Gore, A. C., (2001). Environmental toxicant effects on neuroendocrine function. *Endocrine., 14*, 235–246.

Gore, A. C., Chappell, V. A., Fenton, S. E., Flaws, J. A., Nadal, A., Prins, G. S., & Zoeller, R. T., (2015). EDC-2: The endocrine society's second scientific statement on endocrine-disrupting chemicals. *Endocr. Rev., 36*, 1–150.

Gore, A. C., Crews, D., Doan, L. L., La Merrill, M., Patisaul, H., & Zota, A., (2014). Introduction to endocrine disrupting chemicals (EDCs). *A Guide for Public Interest Organizations and Policy-Makers,* 21, 22.

Gray, L. E Jr., Ostby, J., Cooper, R. L., & Kelce, W. R., (1999). The estrogenic and antiandrogenic pesticide methoxychlor alters the reproductive tract and behavior without affecting pituitary size or LH and prolactin secretion in male rats. *Toxicol. Ind. Health., 15,* 37–47.

Greer, M. A., Goodman, G., Pleus, R. C., & Greer, S. E., (2002). Health effects assessment for environmental perchlorate contamination: The dose response for inhibition of thyroidal radioiodine uptake in humans. *Environ. Health Perspect., 110,* 927–937.

Grünfeld, H. T., & Bonefeld-Jorgensen, E. C., (2004). Effect of *in vitro* estrogenic pesticides on human oestrogen receptor α and β mRNA levels. *Toxicol. Lett., 151,* 467–480.

Guillette, L. J., & Gunderson, M. P., (2001). Alterations in development of reproductive and endocrine systems of wildlife populations exposed to endocrine-disrupting contaminants. *Reprod. Abstr. Ser., 122,* 857–864.

Hallgren, S., Sinjari, T., Håkansson, H., & Darnerud, P. O., (2001). Effects of polybrominated diphenyl ethers (PBDEs) and polychlorinated biphenyls (PCBs) on thyroid hormone and vitamin A levels in rats and mice. *Arch. Toxicol., 75,* 200–208.

Hamers, T., Kamstra, J. H., Sonneveld, E., Murk, A. J., Kester, M. H., Andersson, P. L., & Brouwer, A., (2006). *In vitro* profiling of the endocrine-disrupting potency of brominated flame retardants. *Toxicol. Sci., 92,* 157–173.

Hardell, L., Andersson, S. O., Carlberg, M., Bohr, L., Van, B. B., Lindström, G., Björnfoth, H., & Ginman, C., (2006). Adipose tissue concentrations of persistent organic pollutants and the risk of prostate cancer. *J. Occup. Environ. Med., 48,* 700–707.

Harley, K. G., Schall, R. A., Chevrier, J., Tyler, K., Aguirre, H., Bradman, A., & Eskenazi, B., (2013). Prenatal and postnatal bisphenol A exposure and body mass index in childhood in the CHAMACOS cohort. *Environ. Health Perspect., 121*(4), 514–520.

Harris, C. A., Routledge, E. J., Schaffner, C., Brian, J. V., Giger, W., & Sumpter, J. P., (2007). Benzotriazole is antiestrogenic *in vitro* but not *in vivo*. *Environ. Toxicol. Chem., 26,* 2367–2372.

Harris, R. M., & Waring, R. H., (2012). Diethylstilboestrol—A long-term legacy. *Maturitas, 72,* 108–112.

Harvey, P. W., (2016). Adrenocortical endocrine disruption. *J. Steroid Biochem. Mol. Biol., 155,* 199–206.

Heppell, S. A., Denslow, N. D., Folmar, L. C., & Sullivan, C. V., (1995). Universal assay of vitellogenin as a biomarker for environmental estrogens. *Environ. Health Perspect., 103,* 9–15.

Herbst, A. L., Ulfelder, H., & Poskanzer, D. C., (1971). Adenocarcinoma of the vagina: Association of maternal stilbestrol therapy with tumor appearance in young women. *N. Engl. J. Med., 284,* 878–881.

Herbstman, J. B., Sjödin, A., Kurzon, M., Lederman, S. A., Jones, R. S., Rauh, V., Needham, L. L., et al., (2010). Prenatal exposure to PBDEs and neurodevelopment. *Environ. Health Perspect., 118,* 712–719.

Horiguchi, T., (2006). Masculinization of female gastropod mollusks induced by organotin compounds, focusing on mechanism of actions of tributyltin and triphenyltin for development of imposex. *Environ. Sci., 13,* 77–87.

Howdeshell, K. L., Hotchkiss, A. K., Thayer, K. A., Vandenbergh, J. G., & Vom, S. F. S., (1999). Exposure to bisphenol A advances puberty. *Nature., 401,* 763, 764.

Høyer, A. P., Grandjean, P., Jørgensen, T., Brock, J. W., & Hartvig, H. B., (1998). Organochlorine exposure and risk of breast cancer. *Lancet, 352*, 1816–1820.
Hu, W., Liu, H., Sun, H., Shen, O., Wang, X., Lam, M. H., & Yu, H., (2011). Endocrine effects of methoxylated brominated diphenyl ethers in three *in vitro* models. *Mar. Pollut. Bull., 62*, 2356–2361.
Ikeda, K., Arao, Y., Otsuka, H., Nomoto, S., Horiguchi, H., Kato, S., & Kayama, F., (2002). Terpenoids found in the Umbelliferae family act as agonists/antagonists for ERα and ERβ: Differential transcription activity between ferutinine-liganded ERα and ERβ. *Biochem. Biophys. Res. Commun., 291*, 354–360.
Jobling, S., Nolan, M., Tyler, C. R., Brighty, G., & Sumpter, J. P., (1998). Widespread sexual disruption in wild fish. *Environ. Sci. Technol., 32*, 2498–2506.
Jobling, S., Reynolds, T., White, R., Parker, M. G., & Sumpter, J. P., (1995). A variety of environmentally persistent chemicals, including some phthalate plasticizers, are weakly estrogenic. *Environ. Health Perspect., 103*, 582–587.
Kadhel, P., Monnier, P., Boucoiran, I., Chaillet, N., & Fraser, W. D., (2012). Organochlorine pollutants and female fertility: A systematic review focusing on *in vitro* fertilization studies. *Reprod. Sci., 19*, 1246–1259.
Kamstra, J. H., Hruba, E., Blumberg, B., Janesick, A., Mandrup, S., Hamers, T., & Legler, J., (2014). Transcriptional and epigenetic mechanisms underlying enhanced *in vitro* adipocyte differentiation by the brominated flame retardant BDE-47. *Environ. Sci. Technol., 48*, 4110–4119.
Kay, V. R., Chambers, C., & Foster, W. G., (2013). Reproductive and developmental effects of phthalate diesters in females. *Crit. Rev. Toxicol., 43*, 200–219.
Kelce, W. R., Stone, C. R., Laws, S. C., Gray, L. E., Kemppainen, J. A., & Wilson, E. M., (1995). Persistent DDT metabolite p,p'-DDE is a potent androgen receptor antagonist. *Nature, 375*, 581–585.
Kiliç, N., Sandal, S., Çolakoglu, N., Kutlu, S., Seyran, A., & Yilmaz, B., (2005). Endocrine disruptive effects of polychlorinated biphenyls on the thyroid gland in female rats. *Tohoku J. Exp. Med., 206*, 327–332.
Kim, S. S., Lee, R. D., Lim, K. J., Kwack, S. J., Rhee, G. S., Seok, J. H., & Park, K. L., (2004). Potential estrogenic and anti-androgenic effects of permethrin in rats. *J. Reprod. Dev., 51*, 201–210.
Kim, Y., Ha, E. H., Kim, E. J., et al., (2011). Prenatal exposure to phthalates and infant development at 6 months: Prospective mothers and children's environmental health (MOCEH) study. *Environ. Health Perspect., 119*, 1495–1500.
Kishta, O., Adeeko, A., Li, D., et al., (2007). In utero exposure to tributyltin chloride differentially alters male and female fetal gonad morphology and gene expression profiles in the Sprague-Dawley rat. *Reprod. Toxicol., 23*, 1–11.
Kitamura, S., Suzuki, T., Ohta, S., & Fujimoto, N., (2003). Antiandrogenic activity and metabolism of the organophosphorus pesticide fenthion and related compounds. *Environ. Health Perspect., 111*, 503–508.
Kjærstad, M. B., Taxvig, C., Nellemann, C., Vinggaard, A. M., & Andersen, H. R., (2010). Endocrine disrupting effects *in vitro* of conazole antifungals used as pesticides and pharmaceuticals. *Reprod. Toxicol., 30*, 573–582.
Kojima, H., Katsura, E., Takeuchi, S., Niiyama, K., & Kobayashi, K., (2004). Screening for estrogen and androgen receptor activities in 200 pesticides by *in vitro* reporter gene assays using Chinese hamster ovary cells. *Environ. Health Perspect., 112*, 524–531.

Kristensen, D. M., Hass, U., Lesné, L., Lottrup, G., Jacobsen, P. R., Desdoits-Lethimonier, C., Boberg, J., et al., (2011). Intrauterine exposure to mild analgesics is a risk factor for the development of male reproductive disorders in humans and rats. *Hum. Reprod., 26*, 235–244.

Kuiper, G. G., Lemmen, J. G., Carlsson, B. O., Corton, J. C., Safe, S. H., Van, D. S. P. T., Van, D. B. B., & Gustafsson, J. A., (1998). Interaction of estrogenic chemicals and phytoestrogens with estrogen receptor β. *Endocrinology, 139*, 4252–4263.

Kunz, P. Y., & Fent, K., (2006). Multiple hormonal activities of UV filters and comparison of *in vivo* and *in vitro* estrogenic activity of ethyl-4-aminobenzoate in fish. *Aquat. Toxicol., 79*, 305–324.

Kurita, H., Yoshioka, W., Nishimura, N., Kubota, N., Kadowaki, T., & Tohyama, C., (2009). Aryl hydrocarbon receptor-mediated effects of 2, 3, 7, 8-tetrachlorodibenzo-p-dioxin on glucose-stimulated insulin secretion in mice. *J. Appl. Toxicol., 29*, 689–694.

Kuriyama, S. N., Wanner, A., Fidalgo-Neto, A. A., Talsness, C. E., Koerner, W., & Chahoud, I., (2007). Developmental exposure to low-dose PBDE-99: Tissue distribution and thyroid hormone levels. *Toxicol., 242*, 80–90.

Lai, Y., Lu, M., Lin, S., & Cai, Z., (2012). Glucuronidation of hydroxylated polybrominated diphenyl ethers and their modulation of estrogen UDP-glucuronosyltransferases. *Chemosphere, 86*, 727–734.

Le Guevel, R., & Pakdel, F., (2001). Assessment of oestrogenic potency of chemicals used as growth promoter by *in-vitro* methods. *Hum. Reprod., 16*, 1030–1036.

Lemaire, G., Mnif, W., Mauvais, P., Balaguer, P., & Rahmani, R., (2006). Activation of α-and β-estrogen receptors by persistent pesticides in reporter cell lines. *Life Sci., 79*, 1160–1169.

Lemaire, G., Terouanne, B., Mauvais, P., Michel, S., & Rahmani, R., (2004). Effect of organochlorine pesticides on human androgen receptor activation *in vitro*. *Toxicol. Appl. Pharmacol., 196*, 235–246.

Lewis, D. R., Southwick, J. W., Ouellet-Hellstrom, R., Rench, J., & Calderon, R. L., (1999). Drinking water arsenic in Utah: A cohort mortality study. *Environ. Health Perspect., 107*, 359–365.

Li, Z. M., Hernandez-Moreno, D., Main, K. M., Skakkebæk, N. E., Kiviranta, H., Toppari, J., Feldt-Rasmussen, U., et al., (2018). Association of in utero persistent organic pollutant exposure with placental thyroid hormones. *Endocrinol., 159*, 3473–3481.

Lin, Y., Wei, J., Li, Y., Chen, J., Zhou, Z., Song, L., Wei, Z., et al., (2011). Developmental exposure to di(2- Ethylhexyl) phthalate impairs endocrine pancreas and leads to long-term adverse effects on glucose homeostasis in the rat. *Am. J. Physiol Endocrinol Metab., 301*, 527–538.

Liu, H., Hu, W., Sun, H., Shen, O., Wang, X., Lam, M. H., Giesy, J. P., et al., (2011). *In vitro* profiling of endocrine disrupting potency of 2, 2′, 4, 4′-tetrabromodiphenyl ether (BDE47) and related hydroxylated analogs (HO-PBDEs). *Mar. Pollut. bull., 63*, 287–296.

Longnecker, M. P., Klebanoff, M. A., Brock, J. W., Zhou, H., Gray, K. A., Needham, L. L., & Wilcox, A. J., (2002). Maternal serum level of 1,1-dichloro-2,2-bis(p-chlorophenyl-) ethylene and risk of cryptorchidism, hypospadias, and polythelia among male offspring. *Am. J. Epidemiol., 155*, 313–322.

López-Carrillo, L., Hernández-Ramírez, R. U., Calafat, A. M., Torres-Sánchez, L., Galván-Portillo, M., Needham, L. L., Ruiz-Ramos, R., & Cebrián, M. E., (2010). Exposure to phthalates and breast cancer risk in northern Mexico. *Environ. Health Perspect., 118*, 539–544.

Lymperi, S., & Giwercman, A., (2018). Endocrine disruptors and testicular function. *Metabol., 86*, 79–90.

Ma, Y., Xia, W., Wang, D. Q., Wan, Y. J., Xu, B., Chen, X., Li, Y. Y., & Xu, S. Q., (2013). Hepatic DNA methylation modifications in early development of rats resulting from perinatal BPA exposure contribute to insulin resistance in adulthood. *Diabetologia., 56*, 2059–2067.
Mably, T. A., Bjerke, D. L., Moore, R. W., Gendron-Fitzpatrick, A., & Peterson, R. E., (1992). In utero and lactational exposure of male rats to 2, 3, 7, 8-tetrachlorodibenzo-p-dioxin. 3. Effects on spermatogenesis and reproductive capability. *Toxicol. Appl. Pharmacol., 114*, 118–126.
Martino-Andrade, A. J., & Chahoud, I., (2010). Reproductive toxicity of phthalate esters. *Mol. Nutr. Food Res., 54*, 148–157.
Marty, M. S., & O'Connor, J. C., (2014). Key learnings from the endocrine disruptor screening program (EDSP) tier 1 rodent uterotrophic and Hershberger assays. *Birth Defects Research Part B: Reprod. Toxicol., 101*, 63–79.
Matozzo, V., Gagné, F., Marin, M. G., Ricciardi, F., & Blaise, C., (2008). Vitellogenin as a biomarker of exposure to estrogenic compounds in aquatic invertebrates: A review. *Environ. Int., 34*, 531–545.
McElroy, J. A., Shafer, M. M., Trentham-Dietz, A., Hampton, J. M., & Newcomb, P. A., (2006). Cadmium exposure and breast cancer risk. *J. Natl. Cancer Inst., 98*, 869–873.
McGlynn, K. A., Guo, X., Graubard, B. I., Brock, J. W., Klebanoff, M. A., & Longnecker, M. P., (2009). Maternal pregnancy levels of polychlorinated biphenyls and risk of hypospadias and cryptorchidism in male offspring. *Environ. Health Perspect., 117*, 1472–1476.
McIntyre, B. S., Barlow, N. J., & Foster, P. M., (2001). Androgen-mediated development in male rat offspring exposed to flutamide in utero: Permanence and correlation of early postnatal changes in anogenital distance and nipple retention with malformations in androgen-dependent tissues. *Toxicol Sci., 62*, 236–249.
McKinlay, R., Plant, J. A., Bell, J. N. B., & Voulvoulis, N., (2008). Endocrine disrupting pesticides: Implications for risk assessment. *Environ. Int., 34*, 168–183.
Meeker, J. D., (2012). Exposure to environmental endocrine disruptors and child development. *Arch Pediatr. Adolesc. Med., 166*, 952–958.
Meerts, I. A., Letcher, R. J., Hoving, S., Marsh, G., Bergman, A., Lemmen, J. G., Van, D. B. B., & Brouwer, A., (2001). *In vitro* estrogenicity of polybrominated diphenyl ethers, hydroxylated PDBEs, and polybrominated bisphenol A compounds. *Environ. Health Perspect., 109*, 399–407.
Mills, P. K., & Yang, R., (2006). Regression analysis of pesticide use and breast cancer incidence in California Latinas. *J. Environ. Health, 68*.
Miyawaki, J., Sakayama, K., Kato, H., Yamamoto, H., & Masuno, H., (2007). Perinatal and postnatal exposure to bisphenol A increases adipose tissue mass and serum cholesterol level in mice. *J. Atheroscler Thromb., 14*, 245–252.
Mnif, W., Pascussi, J. M., Pillon, A., Escande, A., Bartegi, A., Nicolas, J. C., Cavaillès, V., Duchesne, M. J., & Balaguer, P., (2007). Estrogens and antiestrogens activate hPXR. *Toxicol Lett., 170*, 19–29.
Morgan, M., Deoraj, A., Felty, Q., & Roy, D., (2017). Environmental estrogen-like endocrine disrupting chemicals and breast cancer. *Mol. Cell. Endocrinol., 457*, 89–102.
Mughal, B. B., Demeneix, B. A., & Fini, J. B., (2018). Evaluating thyroid disrupting chemicals *in vivo* using *Xenopus laevis*. In: *Thyroid Hormone Nuclear Receptor* (pp. 183–192).
Naville, D., Pinteur, C., Vega, N., Menade, Y., Vigier, M., Le Bourdais, A., Labaronne, E., et al., (2013). Low-dose food contaminants trigger sex-specific, hepatic metabolic changes in the progeny of obese mice. *FASEB J., 27*, 3860–3870.

Naylor, L. H., (1999). Reporter gene technology: The future looks bright. *Biochem. Pharmacol., 58*, 749–757.
Nie, J., Beyea, J., Bonner, M. R., Han, D., Vena, J. E., Rogerson, P., Vito, D., et al., (2007). Exposure to traffic emissions throughout life and risk of breast cancer: The Western New York Exposures and Breast Cancer (WEB) study. *Cancer Cause Control, 18*, 947–955.
No, E. R., (2012). *Methods of Sampling and Analysis for the Official Control of Levels of Dioxins, Dioxin- Like PCBs and Non-Dioxin-Like PCBs in Certain Foodstuffs and Repealing Regulation (EC) No 1883/2006*.
Novelli, M., Piaggi, S., & De Tata, V., (2005). 2, 3, 7, 8-Tetrachlorodibenzo-p-dioxin-induced impairment of glucose-stimulated insulin secretion in isolated rat pancreatic islets. *Toxicol Lett., 156*, 307–314.
OECD Test No. 440., (2007). *Uterotrophic Bioassay in Rodents*.
Oh, S. M., Kim, H. R., & Chung, K. H., (2009). *In vitro* estrogenic and antiestrogenic potential of chlorostyrenes. *Toxicol. in Vitro, 23*, 1242–1248.
Oh, S. M., Ryu, B. T., Lee, S. K., & Chung, K. H., (2007). Antiestrogenic potentials of ortho-PCB congeners by single or complex exposure. *Arch. Pharm. Res., 30*, 199–209.
Ohsako, S., Miyabara, Y., Nishimura, N., Kurosawa, S., Sakaue, M., Ishimura, R., Sato, M., et al., (2001). Maternal exposure to a low dose of 2,3,7,8-tetrachlorodibenzo-pdioxin (TCDD) suppressed the development of reproductive organs of male rats: Dose-dependent increase of mRNA levels of 5-reductase type 2 in contrast to decrease of androgen receptor in the pubertal ventral prostate. *Toxicol. Sci., 60*, 132–143.
Orton, F., Rosivatz, E., Scholze, M., & Kortenkamp, A., (2011). Widely used pesticides with previously unknown endocrine activity revealed as *in vitro* antiandrogens. *Environ. Health Perspect., 119*, 794–800.
Pasqualini, C., Sarrieau, A., Dussaillant, M., Corbani, M., Bojda-Diolez, F., Rostène, W., & Kerdelhué, B., (1990). Estrogen-like effects of 7, 12-dimethylbenz (a) anthracene on the female rat hypothalamic-pituitary axis. *J. Steroid Biochem., 36*, 485–491.
Patisaul, H. B., & Bateman, H. L., (2008). Neonatal exposure to endocrine active compounds or an ER agonist increases adult anxiety and aggression in gonadally intact male rats. *Horm. Behav., 53*, 580–588.
Pereira-Fernandes, A., Demaegdt, H., Vandermeiren, K., Hectors, T. L., Jorens, P. G., Blust, R., & Vanparys, C., (2013). Evaluation of a screening system for obesogenic compounds: Screening of endocrine disrupting compounds and evaluation of the PPAR dependency of the effect. *PloS One, 8*, 77481.
Pereira-Fernandes, A., Vanparys, C., Vergauwen, L., Knapen, D., Jorens, P. G., & Blust, R., (2014). Toxicogenomics in the 3T3-L1 cell line, a new approach for screening of obesogenic compounds. *Toxicol. Sci., 140*, 352–363.
Peretz, J., Vrooman, L., Ricke, W. A., Hunt, P. A., Ehrlich, S., Hauser, R., Padmanabhan, V., et al., (2014). Bisphenol A and reproductive health: Update of experimental and human evidence, 2007–2013. *Environ. Health Perspect., 122*, 775–786.
Pesatori, A. C., Consonni, D., Rubagotti, M., Grillo, P., & Bertazzi, P. A., (2009). Cancer incidence in the population exposed to dioxin after the "Seveso accident": Twenty years of follow-up. *Environ. Health, 8*.
Pierik, F. H., Klebanoff, M. A., Brock, J. W., & Longnecker, M. P., (2007). Maternal pregnancy serum level of heptachlor epoxide, hexachlorobenzene, and -hexachlorocyclohexane and risk of cryptorchidism in offspring. *Environ Res., 105*, 364–369.

Pillon, A., Boussioux, A. M., Escande, A., Aït-Aïssa, S., Gomez, E., Fenet, H., Ruff, M., et al., (2005). Binding of estrogenic compounds to recombinant estrogen receptor-α: Application to environmental analysis. *Environ. Health Perspect., 113*, 278–284.

Pinto, C. L., Mansouri, K., Judson, R., & Browne, P., (2016). Prediction of estrogenic bioactivity of environmental chemical metabolites. *Chem. Res. Toxicol., 29*, 1410–1427.

Pocock, V. J., Sales, G. D., Wilson, C. A., & Milligan, S. R., (2002). Effects of perinatal octylphenol on ultrasound vocalization, behavior and reproductive physiology in rats. *Physiol. Behav., 76*, 645–653.

Preuss, T. G., Gurer-Orhan, H., Meerman, J., & Ratte, H. T., (2010). Some nonylphenol isomers show antiestrogenic potency in the MVLN cell assay. *Toxicol. in Vitro, 24*, 129–134.

Rachoń, D., (2015). Endocrine disrupting chemicals (EDCs) and female cancer: Informing the patients. *Rev. Endocr. Metab. Disord., 16*, 359–364.

Rauh, V., Arunajadai, S., Horton, M., Perera, F., Hoepner, L., Barr, D. B., & Whyatt, R., (2011). Seven-year neurodevelopmental scores and prenatal exposure to chlorpyrifos, a common agricultural pesticide. *Environ. Health Perspect., 119*, 1196–1201.

Recio-Vega, R., Velazco-Rodriguez, V., Ocampo-Gómez, G., Hernandez-Gonzalez, S., Ruiz-Flores, P., & Lopez-Marquez, F., (2011). Serum levels of polychlorinated biphenyls in Mexican women and breast cancer risk. *J. Appl. Toxicol., 31*, 270–278.

Regnier, S. M., & Sargis, R. M., (2014). Adipocytes under assault: Environmental disruption of adipose physiology. *Biochim. Biophys. Acta., 1842*, 520–533.

Rubin, B. S., & Soto, A. M., (2009). Bisphenol A: Perinatal exposure and body weight. *Mol. Cell. Endocrinol., 304*, 55–62.

Sanchez, B. C., Carter, B., Hammers, H. R., & Sepúlveda, M. S., (2011). Transcriptional response of hepatic largemouth bass (*Micropterus salmoides*) mRNA upon exposure to environmental contaminants. *J. App. Toxicol., 31*, 108–116.

Schlumpf, M., Cotton, B., Conscience, M., Haller, V., Steinmann, B., & Lichtensteiger, W., (2001). *In vitro* and *in vivo* estrogenicity of UV screens. *Environ. Health Perspect., 109*, 239–244.

Schneider, S. L., Alks, V., Morreal, C. E., Sinha, D. K., & Dao, T. L., (1976). Estrogenic properties of 3, 9-dihydroxybenz [a] anthracene, a potential metabolite of Benz [a] anthracene. *J. Natl. Cancer Inst. Monogr., 57*, 1351–1354.

Schug, T. T., Janesick, A., Blumberg, B., & Heindel, J. J., (2011). Endocrine disrupting chemicals and disease susceptibility. *J. Steroid Biochem. Mol. Biol., 127*, 204–215.

Schulte-Oehlmann, U., (2005). The Prague declaration on endocrine disruption. *Environ. Sci. Pollut. Res. Int., 12*, 188.

Shankar, A., Teppala, S., & Sabanayagam, C., (2012). Urinary bisphenol a levels and measures of obesity: Results from the national health and nutrition examination survey 2003–2008. *Int. Scholarly Res. Notices*.

Sharpe, R. M., & Drake, A. J., (2013). Obesogens and obesity—An alternative view. *Obesity, 21*, 1081–1083.

Shelby, M. D., Newbold, R. R., Tully, D. B., Chae, K., & Davis, V. L., (1996). Assessing environmental chemicals for estrogenicity using a combination of *in vitro* and *in vivo* assays. *Environ. Health Perspect., 104*, 1296–1300.

Shelton, J. F., Geraghty, E. M., Tancredi, D. J., Delwiche, L. D., Schmidt, R. J., Ritz, B., Hansen, R. L., & Hertz-Picciotto, I., (2014). Neurodevelopmental disorders and prenatal residential proximity to agricultural pesticides: The CHARGE study. *Environ. Health Perspect., 122*, 1103–1109.

Simanainen, U., Haavisto, T., Tuomisto, J. T., Paranko, J., Toppari, J., Tuomisto, J., Peterson, R. E., & Viluksela, M., (2004). Pattern of male reproductive system effects after in utero and lactational 2,3,7,8-tetrachlorodibenzo-p-dioxin (TCDD) exposure in three differentially TCDD-sensitive rat lines. *Toxicol Sci., 80*, 101–108.

Solomon, K. R., Giesy, J. P., LaPoint, T. W., Giddings, J. M., & Richards, R. P., (2013). Ecological risk assessment of atrazine in North American surface waters. *Environ. Toxicol. Chem., 32*, 10–11.

Somm, E., Schwitzgebel, V. M., Toulotte, A., Cederroth, C. R., Combescure, C., Nef, S., Aubert, M. L., & Hüppi, P. S., (2009). Perinatal exposure to bisphenol A alters early adipogenesis in the rat. *Environ. Health Perspect., 117*, 1549–1555.

Ssempebwa, J. C., Carpenter, D. O., Yilmaz, B., DeCaprio, A. P., O'Hehir, D. J., & Arcaro, K. F., (2004). Waste crankcase oil: An environmental contaminant with potential to modulate estrogenic responses. *J. Toxicol. Environ. Health Part A, 67*, 1081–1094.

Steinmaus, C., Miller, M. D., Cushing, L., Blount, B. C., & Smith, A. H., (2013). Combined effects of perchlorate, thiocyanate, and iodine on thyroid function in the national health and nutrition examination survey 2007–08. *Environ. Res., 123,* 17–24.

Storey, N. M., Gentile, S., Ullah, H., Russo, A., Muessel, M., Erxleben, C., & Armstrong, D. L., (2006). Rapid signaling at the plasma membrane by a nuclear receptor for thyroid hormone. *Proc. Natl. Acad. Sci., 103*, 5197–5201.

Sun, K. S., Jun, K. S., Da Lee, R., Jo Lim, K., Seek, R., G., Hyun, S. J., Ho Kim, B., et al., (2005). Assessment of estrogenic and androgenic activities of tetramethrin *in vitro* and *in vivo* assays. *J. Toxicol. Environ. Health Part A, 68*, 2277–2289.

Suzuki, T., Kitamura, S., Khota, R., Sugihara, K., Fujimoto, N., & Ohta, S., (2005). Estrogenic and antiandrogenic activities of 17 benzophenone derivatives used as UV stabilizers and sunscreens. *Toxicol. Appl. Pharmacol., 203*, 9–17.

Svobodová, K., Plačková, M., Novotná, V., & Cajthaml, T., (2009). Estrogenic and androgenic activity of PCBs, their chlorinated metabolites and other endocrine disruptors estimated with two *in vitro* yeast assays. *Sci. Total Environ., 407*, 5921–5925.

Swan, S. H., Liu, F., Hines, M., Kruse, R. L., Wang, C., Redmon, J. B., Sparks, A., & Weiss, B., (2010). Prenatal phthalate exposure and reduced masculine play in boys. *Int. J. Androl., 33*, 259–269.

Takeuchi, S., Shiraishi, F., Kitamura, S., Kuroki, H., Jin, K., & Kojima, H., (2011). Characterization of steroid hormone receptor activities in 100 hydroxylated polychlorinated biphenyls, including congeners identified in humans. *Toxicol., 289*, 112–121.

Trabert, B., Longnecker, M. P., Brock, J. W., Klebanoff, M. A., & McGlynn, K. A., (2012). Maternal pregnancy levels of trans-nonachlor and oxychlordane and prevalence of cryptorchidism and hypospadias in boys. *Environ Health Perspect., 120*, 478–482.

Trasande, L., Attina, T. M., & Blustein, J., (2012). Association between urinary bisphenol A concentration and obesity prevalence in children and adolescents. *JAMA, 308*, 1113–1121.

Turque, N., Palmier, K., Le Mével, S., Alliot, C., & Demeneix, B. A., (2005). A rapid, physiologic protocol for testing transcriptional effects of thyroid-disrupting agents in premetamorphic Xenopus tadpoles. *Environ. Health Perspect., 113*, 1588–1593.

Uslu, U., Sandal, S., Cumbul, A., Yildiz, S., Aydin, M., & Yilmaz, B., (2013). Evaluation of estrogenic effects of polychlorinated biphenyls and organo-chlorinated pesticides using immature rat uterotrophic assay. *Hum. Exp. Toxicol, 32*, 476–482.

Uzumcu, M., Zama, A. M., & Oruc, E., (2012). Epigenetic mechanisms in the actions of endocrine-disrupting chemicals: Gonadal effects and role in female reproduction. *Reprod Domest Anim., 47*, 338–347.

Van, D. B. B., Schreurs, R., Van, D. L. S., Seinen, W., Brouwer, A., & Sonneveld, E., (2008). Endocrine effects of polycyclic musks: Do we smell a rat. *Int. J. Androl., 31*, 188–193.

Viel, J. F., Clément, M. C., Hägi, M., Grandjean, S., Challier, B., & Danzon, A., (2008). Dioxin emissions from a municipal solid waste incinerator and risk of invasive breast cancer: a population-based case-control study with GIS-derived exposure. *Int. J. Health Geogr., 7*, 1–8.

Viel, J. F., Daniau, C., Goria, S., Fabre, P., De Crouy-Chanel, P., & Sauleau, E. A., (2008). Empereur-Bissonnet, P. risk for non-Hodgkin's lymphoma in the vicinity of French municipal solid waste incinerators. *Environ. Health., 7*, 1–9.

Viel, J. F., Floret, N., Deconinck, E., Focant, J. F., De Pauw, E., & Cahn, J. Y., (2011). Increased risk of non-Hodgkin lymphoma and serum organochlorine concentrations among neighbors of a municipal solid waste incinerator. *Environ. Int., 37*, 449–453.

Viglietti-Panzica, C., Mura, E., & Panzica, G., (2007). Effects of early embryonic exposure to genistein on male copulatory behavior and vasotocin system of Japanese quail. *Horm. Behav., 51*, 355–363.

Villeneuve, S., Cyr, D., Lynge, E., Orsi, L., Sabroe, S., Merletti, F., Gorini, G., et al., (2010). Occupation and occupational exposure to endocrine disrupting chemicals in male breast cancer: A case-control study in Europe. *Occup. Environ. Med., 67*, 837–844.

Vom, S. F. S., Akingbemi, B. T., Belcher, S. M., Birnbaum, L. S., Crain, D. A., Eriksen, M., Farabollini, F., et al., (2007). Chapel hill bisphenol A expert panel consensus statement: Integration of mechanisms, effects in animals and potential to impact human health at current levels of exposure. *Reprod. Toxicol., 24*, 131.

Vom, S. F. S., Nagel, S. C., Coe, B. L., Angle, B. M., & Taylor, J. A., (2012). The estrogenic endocrine disrupting chemical bisphenol A (BPA) and obesity. *Mol. Cell. Endocrinol., 354*, 74–84.

Wang, T., Li, M., Chen, B., Xu, M., Xu, Y., Huang, Y., Lu, J., et al., (2012). Urinary bisphenol A (BPA) concentration associates with obesity and insulin resistance. *J. Clin. Endocrinol. Metab., 97*, 223–227.

Wei, J., Lin, Y., Li, Y., Ying, C., Chen, J., Song, L., Zhou, Z., et al., (2011). Perinatal exposure to bisphenol A at reference dose predisposes offspring to metabolic syndrome in adult rats on a high-fat diet. *Endocrinol., 152*, 3049–3061.

Weinhold, B., (2012). More chemicals show epigenetic effects across generations. *Environ. Health Perspect., 120*, 228.

White, R., Jobling, S., Hoare, S. A., Sumpter, J. P., & Parker, M. G., (1994). Environmentally persistent alkylphenolic compounds are estrogenic. *Endocrinol., 135*, 175–182.

Winneke, G., Walkowiak, J., & Lilienthal, H., (2002). PCB-induced neurodevelopmental toxicity in human infants and its potential mediation by endocrine dysfunction. *Toxicology, 181*, 161–165.

Wolf, C., Lambright, C., Mann, P., Price, M., Cooper, R. L., Ostby, J., & Gray, Jr. L. E., (1999). Administration of potentially antiandrogenic pesticides (procymidone, linuron, iprodione, chlozolinate, p,p'-DDE, and ketoconazole) and toxic substances (dibutyl- and diethylhexyl phthalate, PCB 169, and ethane dimethane sulphonate) during sexual differentiation produces diverse profiles of reproductive malformations in the male rat. *Toxicol. Ind. Health., 15*, 94–118.

Woods, H. F., (2003). *Phytoestrogens and Health.* Crown copyright UK.

Yamasaki, K., Takeyoshi, M., Sawaki, M., Imatanaka, N., Shinoda, K., & Takatsuki, M., (2003). Immature rat uterotrophic assay of 18 chemicals and Hershberger assay of 30 chemicals. *Toxicol., 183*, 93–115.

You, L., Casanova, M., Archibeque-Engle, S., Sar, M., Fan, L. Q., & Heck, H. A., (1998). Impaired male sexual development in perinatal Sprague-Dawley and long-Evans hooded rats exposed in utero and lactationally top, p'-DDE. *Toxicol. Sci., 45*, 162–173.

Yu, Y., Zhou, Q., Hang, Y., Bu, X., & Jia, W., (2007). Antiestrogenic effect of 20S-protopanaxadiol and its synergy with tamoxifen on breast cancer cells. *Cancer, 109*, 2374–2382.

Zucchi, S., Blüthgen, N., Ieronimo, A., & Fent, K., (2011). The UV-absorber benzophenone-4 alters transcripts of genes involved in hormonal pathways in zebrafish (*Danio rerio*) eleuthero-embryos and adult males. *Toxicol. Appl. Pharmacol., 250*, 137–146.

Index

B

C

H

I

N

O

P

R

T

U

For Product Safety Concerns and Information please contact our EU representative GPSR@taylorandfrancis.com Taylor & Francis Verlag GmbH, Kaufingerstraße 24, 80331 München, Germany